AF356134

Aroma Compound Evolution during Fermentation

Editor

Niël van Wyk

Basel • Beijing • Wuhan • Barcelona • Belgrade • Novi Sad • Cluj • Manchester

Editor
Niël van Wyk
Hochschule Geisenheim
University
Geisenheim
Germany

Editorial Office
MDPI AG
Grosspeteranlage 5
4052 Basel, Switzerland

This is a reprint of articles from the Special Issue published online in the open access journal *Fermentation* (ISSN 2311-5637) (available at: https://www.mdpi.com/journal/fermentation/special_issues/3624791341).

For citation purposes, cite each article independently as indicated on the article page online and as indicated below:

Lastname, A.A.; Lastname, B.B. Article Title. *Journal Name* **Year**, *Volume Number*, Page Range.

ISBN 978-3-7258-2387-1 (Hbk)
ISBN 978-3-7258-2388-8 (PDF)
doi.org/10.3390/books978-3-7258-2388-8

Contents

About the Editor

Niël van Wyk

Niël van Wyk is a former post-doctoral fellow at the Macquarie University in Sydney, Australia, and Geisenheim University in Germany, where he focused on yeast's impact on fermented beverages' aroma. He was a post-doctoral fellow at the University of Stellenbosch, South Africa, and CNRS in Montpellier, France, working primarily on novel drug targets for mycobacterial infections. He received his Ph.D. in Microbiology in 2010 at Stellenbosch University. He is a researcher at the BRAIN Biotech AG, a listed biotechnology company headquartered in Zwingenberg, Germany.

 *fermentation*

Editorial

Aroma Compound Evolution during Fermentation

Niël van Wyk [1,2,†]

[1] Department of Microbiology and Biochemistry, Hochschule Geisenheim University, Von-Lade-Strasse 1, 65366 Geisenheim, Germany; niel.wyk@hs-gm.de
[2] ARC Centre of Excellence in Synthetic Biology, Department of Molecular Sciences, Macquarie University, Sydney, NSW 2113, Australia
[†] Current address: Brain-Biotech AG, 34-36 Darmstädter Strasse, 64673 Zwingenberg, Germany.

Citation: van Wyk, N. Aroma Compound Evolution during Fermentation. *Fermentation* 2023, 9, 797. https://doi.org/10.3390/fermentation9090797

Received: 9 August 2023
Accepted: 25 August 2023
Published: 29 August 2023

Microorganisms involved in the fermentation process play a significant role in shaping the aromatic characteristics of the final food product. During fermentation, various precursor compounds undergo transformation into aroma-active compounds due to microbial activity. Understanding the timing, mechanisms, and reasons behind the release of these compounds is essential to ensuring the consistent development of fermentation products, such as wine or beer, with well-defined aroma profiles.

The aim of the Special Edition titled "Aroma Compound Evolution during Fermentation" was to present new investigations conducted in this field. The edition featured ten publications (nine original research papers and one review) covering diverse topics. Special mention should be given to papers focusing on the aroma profiles of non-traditional foodstuffs [1–3].

Wine-related studies included:

Badura et al. [4] investigated apiculate yeasts belonging to the *Hanseniaspora* genus in simultaneous inoculations with *Saccharomyces cerevisiae* during grape must fermentation. Aroma profiles were compared, and distinct differences were observed between the two lineages of *Hanseniaspora*, impacting the final product. The study also revealed that specific *Hanseniaspora* species esterify citronellol into citronellyl acetate, which is a significant aroma molecule.

Guittin et al. [5] implemented an online monitoring system to measure acetaldehyde during wine fermentation, emphasizing the role of temperature in the dynamic production of this critical yeast fermentation metabolite, which affects wine quality, including its aroma.

Gottardi et al. [6] compared the performance of seven *S. cerevisiae* strains on a pilot scale (0.9 hL) and highlighted significant differences among the strains. The study aimed to simulate wine fermentations more closely to real-world scenarios than small-scale laboratory experiments.

Vicente et al. [7] used two popular non-*Saccharomyces* yeasts, *Schizosaccharomyces pombe* and *Lachancea thermotolerans*, in a sequential wine fermentation with *S. cerevisiae*. The study demonstrated that these yeasts not only dramatically influenced the final aroma profile but also converted malic acid into less tart lactic acid, resulting in the expected organic acid modulation.

Agarbati et al. [8] investigated the use of another non-*Saccharomyces* yeast, *Metschnikovia pulcherrima*, in wine fermentation. The addition of this yeast altered the wine aroma and also had a meaningful impact on the biocontrol of the wild microflora population.

Beer-related study:

Roberts et al. [9] explored geraniol, an important hops-derived terpene that imparts a floral aroma to beer. The study used two different measuring techniques and demonstrated that geraniol can undergo transformation into several different compounds during beer fermentation.

Non-conventional fermentations:

Liu et al. [2] reported on the aroma compounds that are produced during the fermentation of lamb liver paste. Hung et al. [1] showed what effect the addition of starter cultures like *Bacillus subtilis* and *S. cerevisiae* can have on the analytical and sensory composition of seaweed. With the addition of the basidiomycete *Laetiporus persicinus* to cocoa pulp, Klis et al. [3] showed that this fungus produced several interesting aromas that imparted tropical notes to the fermented product. Finally, Meneses Queral et al. [10] wrote a review discussing the aroma compounds that are present during the fermentation process of cocoa.

Overall, these studies contribute valuable insights into the evolution of aroma during fermentation, enhancing our understanding of the factors influencing the aromatic characteristics of fermented food and beverage products.

Funding: The research fellowship of N.V.W. was co-funded by Geisenheim University and Macquarie University. The Hesse State Ministry of Higher Education, Research and the Arts for the Hesse initiative for scientific and economic excellence (LOEWE) in the framework of AROMAplus (https://www.hs-geisenheim.de/aromaplus/) (accessed on 28 August 2023) and the Macquarie-led national Centre of Excellence in Synthetic Biology funded by the Australian Government thorough its agency, the Australian Research Council are thanks for funding.

Conflicts of Interest: The author declares no conflict of interest.

References

1. Hung, Y.H.R.; Peng, C.Y.; Huang, M.Y.; Lu, W.J.; Lin, H.J.; Hsu, C.L.; Fang, M.C.; Lin, H.T.V. Monitoring the Aroma Compound Profiles in the Microbial Fermentation of Seaweeds and Their Effects on Sensory Perception. *Fermentation* **2023**, *9*, 135. [CrossRef]
2. Liu, T.; Zhang, T.; Yang, L.; Zhang, Y.; Kang, L.; Yang, L.; Zhai, Y.; Jin, Y.; Zhao, L.; Duan, Y. Effects of Fermentation on the Physicochemical Properties and Aroma of Lamb Liver Paste. *Fermentation* **2022**, *8*, 676. [CrossRef]
3. Klis, V.; Pühn, E.; Jerschow, J.J.; Fraatz, M.A.; Zorn, H. Fermentation of Cocoa (*Theobroma cacao* L.) Pulp by Laetiporus persicinus Yields a Novel Beverage with Tropical Aroma. *Fermentation* **2023**, *9*, 533. [CrossRef]
4. Badura, J.; Kiene, F.; Brezina, S.; Fritsch, S.; Semmler, H.; Rauhut, D.; Pretorius, I.S.; von Wallbrunn, C.; van Wyk, N. Aroma Profiles of *Vitis vinifera* L. cv. Gewürztraminer Must Fermented with Co-Cultures of Saccharomyces cerevisiae and Seven *Hanseniaspora* spp. *Fermentation* **2023**, *9*, 109. [CrossRef]
5. Guittin, C.; Maçna, F.; Picou, C.; Perez, M.; Barreau, A.; Poitou, X.; Sablayrolles, J.M.; Mouret, J.R.; Farines, V. New Online Monitoring Approaches to Describe and Understand the Kinetics of Acetaldehyde Concentration during Wine Alcoholic Fermentation: Access to Production Balances. *Fermentation* **2023**, *9*, 299. [CrossRef]
6. Gottardi, D.; Siesto, G.; Bevilacqua, A.; Patrignani, F.; Campaniello, D.; Speranza, B.; Lanciotti, R.; Capece, A.; Romano, P. Pilot Scale Evaluation of Wild Saccharomyces cerevisiae Strains in Aglianico. *Fermentation* **2023**, *9*, 245. [CrossRef]
7. Vicente, J.; Kelanne, N.; Navascués, E.; Calderón, F.; Santos, A.; Marquina, D.; Yang, B.; Benito, S. Combined Use of Schizosaccharomyces pombe and a Lachancea thermotolerans Strain with a High Malic Acid Consumption Ability for Wine Production. *Fermentation* **2023**, *9*, 165. [CrossRef]
8. Agarbati, A.; Canonico, L.; Ciani, M.; Comitini, F. Metschnikowia pulcherrima in Cold Clarification: Biocontrol Activity and Aroma Enhancement in Verdicchio Wine. *Fermentation* **2023**, *9*, 302. [CrossRef]
9. Roberts, R.; Khomenko, I.; Eyres, G.T.; Bremer, P.; Silcock, P.; Betta, E.; Biasioli, F. Investigation of Geraniol Biotransformation by Commercial Saccharomyces Yeast Strains by Two Headspace Techniques: Solid-Phase Microextraction Gas Chromatography/Mass Spectrometry (SPME-GC/MS) and Proton Transfer Reaction-Time of Flight-Mass Spectrometry. *Fermentation* **2023**, *9*, 294. [CrossRef]
10. Meneses Quelal, O.; Pilamunga Hurtado, D.; Arroyo Benavides, A.; Vidaurre Alanes, P.; Vidaurre Alanes, N. Key Aromatic Volatile Compounds from Roasted Cocoa Beans, Cocoa Liquor, and Chocolate. *Fermentation* **2023**, *9*, 166. [CrossRef]

 *fermentation*

Article

Effects of Fermentation on the Physicochemical Properties and Aroma of Lamb Liver Paste

Ting Liu [1], Taiwu Zhang [1], Lirong Yang [1], Yanni Zhang [1], Letian Kang [1], Le Yang [1], Yujia Zhai [2], Ye Jin [1], Lihua Zhao [1] and Yan Duan [1,*]

[1] College of Food Science and Engineering, Inner Mongolia Agricultural University, 306 Zhaowuda Road, Hohhot 010018, China
[2] Institute of Product Quality Inspection, Inner Mongolia Autonomous Region, Shihua Road, Hohhot 010070, China
* Correspondence: duanyan@imau.edu.cn

Abstract: The probiotic fermentation of lamb liver paste is a new method with which to utilize sheep by-products and address the issue of waste. In this study, a pH meter, chromaticity meter, texture analyzer, and gas chromatograph–mass spectrometer (GC–MS) were used to determine various indicators. The objective was to investigate the effect of fermentation on the physical properties and aroma of lamb liver paste. The results showed that the L* (brightness), a* (redness), and b* (yellowness) of the samples were significantly higher in the starter fermentation group than in the other two groups after storage for 0, 1, 7, 14, 21, and 28 days ($p < 0.05$). In addition, cohesiveness, adhesion, and chewiness were lower in the starter fermentation group after 7 days ($p < 0.05$). TVB-N and fat were lower in the starter fermentation group compared to the sterilization group at 28 days. pH was significantly lower in the starter fermentation group at the beginning of storage, and lactic acid bacteria numbers were significantly higher than in the sterilization groups ($p < 0.05$). Important aroma compounds, such as 2-undecenal, 1-octen-3-ol, and anethole, were significantly higher in the starter fermentation group than in the sterilization group ($p < 0.05$). Fermented lamb liver paste is a new by-product that exhibits a high degree of freshness and a low degree of fat oxidation during storage. This study provides a theoretical basis for future industrial production.

Keywords: lamb liver; by-product; GC–MS; texture; volatile compounds

Citation: Liu, T.; Zhang, T.; Yang, L.; Zhang, Y.; Kang, L.; Yang, L.; Zhai, Y.; Jin, Y.; Zhao, L.; Duan, Y. Effects of Fermentation on the Physicochemical Properties and Aroma of Lamb Liver Paste. *Fermentation* **2022**, *8*, 676. https://doi.org/10.3390/fermentation8120676

Academic Editor: Niel Van Wyk

Received: 21 October 2022
Accepted: 23 November 2022
Published: 25 November 2022

Publisher's Note: MDPI stays neutral with regard to jurisdictional claims in published maps and institutional affiliations.

1. Introduction

Lamb meat is tender, tasty, unique in flavor, rich in nutrients, and loved by consumers. It also contributes to sustainable development in animal husbandry and the meat industry. The high demand for lamb has led to an increase in the production of sheep by-products. However, these by-products have not been efficiently utilized, resulting in great economic losses in animal husbandry and negative impacts on the ecological environment. A low comprehensive utilization rate of sheep by-products, a lack of product variety, and resource waste are problems associated with the process of animal slaughter. To address these issues, animal liver is worthy of attention in terms of converting animal by-products into valuable animal products. The vitamin (vitamin B2, vitamin C, and vitamin A), mineral (trace element selenium) [1,2], and other nutrient contents in animal liver are higher than those in milk, eggs, meat, fish, and other foods. The nutrients in lamb liver are components of many enzymes and coenzymes in human biochemical metabolism, which accelerate the body's metabolism, relieve visual fatigue, and enhance the body's immunity [3–5]. Consequently, fermented products can both alleviate these problems and provide benefits to the body.

Fermentation technology refers to the artificial application of methods (conditions such as temperature and pH) to control microorganisms (bacteria, yeast, etc.) to use organic matter as a medium for fermentation. The addition of probiotics induces host digestive protease and peptidase activity and also improves the absorption of small peptides and amino acids in the organism by improving epithelial absorption and enhancing translocation [6]. Moreover, endogenous meat enzyme activity, microbial growth, and lipid oxidation reactions are partly responsible for the production of many aroma and carrion compounds [7]. Studies have found that, when lactic acid bacteria are added to fermented sausages, their metabolites (lactic acid, formic acid, acetic acid, etc.) give the product a special sour flavor [8,9]. Casaburi used two strains with differing abilities to decompose protein and fat as starter cultures to make fermented sausage. The experiment proved that both strains improved the nutritional quality of fermented sausage [10]. Antara found that after mixing *Lactobacillus plantarum* and *Pediococcus lactis* into sausages, the control of enterotoxin production by metabolizing lactic acid extended the shelf life of fermented sausages [11]. The Xing TK study showed that the addition of lactic acid bacteria and Staphylococcus significantly promoted the accumulation of free amino acids during the fermentation of beef jerky sausage [12]. In summary, the fermentation technology used for meat products is sophisticated, but there are few reports on the fermentation of by-products.

Microorganisms decompose sugars to produce acids, reducing the pH value of meat products. The rapid production of lactic acid, resulting in a decrease in pH, inhibits the growth of pathogenic and spoilage microorganisms and prolongs the shelf life of fermented foods. The pH value is often used to measure the maturity of fermented meat products and indicates the fermentation stage according to the evaluation index. This is used to establish the best fermentation time for the fermented product. In addition, other metabolites, such as lactic acid, acetic acid, propionic acid, benzoic acid, hydrogen peroxide, and bacteriocins, improve food safety [13]. Therefore, with the continuous improvement of comprehensive animal liver processing technology, animal liver will eventually be utilized to its full potential. Therefore, this experiment investigated the effects of fermentation on the probiotic count, texture, and aroma of lamb liver paste. The fermented lamb liver paste in this study had a pleasant aroma and represents a high-value sheep by-product. Moreover, it increases the variety of sheep by-products available, which will act to reduce waste in this industry.

2. Materials and Methods

2.1. Preparation of Fermented Lamb Liver Paste (Basic Recipe and Process Flow)

The preparation of goat liver paste refers to the "Liver Paste" product in the GOST 12319-77 standard report [14]. (1) Trimming: Raw lamb livers were trimmed for membranes, bile ducts, and other inclusions, and rinsed of blood. (2) Maceration: Trimmed lamb livers were macerated in 1% salt water for 3 h. (3) Pickling: According to the recipe (salt 2.5%, sugar 2%, Chinese rice cooking wine 3%, sodium nitrite 50 mg/kg, and vitamin C 0.05%), the livers were pickled at 0–4 °C for 12 h. (4) Cooking: The pickled lamb livers were cooked for 8–10 min (star anise 0.05%, pepper 0.05%, chili 0.1%, cinnamon 0.1%, dahurica 0.1%, garlic 0.5%, and fresh ginger 0.5%), and then placed on a rack to cool at room temperature. (5) Mixing: The cooled lamb livers were cut into pieces, and the small pieces were put into a ZB-40 chopper (Ruiheng Food Machinery Factory) for grinding. Spices (five-spice powder 0.3%, black pepper powder 0.2%, ginger powder 0.2%, salt 1.5%, sugar 1.5%, and corn starch 4%), emulsifier 3% (glycerol monostearate: sodium caseinate = 1:1), thickening agent 5% (sodium carboxymethyl cellulose: β-cyclodextrin = 4:1), ice water 40%, corn germ oil 9%, and starter culture 0.02% were continually added during mixing. (6) Fermentation and storage: The mixed filling was fermented in a humidity chamber at a constant temperature (39 °C, 7 h) to obtain the finished product, which was vacuum packed and stored at 4 °C in a refrigerator.

2.2. Experimental Design Grouping of Mutton Liver Paste during Storage

Starter fermentation group (SF): 0.02% starter F-1 Bactoferm® (*Staphylococcus xylosus* DD-34 and *Pediococcus pentosaceus* PCFF-1, Chr. Hansen Holding A/S, Hørsholm, Denmark) was added to basic formula lamb liver paste before fermenting in a humidity box at a constant temperature (39 °C, 7 h). Natural fermentation group (NF): basic formula lamb liver paste with no starter culture added was fermented in a humidity box at a constant temperature (39 °C, 7 h). Sterilization treatment group (ST): the basic formula was sterilized at a high temperature (121 °C, 10 min) without adding the starter culture.

The changes in microbes and physicochemical indexes of the three lamb liver paste groups were compared on day 0, the 1st day, 7th day, 14th day, 21st day, and 28th day of storage. In order to explore the quality changes in the lamb liver paste during storage, the most suitable storage period was selected to provide a theoretical basis.

2.3. Indicator Measurement

2.3.1. Microbial Analysis

Under sterile conditions on an ultraclean bench (Haier Co., Ltd., Qindao, China), 25 g of the sample was weighed from the middle of the saucer, added to 225 mL of sterile physiological saline, vigorously shaken for 30 min, and then diluted in gradient after mixing. The appropriate dilution was selected for pouring into agar medium plates, and after culturing at 37 °C for 48 h, a count was performed. PCA medium (plate counting agar medium) was used to measure the total number of colonies. PCA medium ingredients: tryptone 5.0 g, yeast extract 2.5 g, glucose 1.0 g, agar 15.0 g, distilled water 1000 mL, and pH 7.0 ± 0.2. MRS medium (de Man, Rogosa, and Sharpe medium) was used to measure the number of lactic acid bacteria. MRS medium: ingredients: peptone 10.0 g, beef paste 10.0 g, yeast paste 5.0 g, diammonium hydrogen citrate [$(NH_4)_2HC_6H_5O_7$] 2.0 g, glucose ($C_6H_{12}O_6$-H_2O) 20.0 g, Tween 80 1.0 mL, sodium acetate (CH_3COONa-$3H_2O$) 5.0 g, dipotassium hydrogen phosphate (K_2HPO_4-$3H_2O$) 2.0 g, magnesium sulfate ($MgSO_4$-$7H_2O$) 0.58 g, manganese sulfate ($MnSO_4$-H_2O) 0.25 g, agar 18.0 g, distilled water 1000 mL, and pH 6.4 ± 0.2.

2.3.2. Determination of Physical and Chemical Indicators

pH value: The pH value was measured following the method described by Perea-Sanz et al. [15]. Using a portable pH meter (PB-10, Zhicheng Analytical Instrument Manufacturing Co., Ltd., Shanghai, China), which was calibrated in buffers with pH 4.60 and 7.00, each sample was measured three times. Triplicates were taken. Each 25 g sample was mixed with 225 mL of normal saline and shaken for 30 min, and the pH value was measured with a pH meter.

Color: Meat color was measured using a CR-410 chromometer (Konica Minolta, Tokyo, Japan). Standard white plate parameters were used for calibration (D65 light source; Y = 92.6, x = 0.3162, y = 0.3324). Lightness (L*), redness (a*), and yellowness (b*) values of each sample were measured, and the mean values were regarded as the product color [16].

Water activity (Aw): The Aw value was determined with an HD-3A intelligent water activity meter (Huake Instrument Co., Ltd., Wuxi, China).

Textural properties: The lamb liver paste was placed in a 4 cm cylindrical model and arranged into a cylindrical shape. We referred to the texture profile analysis method of Omana et al. [17] and used the TA.XT Express Stable Micro Systems Texture Analyser (Stable Micro systems Ltd., Godalming, UK). The setting parameters of the texture analyzer were as follows: "TPA" setting mode, the strain area was 50%, and the aluminum cylinder probe (diameter 36 mm) was used for double compression cycle test. The compression cycle of each sample lasts for 5 s, the test speed was 1.5 mm/s, and the lamb liver paste was measured in three cycles. The hardness, elasticity, cohesion, adhesiveness, and chewiness of each group of samples were assessed in parallel in the three groups. The parameters were quantified when the product regained its original position [16].

A simple summary of the procedure for TVB-N should be included [18]. Three parallel groups were formed for each group of samples. First, 10 g of goat liver paste was transferred into a distillation tube. Then, 75 mL distilled water was added and homogenized for 30 min. Next, 1 g of magnesium oxide was added to the distillation tube containing the treated sample. Finally, TVB-N content was determined by automatic Kjeldahl nitrogen analyzer (Kjeltec 8200, FOSS, Hilleroed, Denmark).

Thiobarbituric acid-reactive substances (TBARS): According to the method described by Gong et al. [19], lipid oxidation was determined by TBARS. Then, 10 g liver samples were collected and ground, 50 mL of 7.5% trichloroacetic acid was added (containing 0.1% EDTA), the mixture was shaken for 30 min, and the resulting product was filtered twice with double-layer filter paper. Thereafter, 5 mL of supernatant was aspirated with a pipette and added to 5 mL of 0.02% MTBA solution. This was then heated in a water bath at 90 °C for 40 min. The heated sample was cooled, centrifuged for 5 min (16,000 r/min), and then, after 1 h, 5 mL of chloroform was added to the supernatant, and it was well shaken. After standing for stratification, the supernatant was assessed by colorimetry at 532 and 600 nm. TBARS content was expressed as mg malonaldehyde (MDA) per kg liver paste.

$$\text{TBARS (mg/kg)} = (A_{532} - A_{600})/155 \times (1/10) \times 72.6 \times 1000$$

Nutritional indicators: The moisture, protein, and fat content of the samples were determined according to the method described by Perea-Sanz et al. [15]. Three groups of parallel tests were carried out for each group of samples.

2.4. Analysis of Volatile Compounds

The vacuum-preserved lamb liver paste was removed from the 4 °C refrigerator, 5 g of the sample was placed into a 20 mL sample bottle, and 5 mL of saturated sodium chloride solution and 1 μL of 2-methyl-3-heptanone solution (0.168 μg/mL) were added. The mixed sample was placed in the rotor, shaken, and placed on a magnetic stirrer. The extraction head was inserted into the sample bottle at a distance of 1 cm from the sample and removed after adsorption at 60 °C for 45 min. Then, the GC injection port was inserted, and the sample underwent desorption at 250 °C for 4 min.

Referring to the method by Luo et al. [20], analyses were performed on a Trace 1300 Series GC gas chromatograph fitted with an ISQ mass spectrometer and a Xcalibur ChemStationa (Thermo Fisher Scientific, Waltham, MA, USA). A DB-5 chromatographic column (TR-5MS 30 m × 0.25 mm, film thickness 0.25 μm) with Helium as the carrier gas was used for the gas chromatograph–mass spectrometer (GC–MS) determination of volatile aroma components. The carrier gas flow rate was 1.0 mL/min, and the temperature of the inlet and the interface was 250 °C. Heating program: The initial temperature was 40 °C. This was maintained for 3 min, increased to 150 °C at 4 °C /min, and then held for 1 min before increasing to 200 °C at 5 °C/min, again increasing to 230 °C at 20 °C/min, and holding for 5 min, with no split injection. The ion source temperature was 250 °C, the transmission line temperature was 250 °C, the mass scanning range was m/z 30–400, and the solvent delay was 1 min. The mass spectrum was qualitatively analyzed with MEANLIB (MathWorks, Inc., R2006a), library database (NIST MS Search Program 2.0), and Wiley Library, and the matching degree was greater than 98% as the identification basis. Additionally, 2-Methyl-3-heptanone was selected to be the internal standard for the determination of volatile flavor compounds. Quantitative analyses were carried out according to the peak area of 2-methyl-3-heptanone of known mass concentration.

2.5. Data Analysis

An analysis of variance (ANOVA) test and Pearson correlation analysis were performed using the Statistical Package for the Social Science (SPSS Inc., version 26.0). Values are expressed as mean ± standard deviation (RSD). p-values less than 0.05 were considered statistically significant. The origin version from 2021 developed by OriginLab was used for mapping. PCA diagrams were drawn using the R Programming Language "factoextra" (R version 4.1.1).

3. Results

3.1. The Total Number of Colonies and the Number of Lactic Acid Bacteria

Changes in the total number of colonies during storage of lamb liver paste are shown in Figure 1A and Table S1. On day 0 of storage, the total number of colonies in the SF group was significantly higher than in the other two groups ($p < 0.05$), while the total number of colonies in the ST group in each storage period was significantly lower ($p < 0.05$). On the 28th day of storage, the total number of colonies in the three groups significantly increased ($p < 0.05$), and the total number of colonies in the ST group was significantly lower than in the other two groups ($p < 0.05$).

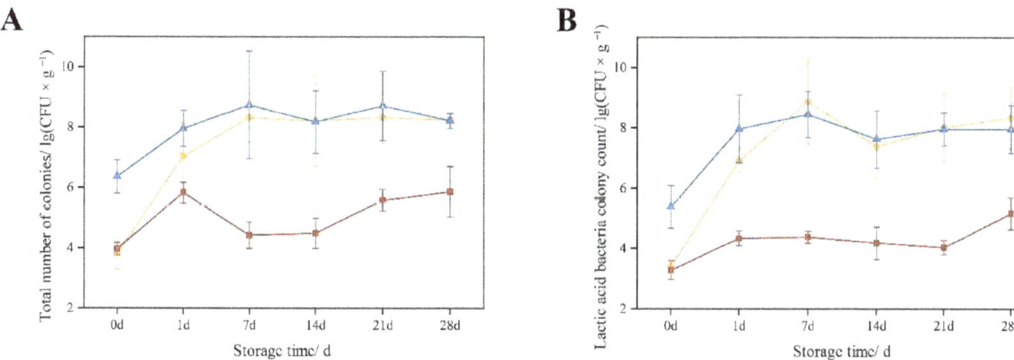

Figure 1. The changing trend of the total number of colonies and the number of lactic acid bacteria in lamb liver paste during the storage period. Note: yellow represents the natural fermentation group (NF); blue represents the starter fermentation group (SF); red represents the sterilization group (ST). (**A**) represents the change of the total colony number of the lamb liver paste with the storage time, and (**B**) represents the change of the lactic acid bacteria colony count of the lamb liver paste with the storage time.

The number of lactic acid bacteria during the storage of lamb liver paste is shown in Figure 1B and Table S2. During the whole storage period, the number of lactic acid bacteria in the ST group was significantly lower than in the other two groups ($p < 0.05$). The high-temperature treatment inhibited the activity of lactic acid bacteria, and the number of lactic acid bacteria in the SF group was significantly higher than in the NF group on the 1st day. The SF group and NF group had more lactic acid bacteria than the ST group.

3.2. pH Value

Acidic metabolites reduce the pH of the fermented product and increase the acidity of the product. As can be seen from Figure 2 and Table S3, after the 1st day of fermentation, the pH value of the SF group was 4.32, which was significantly lower than that of the other two groups ($p < 0.05$). The pH value of the SF group changed less during the storage period. In the NF group, the pH value decreased rapidly from 5.89 to 4.43 on the 7th day and then leveled off. There was no significant change in pH in the ST group ($p > 0.05$).

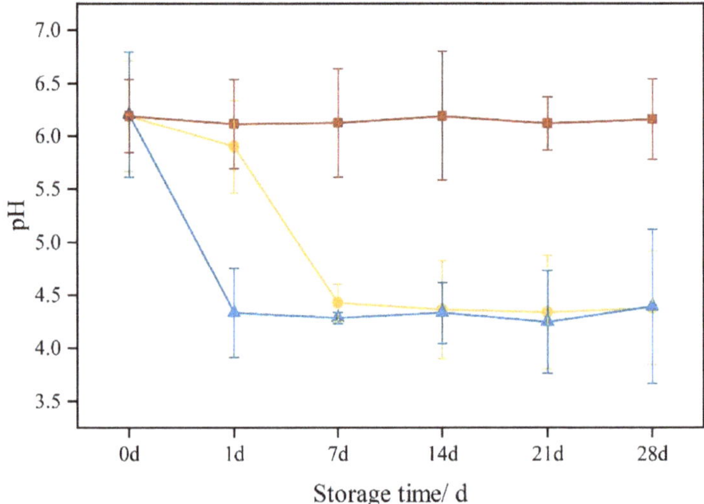

Figure 2. Change trend of pH value of lamb liver paste during storage period. Note: yellow represents the natural fermentation group (NF); blue represents the starter fermentation group (SF); red represents the sterilization group (ST).

3.3. Aw Value

Changes in the Aw value of lamb liver paste during storage are shown in Figure 3 and Table S4. The changing trend of the Aw value of the three groups of lamb liver paste is the same. During the whole storage process, there was no significant change in the Aw value of lamb liver paste in the three groups, and the Aw value was around 0.86.

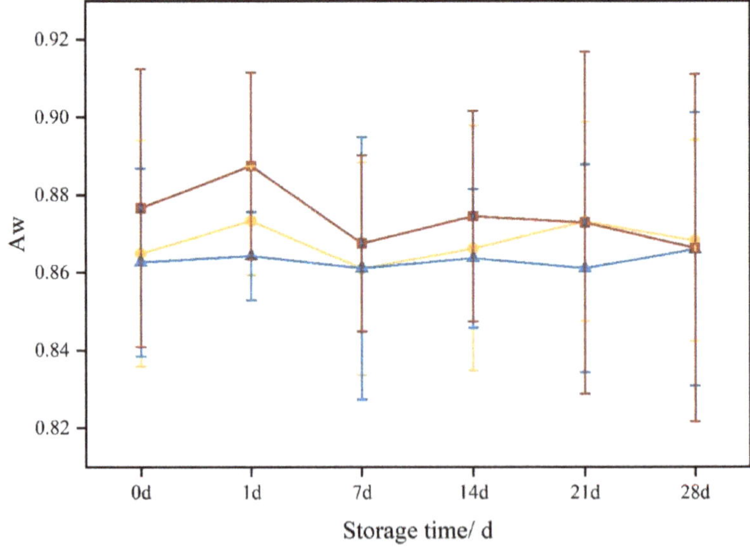

Figure 3. Change trend of Aw value of lamb liver paste during storage period. Note: yellow represents the natural fermentation group (NF); blue represents the starter fermentation group (SF); red represents the sterilization group (ST).

3.4. Chromatic Properties

The changes in L*, a*, and b* during the storage process of lamb liver paste are shown in Figure 4. Figure 4A and Table S5 show the increase in L* in the NF group and the SF group with the prolongation of storage time. The ST group had a significantly lower L* than other groups on the 28th day ($p < 0.05$). Figure 4B and Table S6 show that the a* values of lamb liver paste in the three groups did not change on day 0. The a* of the SF group was significantly higher than those of the other two groups, and the a* of the NF group was significantly higher than that of the ST group after 1 day of storage ($p < 0.05$). With the increase in storage time (Figure 4C and Table S7), the b* of lamb liver paste in the three groups also increased. The b* of the SF group was significantly higher than in the other two groups ($p < 0.05$).

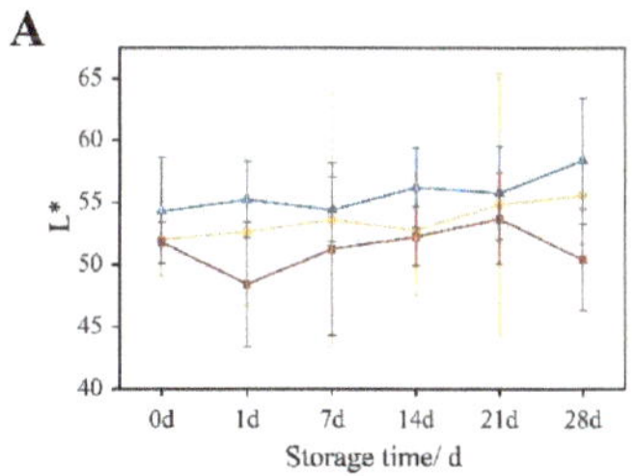 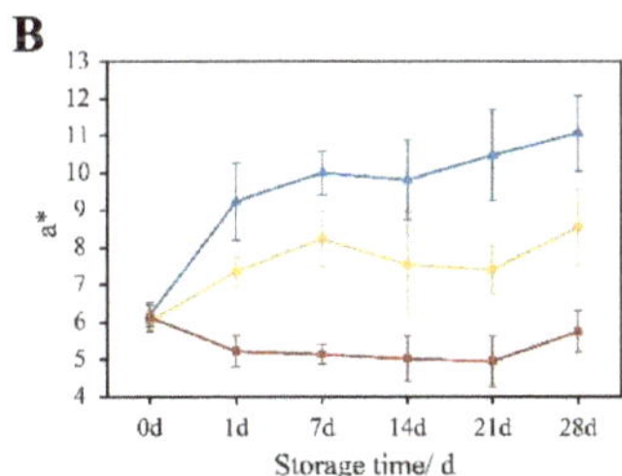 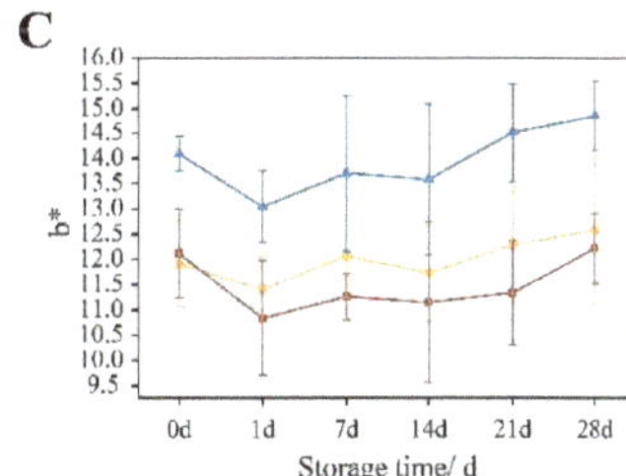

Figure 4. Change trend of color difference (a*, b*, L*) of lamb liver paste during storage period. Note: yellow represents the natural fermentation group (NF); blue represents the starter fermentation group (SF); red represents the sterilization group (ST). (**A**) represents the change of the Lightness value (L*) of the lamb liver paste with the storage time, (**B**) represents the change of the Redness value (a*) of the lamb liver paste with the storage time, and (**C**) represents the change of the Yellowness value (b*) of the lamb liver paste with the storage time.

3.5. Texture

The textural properties reflect the tissue structure and physical state of lamb liver paste. The textural properties of lamb liver paste throughout the storage period are shown in Table 1. The hardness, elasticity, cohesion, adhesiveness, and chewiness of the SF group were significantly higher than the other two groups on day 0 ($p < 0.05$). The adhesion of the SF group was significantly higher than the other two groups on the 1st day ($p < 0.05$), and there was no significant difference in the other indicators ($p > 0.05$). The elasticity, cohesiveness, adhesiveness, and chewiness of the sterilization group were significantly higher than the other two groups on the 7th day and 14th day of storage ($p < 0.05$). With the prolongation of the storage period, there was no significant difference in elasticity among the three groups on the 21st day and 28th day ($p > 0.05$). Cohesion, adhesion, and chewiness were significantly higher in the ST group ($p < 0.05$), but there was no significant difference between the NF group and the SF group.

Table 1. Differences in the texture of lamb liver paste during storage.

	Groups	0 d	1 d	7 d	14 d	21 d	28 d
Hardness/g	NF-group	1770.55 ± 67.60 [Bb]	1963.21 ± 26.12 [Aa]	1621.91 ± 119.71 [Bb]	1592.06 ± 45.13 [Bc]	1367.97 ± 47.51 [Cc]	1353.04 ± 52.84 [Cb]
	SF-group	1963.45 ± 37.48 [ABa]	2054.15 ± 55.98 [Aa]	2017.42 ± 47.64 [Aa]	1870.76 ± 25.70 [Bb]	1472.28 ± 12.69 [Cb]	1409.58 ± 26.61 [Cb]
	ST-group	1844.84 ± 41.44 [Cab]	1964.04 ± 13.45 [BCa]	2097.47 ± 55.96 [Ba]	2056.36 ± 92.52 [Ba]	2262.87 ± 27.78 [Aa]	2348.97 ± 73.99 [Aa]
Elasticity	NF-group	0.18 ± 0.01 [Ab]	0.18 ± 0.01 [Aa]	0.15 ± 0.01 [Ab]	0.16 ± 0.03 [Aab]	0.17 ± 0.05 [Aa]	0.13 ± 0.02 [Aa]
	SF-group	0.30 ± 0.05 [Aa]	0.19 ± 0.02 [Ba]	0.14 ± 0.01 [Bb]	0.12 ± 0.02 [Bb]	0.14 ± 0.04 [Ba]	0.18 ± 0.04 [Ba]
	ST-group	0.18 ± 0.01 [Ab]	0.18 ± 0.02 [Aa]	0.18 ± 0.01 [Aa]	0.17 ± 0.01 [Aa]	0.18 ± 0.01 [Aa]	0.19 ± 0.02 [Aa]
Cohesiveness	NF-group	0.14 ± 0.02 [Ab]	0.12 ± 0.01 [ABa]	0.09 ± 0.01 [BCb]	0.08 ± 0.01 [Cb]	0.08 ± 0.00 [Cb]	0.08 ± 0.01 [Cb]
	SF-group	0.19 ± 0.02 [Aa]	0.12 ± 0.01 [Ba]	0.10 ± 0.01 [Cab]	0.09 ± 0.00 [Cb]	0.09 ± 0.01 [Cb]	0.08 ± 0.00 [Cb]
	ST-group	0.15 ± 0.01 [Ab]	0.12 ± 0.00 [Ba]	0.12 ± 0.00 [Ba]	0.13 ± 0.01 [ABa]	0.12 ± 0.00 [Ba]	0.13 ± 0.01 [ABa]
Adhesion	NF-group	261.05 ± 8.48 [Ab]	227.93 ± 3.04 [Bc]	152.57 ± 10.05 [CDc]	160.30 ± 11.51 [Cb]	127.43 ± 14.35 [DEb]	107.15 ± 8.17 [Eb]
	SF-group	374.55 ± 8.42 [Aa]	268.12 ± 5.58 [Ba]	199.36 ± 4.94 [Cb]	167.51 ± 2.11 [Db]	121.36 ± 1.52 [Eb]	116.24 ± 1.62 [Eb]
	ST-group	256.90 ± 1.39 [CDb]	242.28 ± 5.13 [Db]	251.53 ± 2.21 [Da]	283.36 ± 8.57 [ABa]	274.28 ± 1.60 [BCa]	296.84 ± 14.48 [Aa]
Chewiness	NF-group	49.89 ± 1.76 [Aab]	41.09 ± 1.41 [Aa]	23.47 ± 3.33 [Bb]	23.35 ± 5.25 [Bb]	22.53 ± 8.58 [Bb]	14.85 ± 1.19 [Bb]
	SF-group	58.71 ± 7.13 [Aa]	49.99 ± 7.10 [Aa]	28.51 ± 0.28 [Bab]	21.34 ± 2.02 [BCb]	15.79 ± 3.52 [Cb]	15.68 ± 0.81 [Cb]
	ST-group	44.91 ± 0.92 [Ab]	41.87 ± 1.98 [Aa]	41.24 ± 8.24 [Aa]	51.39 ± 8.47 [Aa]	42.42 ± 2.02 [Aa]	41.85 ± 7.55 [Aa]

Uppercase letters denote significant differences in the same row; lowercase letters denote significant differences in the same column and the same metric.

3.6. TVB-N Value

The TVB-N value during the storage of lamb liver paste is shown in Figure 5 and Table S8. TVB-N values represent the freshness of the food, and a higher TVB-N content indicates higher amino acid destruction. The TVB-N value of the NF group was significantly higher than those of the other two groups on the 1st day of storage ($p < 0.05$). The TVB-N value of the SF group and the ST group were significantly higher than that of the NF group after 7 days ($p < 0.05$). After the 28th day, the TVB-N value of the SF group was lower than those of the other two groups.

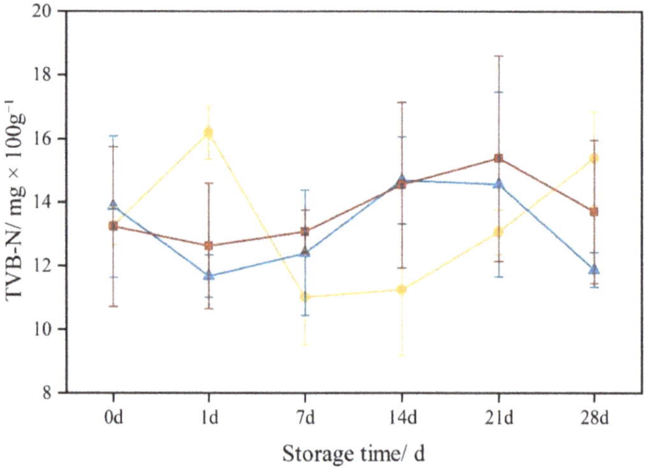

Figure 5. Variation trend of TVB-N value during storage of lamb liver paste. Note: yellow represents the natural fermentation group (NF); blue represents the starter fermentation group (SF); red represents the sterilization group (ST).

3.7. TBARS Value

The TBARS value indicates the degree of fat oxidation according to the amount of malondialdehyde—a secondary product—that is formed by fat oxidation. Fat oxidation causes spoilage and deterioration of meat quality and reduces the storability of meat. The higher the TBARS value, the higher the degree of fat oxidation and the greater the spoilage of the meat. The changes in TBARS values during the storage of lamb liver paste are shown in Figure 6 and Table S9. The TBARS values of the three groups decreased during processing and storage with increasing time. The TBARS values were significantly higher in the NF group than in the other two groups on the 1st day and significantly lower on the 14th day ($p < 0.05$).

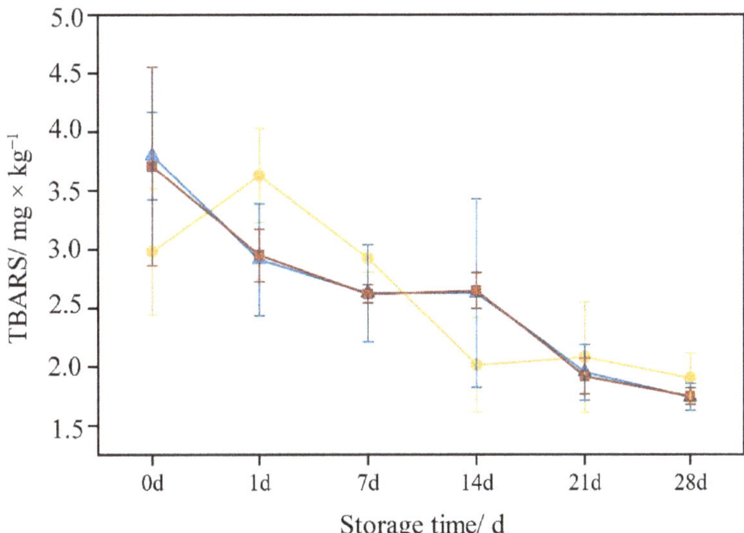

Figure 6. Variation trend of TBARS value during storage of lamb liver paste. Note: yellow represents the natural fermentation group (NF); blue represents the starter fermentation group (SF); red represents the sterilization group (ST).

3.8. Basic Nutritional Indicators

3.8.1. Moisture Content

The moisture content of lamb liver paste during storage is shown in Figure 7A and Table S10. The moisture content of the three groups was similar for all storage days ($p > 0.05$).

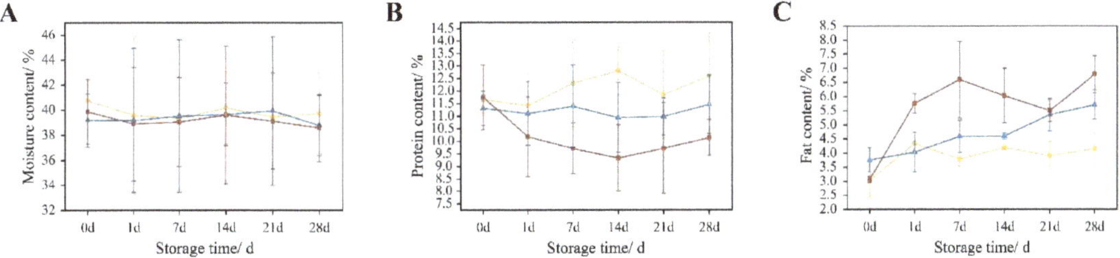

Figure 7. Changes in the basic nutritional quality of lamb liver paste during storage. Note: yellow represents the natural fermentation group (NF); blue represents the starter fermentation group (SF); red represents the sterilization group (ST). (**A**) represents the change of the moisture content of the lamb liver paste with the storage time, (**B**) represents the change of the protein content of the lamb liver paste with the storage time, and (**C**) represents the change of the fat content of the lamb liver paste with the storage time.

3.8.2. Protein Content

The protein content of lamb liver paste during storage is shown in Figure 7B and Table S11. The protein content of the ST group was significantly lower than that of the NF group and the fermentation group throughout the storage period ($p < 0.05$). The protein content of the NF group was significantly higher than that of the SF group and ST group ($p < 0.05$).

3.8.3. Fat Content

The fat content of lamb liver paste during storage is shown in Figure 7C and Table S12. In the ST group, the fat content increased significantly from day 0 to the 1st day, and the fat content was higher than in the other groups ($p < 0.05$). Except for the 1st day, the fat content in the SF group was significantly higher than in the NF group for all the storage times ($p > 0.05$).

3.9. Volatile Profile

The differences between groups were analyzed using the dimensionality reduction method and were ranked by principal component analysis (PCA). It can be seen from the PCA graph that the aroma of the three groups of samples were well-differentiated throughout the storage period (Figure 8). The scattered areas of the sample points are relatively dense, and the differences within the groups are small. The values of PC1 and PC2 accounted for 97.8%, representing the whole sample, indicating that the lamb liver pastes with different treatments were different.

The aroma analysis of lamb liver paste is shown in Table 2. Benzaldehyde in the NF group and SF group was significantly higher than in the ST group ($p < 0.05$). Octanal, decanal, 2,4-decadienal, and tetradecanal were significantly higher in the SF group and NF group as compared to the ST group ($p < 0.05$). The proportions of 2-undecenal and 2-octenal in the SF group and NF group were significantly higher than in the ST group ($p < 0.01$). The content of 1-hexanol in the NF group was significantly higher than in the other two groups ($p < 0.05$). Additionally, 1-Heptanol in the SF group was significantly higher than in the other two groups ($p < 0.05$). The 1-octen-3-ol in the SF group and the NF group was significantly higher than in the ST group ($p < 0.05$). Benzaldehyde, 4-(1-methyl ethyl)- and anethole in the SF group were significantly higher than in the other two groups ($p < 0.05$). Furthermore, 2-Nonanone was significantly higher in the NF group ($p < 0.05$).

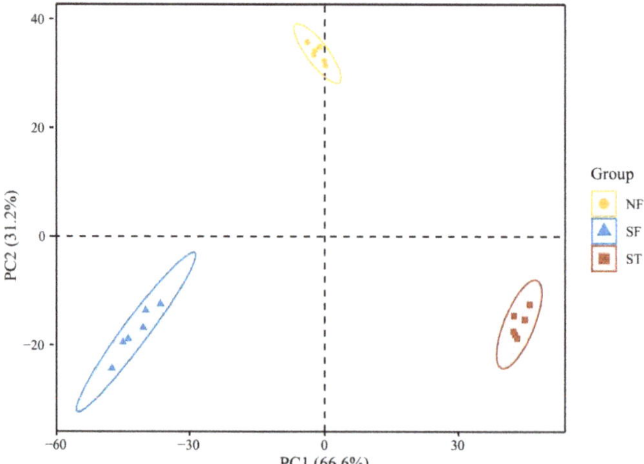

Figure 8. Principal coordinate analysis (PCoA) scatter plot analysis of two groups of samples. Note: yellow represents the natural fermentation group (NF); blue represents the starter fermentation group (SF); red represents the sterilization group (ST).

Table 2. Effects of three processing methods on volatile flavor compounds in lamb liver paste (µg/kg).

Volatile Flavor Compounds	Groups	0 d	1 d	7 d	14 d	21 d	28 d
Pentanal	NF	7.42 ± 0.85 [Db]	24.62 ± 0.82 [Cb]	32.24 ± 3.18 [Ba]	21.44 ± 2.67 [Cb]	31.19 ± 1.55 [Ba]	40.94 ± 3.08 [Aa]
	SF	7.29 ± 0.41 [Cb]	27.01 ± 1.80 [Ba]	34.88 ± 3.03 [Aa]	26.00 ± 3.28 [Ba]	2.36 ± 0.24 [Db]	3.25 ± 0.38 [Db]
	ST	20.08 ± 2.68 [Aa]	4.44 ± 0.57 [Cc]	15.93 ± 1.05 [Bb]	16.51 ± 1.50 [Bc]	1.77 ± 0.10 [Db]	2.16 ± 0.32 [Db]
Hexanal	NF	146.18 ± 15.29 [Cb]	492.83 ± 20.58 [Ab]	84.92 ± 7.92 [Da]	380.33 ± 38.01 [Bb]	18.46 ± 1.22 [Ea]	16.23 ± 1.97 [Eb]
	SF	112.24 ± 8.49 [Dc]	540.75 ± 21.68 [Ba]	662.09 ± 63.30 [Aa]	451.44 ± 39.54 [Ca]	15.76 ± 1.45 [Eb]	18.13 ± 1.78 [Eab]
	ST	413.27 ± 18.15 [Aa]	64.47 ± 2.09 [Cc]	272.52 ± 28.61 [Bb]	274.21 ± 15.69 [Ba]	17.15 ± 2.36 [Dab]	19.36 ± 2.01 [Da]
Heptanal	NF	25.54 ± 1.63 [Cb]	8.44 ± 1.09 [Db]	41.82 ± 5.19 [Aa]	31.50 ± 3.09 [Bb]	5.36 ± 0.58 [Db]	4.18 ± 0.36 [Db]
	SF	21.81 ± 1.48 [Bc]	40.55 ± 2.82 [Aa]	42.22 ± 3.03 [Aa]	39.10 ± 6.23 [Aa]	4.74 ± 0.65 [Cb]	4.78 ± 0.77 [Cb]
	ST	38.99 ± 2.97 [Aa]	9.67 ± 1.10 [Cb]	21.46 ± 1.45 [Bb]	22.01 ± 1.29 [Bc]	7.70 ± 0.52 [Ca]	7.57 ± 1.13 [Ca]
Octanal	NF	43.54 ± 3.30 [Cb]	81.70 ± 4.87 [Bb]	106.83 ± 16.95 [Aa]	91.79 ± 7.27 [Bb]	14.16 ± 2.07 [Da]	18.82 ± 2.26 [Da]
	SF	40.62 ± 5.38 [Bb]	137.87 ± 20.03 [Aa]	125.30 ± 15.41 [Aa]	130.46 ± 10.86 [Aa]	6.44 ± 0.53 [Cb]	8.42 ± 0.87 [Cb]
	ST	72.43 ± 12.41 [Aa]	7.11 ± 1.08 [Cc]	51.21 ± 3.38 [Bb]	49.14 ± 3.46 [Bc]	12.92 ± 1.13 [Ca]	9.94 ± 0.79 [Cb]
Nonanal	NF	126.47 ± 15.31 [Cb]	168.31 ± 17.91 [Aa]	158.63 ± 12.55 [Ba]	145.30 ± 12.91 [Cb]	34.67 ± 1.82 [Db]	30.51 ± 3.23 [Db]
	SF	127.41 ± 10.25 [Cb]	145.92 ± 15.58 [Cb]	165.23 ± 6.47 [Ba]	184.53 ± 18.41 [Aa]	27.10 ± 3.70 [Dc]	30.52 ± 3.48 [Db]
	ST	167.78 ± 13.44 [Aa]	55.53 ± 6.50 [Cc]	91.29 ± 10.52 [Bb]	94.32 ± 7.34 [Bc]	42.44 ± 4.37 [Ca]	43.65 ± 3.68 [Ca]
Benzaldehyde	NF	60.94 ± 4.13 [Cb]	112.34 ± 14.23 [Ba]	130.81 ± 15.94 [Aa]	99.75 ± 5.07 [Bb]	38.25 ± 3.93 [Db]	32.26 ± 2.68 [Db]
	SF	36.13 ± 3.74 [Bc]	118.83 ± 14.76 [Aa]	113.52 ± 5.31 [Ab]	115.86 ± 7.56 [Aa]	34.12 ± 2.09 [Bb]	34.69 ± 4.06 [Bb]
	ST	97.58 ± 9.94 [Aa]	59.12 ± 6.16 [Cb]	83.73 ± 5.10 [Bc]	85.84 ± 5.34 [Bc]	51.45 ± 4.55 [Ca]	48.79 ± 6.37 [Ca]
Tetradecanal	NF	1.77 ± 0.18 [Ec]	4.72 ± 0.40 [Db]	7.30 ± 0.62 [Ba]	5.72 ± 0.25 [Cb]	12.41 ± 0.87 [Aa]	6.25 ± 0.55 [Cb]
	SF	6.88 ± 0.48 [Ba]	7.42 ± 0.67 [Aa]	4.57 ± 0.57 [Cb]	7.61 ± 0.52 [Aa]	4.98 ± 0.52 [Cb]	6.30 ± 0.60 [Bb]
	ST	1.77 ± 0.18 [Db]	5.10 ± 0.66 [Cb]	4.56 ± 0.66 [Cb]	4.35 ± 0.39 [Cc]	12.14 ± 1.54 [Aa]	8.81 ± 1.23 [Ba]
2-Undecenal	NF	10.16 ± 1.00 [Ca]	55.30 ± 5.80 [Bb]	89.24 ± 11.83 [Aa]	65.28 ± 6.48 [Bb]	2.90 ± 0.27 [Ca]	2.36 ± 0.30 [Ca]
	SF	13.65 ± 1.03 [Ca]	92.88 ± 8.53 [Aa]	77.12 ± 4.86 [Bb]	85.07 ± 13.86 [Ba]	2.32 ± 0.13 [Cb]	2.39 ± 0.33 [Ca]
	ST	10.01 ± 1.42 [Ba]	3.00 ± 0.23 [Cc]	18.31 ± 1.68 [Ac]	19.21 ± 2.47 [Ac]	1.86 ± 0.19 [Cc]	1.91 ± 0.20 [Cb]
2-Octenal	NF	45.60 ± 3.01 [Db]	168.06 ± 15.66 [Bb]	212.57 ± 10.47 [Aa]	146.82 ± 10.09 [Cb]	12.18 ± 0.70 [Ea]	14.16 ± 1.19 [Ea]
	SF	37.79 ± 2.31 [Cb]	216.19 ± 22.17 [Aa]	222.56 ± 15.11 [Aa]	168.40 ± 14.09 [Ba]	9.86 ± 0.58 [Db]	11.00 ± 1.84 [Db]
	ST	102.38 ± 9.95 [Aa]	18.30 ± 1.21 [Cc]	93.61 ± 4.96 [Bb]	90.52 ± 8.31 [Bc]	8.77 ± 0.67 [Cc]	11.58 ± 0.34 [Cb]
2-Nonenal	NF	2.85 ± 0.21 [Db]	5.95 ± 0.54 [Bb]	7.05 ± 1.08 [Ab]	6.01 ± 0.49 [Bb]	7.13 ± 0.31 [Aa]	4.04 ± 0.17 [Ca]
	SF	2.60 ± 0.29 [Db]	10.97 ± 0.88 [Ba]	23.27 ± 3.46 [Aa]	6.68 ± 0.52 [Ca]	1.84 ± 0.28 [Db]	3.57 ± 0.26 [Db]
	ST	6.62 ± 0.68 [Ba]	1.49 ± 0.14 [Dc]	2.89 ± 0.24 [Cc]	2.97 ± 0.28 [Cc]	7.44 ± 0.76 [Aa]	1.71 ± 0.21 [Dc]
2-Decenal	NF	14.38 ± 0.84 [Db]	62.14 ± 3.81 [Ca]	117.49 ± 6.83 [Aa]	82.98 ± 8.66 [Bb]	7.39 ± 0.73 [Da]	6.49 ± 0.75 [Da]
	SF	15.83 ± 2.52 [Db]	63.46 ± 2.84 [Ba]	46.77 ± 4.38 [Cb]	99.53 ± 8.95 [Aa]	5.14 ± 0.34 [Eb]	5.70 ± 0.33 [Eb]
	ST	38.45 ± 4.53 [Aa]	8.19 ± 1.00 [Cb]	28.32 ± 1.55 [Bc]	29.78 ± 3.33 [Bc]	5.79 ± 0.49 [Cb]	4.99 ± 0.24 [Cb]
2,4-Decadienal	NF	3.86 ± 0.39 [Dc]	15.30 ± 1.79 [Cb]	26.54 ± 2.53 [Ba]	35.84 ± 4.29 [Aa]	28.55 ± 1.54 [Bb]	25.66 ± 4.09 [Ba]
	SF	4.73 ± 0.57 [Db]	30.18 ± 2.57 [Ca]	25.08 ± 2.05 [Ca]	31.94 ± 2.91 [Ba]	36.25 ± 5.58 [Aa]	29.43 ± 2.05 [Ca]
	ST	9.32 ± 0.39 [Aa]	4.52 ± 0.63 [Cc]	8.22 ± 0.68 [Bb]	7.51 ± 0.43 [Bb]	4.56 ± 0.77 [Cc]	4.04 ± 0.31 [Cb]
2,4-Dodecadienal	NF	18.98 ± 0.98 [Bb]	89.33 ± 7.75 [Aa]	53.79 ± 4.28 [Ba]	82.05 ± 6.75 [Ab]	5.47 ± 0.61 [Da]	3.36 ± 0.21 [Db]
	SF	19.27 ± 1.99 [Bb]	12.24 ± 1.76 [Cb]	11.75 ± 1.50 [Cc]	118.49 ± 8.87 [Aa]	3.27 ± 0.25 [Db]	3.32 ± 0.25 [Db]
	ST	62.12 ± 2.93 [Aa]	9.22 ± 0.84 [Cb]	38.86 ± 5.48 [Bb]	38.38 ± 4.03 [Bc]	5.62 ± 0.75 [Ca]	5.11 ± 0.35 [Ca]

Table 2. *Cont.*

Volatile Flavor Compounds	Groups	0 d	1 d	7 d	14 d	21 d	28 d
1-Hexanol	NF	9.59 ± 0.76 [Ba]	3.78 ± 0.38 [Db]	11.34 ± 1.07 [Aa]	7.47 ± 1.03 [Ca]	2.89 ± 0.20 [Ea]	2.14 ± 0.19 [Ea]
	SF	2.96 ± 0.10 [Cb]	4.40 ± 0.30 [Ba]	3.82 ± 0.36 [Bb]	5.17 ± 0.44 [Ab]	3.08 ± 0.23 [Ca]	0.98 ± 0.46 [Db]
	ST	1.38 ± 0.22 [Bc]	2.06 ± 0.34 [Ac]	2.12 ± 0.22 [Ac]	2.15 ± 0.28 [Ac]	2.34 ± 0.31 [Ab]	1.39 ± 0.20 [Bb]
1-Heptanol	NF	3.75 ± 0.29 [Cc]	19.05 ± 2.20 [Bb]	27.14 ± 2.97 [Aa]	16.08 ± 1.87 [Bb]	3.83 ± 0.43 [Ca]	1.76 ± 0.18 [Cc]
	SF	12.06 ± 1.25 [Ca]	25.44 ± 2.56 [Ba]	23.68 ± 3.67 [Ba]	46.03 ± 3.55 [Aa]	2.88 ± 0.19 [Db]	3.00 ± 0.19 [Db]
	ST	5.27 ± 1.04 [Bb]	17.58 ± 1.56 [Ab]	3.30 ± 0.55 [Cb]	3.26 ± 0.24 [Cc]	3.63 ± 0.37 [Ca]	3.99 ± 0.63 [Ca]
1-Octanol	NF	6.35 ± 0.37 [Db]	15.24 ± 1.63 [Ba]	18.52 ± 2.35 [Aa]	12.58 ± 1.73 [Cb]	7.13 ± 0.21 [Da]	4.78 ± 0.41 [Da]
	SF	6.06 ± 0.66 [Cb]	15.99 ± 1.33 [Ba]	19.37 ± 1.25 [Aa]	15.96 ± 1.71 [Ba]	3.28 ± 0.28 [Db]	2.44 ± 0.22 [Db]
	ST	9.88 ± 0.62 [Aa]	3.41 ± 0.24 [Cb]	7.92 ± 0.93 [Bb]	7.83 ± 0.91 [Bc]	1.85 ± 0.11 [Dc]	1.86 ± 0.19 [Dc]
1-Octen-3-ol	NF	78.13 ± 2.42 [Cab]	109.64 ± 6.39 [Aa]	89.92 ± 5.07 [Ba]	60.79 ± 4.45 [Db]	7.96 ± 0.77 [Eb]	6.77 ± 0.71 [Eb]
	SF	72.55 ± 4.8 [Cb]	91.86 ± 10.26 [Ab]	84.32 ± 8.81 [Ba]	76.48 ± 4.86 [Ca]	10.45 ± 0.67 [Da]	10.66 ± 1.05 [Da]
	ST	85.34 ± 7.21 [Aa]	19.50 ± 1.85 [Cc]	51.85 ± 3.94 [Bb]	47.36 ± 3.87 [Bc]	7.31 ± 0.91 [Db]	9.31 ± 1.31 [Da]
Benzaldehyde, 4-(1-methyl lethyl)-	NF	24.14 ± 2.46 [Ab]	15.39 ± 1.17 [Bb]	11.68 ± 0.73 [Cb]	8.17 ± 1.17 [Dc]	7.56 ± 0.35 [Db]	6.54 ± 0.58 [Dc]
	SF	31.69 ± 2.50 [Ca]	101.55 ± 10.90 [Aa]	83.26 ± 11.65 [Ba]	16.57 ± 2.12 [Da]	15.98 ± 2.35 [Da]	15.48 ± 1.19 [Da]
	ST	15.34 ± 0.66 [Bc]	19.08 ± 0.90 [Ab]	14.19 ± 0.80 [Bb]	13.75 ± 1.24 [Bb]	18.45 ± 2.00 [Aa]	15.89 ± 1.82 [Ba]
2-Heptanone	NF	2.82 ± 0.16 [Db]	6.97 ± 0.65 [Aa]	4.99 ± 0.44 [Ba]	3.19 ± 0.44 [Db]	2.56 ± 0.24 [Db]	3.50 ± 0.24 [Cb]
	SF	2.86 ± 0.20 [Cb]	6.10 ± 0.86 [Aa]	5.45 ± 0.71 [Aa]	4.43 ± 0.48 [Ba]	2.86 ± 0.19 [Cab]	2.62 ± 0.34 [Cc]
	ST	4.42 ± 0.53 [Ba]	3.86 ± 0.55 [Cb]	4.16 ± 0.34 [Bb]	3.79 ± 0.38 [Cab]	3.33 ± 0.45 [Ca]	5.51 ± 0.32 [Aa]
2-Nonanone	NF	4.21 ± 0.35 [Cb]	10.79 ± 0.40 [Ba]	4.37 ± 0.38 [Cb]	37.34 ± 3.70 [Aa]	3.01 ± 0.31 [Ca]	3.64 ± 0.38 [Ca]
	SF	4.73 ± 0.68 [Bb]	5.78 ± 0.33 [Ab]	5.44 ± 0.32 [Aa]	3.86 ± 0.34 [Cb]	1.63 ± 0.19 [Eb]	2.36 ± 0.27 [Db]
	ST	13.77 ± 1.44 [Aa]	3.70 ± 0.33 [Bc]	2.41 ± 0.21 [Ca]	2.50 ± 0.15 [Cb]	2.46 ± 0.25 [Cb]	2.47 ± 0.17 [Cb]
2-Tridecanone	NF	16.31 ± 2.13 [Aa]	10.73 ± 1.03 [Ca]	11.46 ± 0.89 [Ca]	12.15 ± 1.02 [Bb]	9.58 ± 0.58 [Ca]	9.40 ± 0.94 [Ca]
	SF	17.94 ± 3.11 [Aa]	8.72 ± 0.32 [Bb]	6.64 ± 0.57 [Bb]	17.60 ± 0.69 [Aa]	1.76 ± 0.16 [Cc]	2.39 ± 0.25 [Cb]
	ST	3.92 ± 0.28 [Ab]	2.60 ± 0.13 [Bc]	2.09 ± 0.21 [Dc]	2.06 ± 0.18 [Dc]	2.65 ± 0.13 [Bb]	2.44 ± 0.24 [Cb]
Ethyl octanoate	NF	6.89 ± 0.56 [Db]	10.40 ± 1.07 [Ab]	8.71 ± 0.88 [Cb]	7.66 ± 0.76 [Db]	10.02 ± 1.00 [Ba]	10.05 ± 1.19 [Bb]
	SF	11.31 ± 1.00 [Ba]	116.18 ± 9.85 [Aa]	106.08 ± 14.01 [Aa]	12.17 ± 0.85 [Ba]	10.29 ± 0.76 [Ba]	12.54 ± 1.65 [Ba]
	ST	12.36 ± 1.30 [Aa]	5.14 ± 0.48 [Cb]	3.09 ± 0.29 [Bb]	3.13 ± 0.17 [Bc]	3.97 ± 0.39 [Bb]	6.41 ± 0.62 [Bc]
Ethyl nonanoate	NF	5.73 ± 0.51 [Bc]	3.77 ± 0.31 [Cc]	11.17 ± 0.70 [Aa]	5.38 ± 0.36 [Bc]	2.25 ± 0.13 [Dc]	2.13 ± 0.21 [Da]
	SF	6.73 ± 0.52 [Db]	11.74 ± 1.08 [Ba]	7.13 ± 1.04 [Db]	14.10 ± 1.33 [Aa]	7.64 ± 0.60 [Db]	9.93 ± 0.81 [Ca]
	ST	9.42 ± 0.57 [Ba]	8.42 ± 0.68 [Db]	7.26 ± 0.61 [Db]	7.05 ± 0.56 [Db]	10.47 ± 1.00 [Aa]	8.50 ± 1.18 [Cb]
Ethyl tridecanoate	NF	17.15 ± 1.52 [Cb]	18.58 ± 2.26 [Cb]	23.86 ± 1.31 [Ba]	20.60 ± 1.95 [Ca]	26.21 ± 2.73 [Aa]	22.42 ± 2.48 [Ba]
	SF	30.55 ± 3.33 [Aa]	23.34 ± 1.86 [Ca]	18.94 ± 0.75 [Cb]	22.21 ± 2.14 [Ca]	22.84 ± 3.34 [Cab]	25.11 ± 3.45 [Ba]
	ST	14.43 ± 1.02 [Cb]	18.26 ± 1.68 [Bb]	12.05 ± 1.39 [Dc]	12.18 ± 0.85 [Db]	20.44 ± 1.18 [Bb]	21.20 ± 1.61 [Aa]
Phenol	NF	10.92 ± 0.82 [Cb]	3.60 ± 0.11 [Dc]	67.85 ± 7.38 [Aa]	16.75 ± 0.90 [Bb]	7.51 ± 0.84 [Db]	11.42 ± 1.12 [Ca]
	SF	27.03 ± 2.96 [Ba]	15.13 ± 0.90 [Cb]	17.51 ± 0.97 [Cb]	20.69 ± 2.92 [Ca]	69.03 ± 6.81 [Aa]	7.31 ± 0.67 [Db]
	ST	8.49 ± 0.89 [Bb]	19.91 ± 2.43 [Aa]	5.49 ± 0.75 [Cc]	5.61 ± 0.55 [Cc]	7.73 ± 0.38 [Bb]	5.33 ± 0.88 [Cc]
D-Limonene	NF	20.07 ± 2.83 [Ab]	16.36 ± 1.69 [Cb]	13.27 ± 1.24 [Cb]	13.52 ± 1.12 [Cb]	18.83 ± 2.10 [Bb]	20.61 ± 1.55 [Ab]
	SF	30.01 ± 3.60 [Aa]	14.45 ± 0.57 [Db]	11.73 ± 0.42 [Db]	15.89 ± 0.63 [Ca]	23.65 ± 1.49 [Ba]	23.32 ± 1.25 [Bb]
	ST	15.90 ± 1.58 [Cb]	28.29 ± 2.23 [Aa]	16.29 ± 1.70 [Ca]	16.50 ± 2.23 [Ca]	22.94 ± 2.51 [Ba]	29.50 ± 3.64 [Aa]

Table 2. *Cont.*

Volatile Flavor Compounds	Groups	0 d	1 d	7 d	14 d	21 d	28 d
Caryophyllene	NF	37.72 ± 5.8 [Cb]	34.58 ± 3.23 [Db]	35.88 ± 1.94 [Da]	29.96 ± 2.68 [Db]	48.19 ± 5.27 [Aa]	44.27 ± 3.31 [Bb]
	SF	69.13 ± 5.06 [Aa]	33.70 ± 2.69 [Db]	19.17 ± 2.87 [Eb]	43.10 ± 4.25 [Ca]	32.26 ± 3.37 [Db]	51.46 ± 6.40 [Ba]
	ST	33.36 ± 2.74 [Bb]	48.43 ± 5.83 [Aa]	21.31 ± 3.01 [Cb]	22.20 ± 1.49 [Cc]	44.45 ± 4.27 [Aa]	49.58 ± 3.63 [Aab]
3-Ethyl-2-methyl-1,3-hexadiene	NF	7.15 ± 0.73 [Cb]	21.53 ± 1.69 [Aa]	23.63 ± 2.41 [Aa]	17.29 ± 1.82 [Bb]	16.85 ± 1.64 [Bb]	1.99 ± 0.12 [Db]
	SF	7.77 ± 0.35 [Cb]	20.54 ± 3.02 [Ba]	24.07 ± 3.43 [Ba]	20.59 ± 1.98 [Ba]	27.28 ± 1.44 [Aa]	2.56 ± 0.26 [Da]
	ST	13.71 ± 1.55 [Ba]	12.19 ± 0.60 [Bb]	11.67 ± 1.69 [Bb]	12.87 ± 1.29 [Bc]	14.85 ± 1.06 [Ab]	2.54 ± 0.33 [Ca]
Oxime-, methoxy-phenyl-	NF	5.15 ± 0.25 [Da]	20.62 ± 10.00 [Bb]	33.58 ± 5.09 [Ab]	19.75 ± 1.10 [Bb]	12.80 ± 1.27 [Cc]	2.02 ± 0.16 [Da]
	SF	5.48 ± 0.51 [Ea]	31.65 ± 3.70 [Aa]	20.60 ± 2.20 [Bc]	11.00 ± 1.24 [Dc]	16.16 ± 0.47 [Cb]	2.04 ± 0.30 [Fa]
	ST	2.47 ± 0.14 [Db]	14.82 ± 0.86 [Ca]	51.84 ± 10.70 [Aa]	48.70 ± 7.11 [Aa]	25.12 ± 3.61 [Ba]	1.54 ± 0.18 [Db]
Anethole	NF	25.04 ± 3.31 [Db]	27.23 ± 3.86 [Da]	32.26 ± 2.46 [Ca]	33.09 ± 2.01 [Bb]	31.64 ± 2.35 [Cab]	39.78 ± 3.98 [Aa]
	SF	50.17 ± 7.23 [Aa]	10.29 ± 1.20 [Cb]	6.65 ± 0.65 [Cc]	39.67 ± 4.36 [Ba]	35.05 ± 2.42 [Ba]	37.36 ± 2.72 [Ba]
	ST	26.89 ± 2.45 [Bb]	26.06 ± 1.91 [Ba]	17.86 ± 2.10 [Cb]	18.89 ± 1.60 [Cc]	30.51 ± 2.28 [Ab]	27.59 ± 1.38 [Bb]
Pentadecane	NF	6.52 ± 0.78 [Aa]	4.25 ± 0.39 [Ca]	4.17 ± 0.48 [Ca]	3.89 ± 0.19 [Cb]	5.78 ± 0.85 [Bb]	5.33 ± 0.61 [Ba]
	SF	7.30 ± 0.68 [Aa]	4.94 ± 0.47 [Ca]	4.57 ± 0.24 [Ca]	5.13 ± 0.50 [Ca]	5.48 ± 0.60 [Bb]	5.51 ± 0.43 [Ba]
	ST	4.50 ± 0.62 [Bb]	4.65 ± 0.60 [Ba]	3.30 ± 0.33 [Cb]	3.54 ± 0.25 [Cb]	7.03 ± 0.83 [Aa]	5.25 ± 0.31 [Ba]

Different capital letters in the same row indicate significant differences in the same volatile flavor compound under different storage times within the same group; lowercase letters in the same column indicate significant differences among the three groups for the same volatile flavor at the same storage time.

4. Discussion

Fermentation is a key process in the production of fermented lamb liver paste and can give the product a special fermentation aroma and play a certain role in removing the muttony taste. There are few reports on the application of fermentation technology to liver products, and systematic research on the formulation and process is scarce. Mokhtar [21] compounded *Lactobacillus plantarum*, *Bifidobacterium lactis*, and *Bifidobacterium bifidum* into a starter culture, applied it to the production of sausages, and found that the contents of tyramine, putrescine, cadaverine, and tryptamine were significantly reduced. Zang [22] studied the mixed starter culture to help improve the flavor quality of traditional Chinese fermented fish. The Cenci-Goga [23] study found that the addition of dairy starter cultures and commercial probiotics inhibited the growth of undesirable microorganisms in salami and improved its sensory properties. Wang [24] used high-throughput sequencing technology to determine the bacterial community in sausages, dry-cured sausages, and smoked sausages and found that the main microorganisms in fermented sausages were Staphylococcus and Lactobacillus. Cano Garxia [25] found that yeast decomposes proteins and fats in dry fermented sausages to produce phenols and alcohols. Alcohols can react with lactic acid produced by lactic acid bacteria to give sausages a certain ester flavor. Antara [11] found that, after mixing *Lactobacillus plantarum* and *Pediococcus lactis* into sausages, the shelf life of fermented sausages was prolonged by metabolizing lactic acid to control enterotoxins. The main fermenting bacteria in fermented lamb liver paste are lactic acid bacteria, which participate in the metabolism of various substances during the fermentation process.

In the storage of lamb liver paste, the total number of colonies and lactic acid bacteria were higher in the NF and SF groups than in the ST group. On the one hand, this may have been due to the higher protein content of the NF group and SF group providing sufficient nitrogen sources for microbial growth and reproduction, and on the other, it may have been due to the addition of the starter culture increasing the total number of initial microorganisms [26]. With the increase in storage time, the total number of colonies in each group exhibited an upward trend. The NF group and the SF group exhibited the same trend.

The ST group underwent high-temperature treatment, which inactivated most of the heat-labile microorganisms. The microorganisms in the NF group and the SF group both grew and reproduced well under suitable fermentation conditions, increasing the number of microorganisms. After the 7th day of storage, the total number of colonies in the NF and SF groups was similar. However, after 14 days of storage, the TVB-N of the NF group continued to increase, and the highest degrees of amino acid destruction and lowest nutritional value were found in the lamb liver paste of the NF group at 28 days of storage. In conclusion, starter culture is of great significance to the quality of fermented meat products. In this research, the fermented lamb liver paste was made with lamb liver as the main raw material, and the paste products were made by adding an emulsifier, thickener, seasoning, other auxiliary materials, and starter culture. Probiotic products promote nutrient absorption in the small intestine, affect immune homeostatic cell signaling pathways in the intestinal mucosa, inhibit pathogenic bacteria, and improve intestinal health [27]. Therefore, a new type of lamb liver fermented product was developed in this experiment, which provides a theoretical basis for the high-value utilization of lamb liver.

Microorganisms decompose sugars to produce acid, which reduces the pH of meat products and prevents the growth of spoilage bacteria [13]. The pH value of the ST group was stable between 6.1 and 6.2. After the product is sterilized, a large number of microorganisms are inactivated, which does not affect the pH value of the product. The product was vacuum packed and stored at 4 °C. The environmental humidity of lamb liver paste storage was unchanged, and the Aw value of the three groups of lamb liver paste was not significantly different at any point. The L*, a*, and b* in the SF group were higher than in the ST group, which may be related to the effect of microorganisms in the fermentation process. With the increase in storage time, an upward trend was observed, with the SF group exhibiting significantly higher values than the other two groups. This situation shows that the use of leavening agents promotes the formation of product color. The microbes in the starter culture break down the proteins in the liver paste. The decreased adhesiveness of liver paste during storage is the result of microbial action. The ST group was treated at high temperature and pressure, and most of the microorganisms were inactivated, which also inactivated and denatured most of the proteins. This ultimately led to increased interstructural forces in the liver paste and the increased adhesion of the paste in the ST group.

The TVB-N and TBARS values in the SF group were lower than in the NF groups on the 1st day and 28th day, indicating that the fermentation technology ensures the quality of the product within 1 month of storage. The change in the TVB-N value in the NF group may have been caused by the mutual inhibition of bacteria in the early stage and the decrease in bacterial activity in the later stage. The analysis of the results on the 28th day showed that the addition of starter culture reduced the TVB-N value and improved the safety of lamb liver paste. There was no significant change in the ST group, indicating that the oxidation of fat was inhibited by autoclaving. In addition, lactic acid bacteria in the starter culture may also inhibit the production of malondialdehyde from peroxides and reduce fat oxidation. The addition of starter culture improved the freshness of lamb liver paste.

The protein content of the three groups of lamb liver paste was generally stable, i.e., remaining between 10 and 12%. This indicated that, as compared with fresh lamb liver (23.26%), the effect of 1-month storage time on protein content was relatively small. It is possible that the protein content of the lamb liver paste was reduced after the addition of water, emulsifiers, and thickeners. Studies have shown that under the dual decomposition of endogenous enzymes and microbial enzymes, protein degradation produces free amino acids as a taste substance [28]. The protein content of animal liver is lower as compared to the muscle [29]. Therefore, fewer proteins were available for microbial decomposition during the production of fermented products, resulting in insignificant changes in the protein content of the fermented liver paste. The low protein content of the ST group may have been caused by the destruction of the protein structure by autoclaving. The higher content of naturally fermented histones may have been due to the catabolism of proteins by miscellaneous bacteria. The fat content of the NF group, the SF group, and the ST group changed from 3% to 4%, from 3.7% to 5.5%, and from 3% to 7%, respectively. Except for the 21st day, the fat content of the NF group and SF group was significantly lower than that of the ST group, probably due to the consumption of the carbon source by microorganisms during fermentation.

Liver-endogenous enzymes degrade fat and protein to produce free fatty acids, amino acids, and volatile flavor substances, which not only improve the nutritional value but also give fermented meat products a unique flavor [2,30]. Lactic acid bacteria are the focus of microbial starter culture screening for fermented meat products. They use carbohydrate fermentation to produce by-products, such as acetic acid, formic acid, and succinic acid, which have a certain positive effect on the flavor of fermented meat products. The results of the flavor analysis revealed several flavor compounds that we were concerned about during the storage period [8,31]. Aldehydes give lamb liver paste an almond and sweet aroma. Benzaldehyde is one of the representatives of the aromatic aldehydes that has cherry and nut aromas. Octanal, decanal, 2,4-decadienal, and tetradecanal have a gentle oily, slightly citrus, and iris-like aroma, with an aroma strength value of 2 but with a short duration [32]. The key aroma compounds in pork soup—2-undecenal and 2-octenaland hexanal—were characterized in the directional aroma analysis, which showed that the meat aroma was stronger in the SF group. Bai Shuang [33] found in an experiment on the formation mechanism of volatile compounds in fried mutton that aldehydes, alcohols, and esters produced by fat oxidation during the frying process were the main sources of volatile compounds. Alcohols are one of the main aroma compounds of fermented lamb liver paste. In addition, 1-Hexanol has an herbaceous aroma and also affects actin interactions [34]. Consistent with changes in hydrophobic interactions at the interface of myosin and actin in transition from a weakly to strongly bound state, 1-hexanol accelerated Pi release from myosin [35]. In addition, 1-heptanol also has a relatively strong fruit aroma and exhibits a binding ability to the myofibrillar protein, which can make the meat have a stronger flavor [36]. Furthermore, 1-Octen-3-ol has an important contribution to meat aroma. Mevalonate, a key substance in cholesterol synthesis, leads to a decrease in corticosterone content, affects ZNF414 and KLF15 gene expression, and positively regulates 1-octen-3-ol production in chicken [37]. The platycodon grandiflorum extracts anethole (40.27%) and 4-methoxy benzaldehyde (4.25%) can activate the acquired immune response and are beneficial to the human body [38]. Ketones are also part of the aroma of lamb liver paste. The fermented aroma compound 2-nonanone has a pleasing aroma and acts as a pheromone component to improve olfactory learning through persistent modulation of appetite motivation [39]. Limonene is the most common terpene in nature and a major constituent of several citrus oils (orange, lemon, mandarin, lime, and grapefruit). As a solvent for cholesterol, limonene has been clinically used to dissolve cholesterol-containing gallstones [40]. The limonene in the ST group was significantly higher than in the other two groups ($p < 0.05$). Caryophyllene is a class of bicyclic sesquiterpenes with cloves and turpentine aroma notes as functional food factors [41]. It is naturally found in lemon, nutmeg, and cinnamon leaf oils [42]. Studies have shown that exposure to volatile BCP in mice is detectable in the lung, olfactory bulb, brain, serum, heart, liver, kidney, epididymal adipose, and brown adipose tissue. Furthermore, inhaled volatile BCP is widely distributed in mouse tissues and affects the kinetics of metabolites in the liver [43]. Caryophyllene in the SF group and NF group was significantly higher than in the ST group ($p < 0.05$). As a new type of liver paste product, fermented lamb liver paste has a long storage period. In future research, the effect of functional substances on fermented lamb liver paste can be explored to improve its quality and nutritional value.

5. Conclusions

In conclusion, the probiotic fermented lamb liver paste in this study represents a high-value sheep by-product. The freshness, the number of lactic acid bacteria, and the L*, a*, and b* values were higher in the starter fermentation group, and the fat content was lower. The lower pH in the early storage period of the starter fermentation group acted to inhibit the growth of microorganisms to a certain extent and prolong the storage period of fermented lamb liver paste. Volatile aroma compounds, such as aldehydes, 1-octen-3-ol, anethole, and 2-nonanone, as detected by GC–MS, contributed to the aroma composition of the fermented lamb liver paste.

Supplementary Materials: The following supporting information can be downloaded at: https://www.mdpi.com/article/10.3390/fermentation8120676/s1, Table S1: Changes of the total colony number in lamb liver paste with three different treatments during storage period; Table S2: Changes of the lactic acid bacteria colony count in lamb liver paste with three different treatments during storage; Table S3: Changes of pH value of lamb liver paste with three different treatments during storage period; Table S4: Changes of Aw value of lamb liver paste with three different treatments during storage period; Table S5: Changes of L* value of lamb liver paste with three different treatments during storage period; Table S6: Changes of a* value of lamb liver paste with three different treatments during storage period; Table S7: Changes of b* value of lamb liver paste with three different treatments during storage period; Table S8: Changes of TVB-N value in lamb liver paste with three different treatments during storage period; Table S9: Changes of TBARS value in lamb liver paste with three different treatments during storage period; Table S10: Changes of moisture content in lamb liver paste with three different treatments during storage period; Table S11: Changes of protein content in lamb liver paste with three different treatments during storage period; Table S12: Changes of fat content in lamb liver paste with three different treatments during storage period.

Author Contributions: Conceptualization, T.L. and Y.D.; formal analysis, L.Y. (Le Yang), T.Z. and Y.Z. (Yanni Zhang); investigation, L.K., Y.Z. (Yuji Zhai) and L.Y. (Lirong Yang); resources, Y.D.; data curation, Y.J.; writing—original draft preparation, T.L. and L.Y. (Lirong Yang); writing—review and editing, T.L., T.Z. and L.Z.; project administration, Y.D.; funding acquisition, Y.D. All authors have read and agreed to the published version of the manuscript.

Funding: This research was funded by the National Natural Science Foundation of China (32160589); the Inner Mongolia Natural Science Foundation Program of China (No. 2021MS03012).

Institutional Review Board Statement: Not applicable.

Informed Consent Statement: Not applicable.

Data Availability Statement: Data are contained within the article and Supplementary Materials.

Conflicts of Interest: The authors declare no conflict of interest.

References

1. Kozeniecki, M.; Ludke, R.; Kerner, J.; Patterson, B. Micronutrients in Liver Disease: Roles, Risk Factors for Deficiency, and Recommendations for Supplementation. *Nutr. Clin. Pract.* **2020**, *35*, 50–62. [CrossRef] [PubMed]
2. Hou, Y.; Hu, S.; Li, X.; He, W.; Wu, G. Amino Acid Metabolism in the Liver: Nutritional and Physiological Significance. *Adv. Exp. Med. Biol.* **2020**, *1265*, 21–37. [PubMed]
3. Tanumihardjo, S.A. Vitamin A: Biomarkers of nutrition for development. *Am. J. Clin. Nutr.* **2011**, *94*, 658S–665S. [CrossRef] [PubMed]
4. Trefts, E.; Gannon, M.; Wasserman, D.H. The liver. *Curr. Biol.* **2017**, *27*, 1147–1151. [CrossRef] [PubMed]
5. Cornelius, C.E. A review of new approaches to assessing hepatic function in animals. *Vet. Res. Commun.* **1987**, *11*, 423–441. [CrossRef] [PubMed]
6. Wang, J.; Ji, H. Influence of Probiotics on Dietary Protein Digestion and Utilization in the Gastrointestinal Tract. *Curr. Protein Pept. Sci.* **2019**, *20*, 125–131. [CrossRef]
7. Ordóñez, J.A.; Hierro, E.M.; Bruna, J.M.; de la Hoz, L. Changes in the components of dry-fermented sausages during ripening. *Crit. Rev. Food Sci. Nutr.* **1999**, *39*, 329–367. [CrossRef]
8. Berdagué, J.L.; Monteil, P.; Montel, M.C.; Talon, R. Effect of starter cultures on the formation of flavor compounds in dry sausage. *Meat Sci.* **1993**, *35*, 275–287. [CrossRef]
9. Ardö, Y. Flavour formation by amino acid catabolism. *Biotechnol. Adv.* **2006**, *24*, 238–242. [CrossRef]
10. Casaburi, A.; Aristoy, M.C.; Cavella, S.; Di Monaco, R.; Ercolini, D.; Toldrá, F.; Villani, F. Biochemical and sensory characteristics of traditional fermented sausages of Vallo di Diano (Southern Italy) as affected by the use of starter cultures. *Meat Sci.* **2007**, *76*, 295–307. [CrossRef]

11. Antara, N.S.; Sujaya, I.N.; Yokota, A.; Asano, K.; Tomita, F. Effects of indigenous starter cultures on the microbial and physicochemical characteristics of Urutan, a Balinese fermented sausage. *J. Biosci. Bioeng.* **2004**, *98*, 92–98. [CrossRef] [PubMed]
12. Xing, T.K.; Shi, T.; Michael, G. Effect of starter cultures on taste-active amino acids and survival of *Escherichia coli* in dry fermented beef sausages. *Eur. Food Res. Technol.* **2018**, *244*, 2203–2212.
13. García-Díez, J.; Saraiva, C. Use of Starter Cultures in Foods from Animal Origin to Improve Their Safety. *Int. J. Environ. Res. Public Health* **2021**, *18*, 2544. [CrossRef]
14. *GOST 12319-77*; The Standard Applies to Canned Meat "Liver Pate", Packed in Cans, Hermetically Sealed and Sterilized. Russian Gost. 2012. Available online: https://www.russiangost.com/ (accessed on 9 October 2022).
15. Perea-Sanz, L.; Peris, D.; Belloch, C.; Flores, M. *Debaryomyces hansenii* metabolism of sulfur amino acids as precursors of volatile sulfur compounds of interest in meat products. *J. Agric. Food Chem.* **2019**, *67*, 9335–9343. [CrossRef] [PubMed]
16. Luo, Y.; Zhao, L.; Xu, J.; Su, L.; Jin, Z.; Su, R.; Jin, Y. Effect of fermentation and postcooking procedure on quality parameters and volatile compounds of beef jerky. *Food Sci. Nutr.* **2020**, *8*, 2316–2326. [CrossRef]
17. Omana, D.A.; Moayedi, V.; Xu, Y.; Betti, M. Alkali-aided protein extraction from chicken dark meat: Textural properties and color characteristics of recovered proteins. *Poult. Sci.* **2010**, *89*, 1056–1064. [CrossRef]
18. Liang, C.; Zhang, D.; Zheng, X.; Wen, X.; Yan, T.; Zhang, Z.; Hou, C. Effects of Different Storage Temperatures on the Physicochemical Properties and Bacterial Community Structure of Fresh Lamb Meat. *Food Sci. Anim. Resour.* **2021**, *41*, 509–526. [CrossRef]
19. Gong, S.; Jiao, C.; Guo, L. Antibacterial mechanism of beetroot (*Beta vulgaris*) extract against *Listeria monocytogenes* through apoptosis-like death and its application in cooked pork. *LWT-Food Sci. Technol.* **2022**, *165*, 113711. [CrossRef]
20. Luo, Y.; Wang, B.; Liu, C.; Su, R.; Hou, Y.; Yao, D.; Zhao, L.; Su, L.; Jin, Y. Meat quality, fatty acids, volatile compounds, and antioxidant properties of lambs fed pasture versus mixed diet. *Food Sci. Nutr.* **2019**, *7*, 2796–2805. [CrossRef]
21. Mokhtar, S.; Mostafa, G.; Taha, R.; Eldeep, G.S.S. Effect of different starter cultures on the biogenic amines production as a critical control point in fresh fermented sausages. *Eur. Food Res. Technol.* **2012**, *235*, 527–535. [CrossRef]
22. Zang, J.; Xu, Y.; Xia, W.; Regenstein, J.M.; Yu, D.; Yang, F.; Jiang, Q. Correlations between microbiota succession and flavor formation during fermentation of Chinese low-salt fermented common carp (*Cyprinus carpio* L.) inoculated with mixed starter cultures. *Food Microbiol.* **2020**, *90*, 103487. [CrossRef] [PubMed]
23. Cenci-Goga, B.T.; Karama, M.; Sechi, P.; Iulietto, M.F.; Novelli, S.; Selvaggini, R.; Barbera, S. Effect of a novel starter culture and specific ripening conditions on microbiological characteristics of nitrate-free dry-cured pork sausages. *J. Anim. Sci.* **2016**, *15*, 358–374. [CrossRef]
24. Wang, X.; Zhang, Y.; Ren, H.; Zhan, Y. Comparison of bacterial diversity profiles and microbial safety assessment of salami, Chinese dry-cured sausage and Chinese smoked-cured sausage by high-through put sequencing. *LWT-Food Sci. Technol.* **2018**, *90*, 108–115. [CrossRef]
25. Cano-Garcia, L.; Belloch, C.; Flores, M. Impact of Debaryomyces hansenii strains inoculation on the quality of slow dry-cured fermented sausages. *Meat Sci.* **2014**, *96*, 1469–1477. [CrossRef]
26. Kim, S.; Lee, J.Y.; Jeong, Y.; Kang, C.-H. Antioxidant Activity and Probiotic Properties of Lactic Acid Bacteria. *Fermentation* **2022**, *8*, 29. [CrossRef]
27. Yan, F.; Polk, D.B. Probiotics and Probiotic-Derived Functional Factors-Mechanistic Insights into Applications for Intestinal Homeostasis. *Front. Immunol.* **2020**, *11*, 1428. [CrossRef]
28. Rui, L. Energy metabolism in the liver. *Compr. Physiol.* **2014**, *4*, 177–197.
29. Koshland, D.E.; Haurowitz, F. Protein. Encyclopedia Britannica. 2022. Available online: https://www.britannica.com/science/protein (accessed on 18 October 2022).
30. Stadtman, E.R.; Levine, R.L. Free radical-mediated oxidation of free amino acids and amino acid residues in proteins. *Amino Acids* **2003**, *25*, 207–218. [CrossRef]
31. Yu, H.; Xie, T.; Xie, J.; Chen, C.; Ai, L.; Tian, H. Aroma perceptual interactions of benzaldehyde, furfural, and vanillin and their effects on the descriptor intensities of Huangjiu. *Food Res. Int.* **2020**, *129*, 108808. [CrossRef]
32. Högnadóttir, A.; Rouseff, R.L. Identification of aroma active compounds in orange essence oil using gas chromatography-olfactometry and gas chromatography-mass spectrometry. *J. Chromatogr. A* **2003**, *998*, 201–211. [CrossRef]
33. Bai, S.; Wang, Y.; Luo, R.; Shen, F.; Bai, H.; Ding, D. Formation of flavor volatile compounds at different processing stages of household stir-frying mutton Sao Zi in the northwest of China. *LWT-Food Sci. Technol.* **2020**, *139*, 110735. [CrossRef]
34. Petronilho, S.; Lopez, R.; Ferreira, V.; Coimbra, M.A.; Rocha, S.M. Revealing the Usefulness of Aroma Networks to Explain Wine Aroma Properties: A Case Study of Portuguese Wines. *Molecules* **2020**, *25*, 272. [CrossRef] [PubMed]
35. Kołakowski, J.; Karkucińska, A.; Dabrowska, R. Calponin inhibits actin-activated MgA.TPase of myosin subfragment 1 (S1) without displacing S1 from its binding site on actin. *Eur. J. Biochem.* **1997**, *243*, 624–629. [CrossRef] [PubMed]
36. Wang, H.; Xia, X.; Yin, X.; Liu, H.; Chen, Q.; Kong, B. Investigation of molecular mechanisms of interaction between myofibrillar proteins and 1-heptanol by multiple spectroscopy and molecular docking methods. *Int. J. Biol. Macromol.* **2021**, *193 Pt A*, 672–680. [CrossRef]
37. Jin, Y.; Yuan, X.; Liu, J.; Wen, J.; Cui, H.; Zhao, G. Inhibition of cholesterol biosynthesis promotes the production of 1-octen-3-ol through mevalonic acid. *Food Res. Int.* **2022**, *158*, 111392. [CrossRef] [PubMed]

38. Peng, W.; Lin, Z.; Wang, L.; Chang, J.; Gu, F.; Zhu, X. Molecular characteristics of *Illicium verum* extractives to activate acquired immune response. *Saudi J. Biol. Sci.* **2016**, *23*, 348–352. [CrossRef]
39. Yang, W.; Wu, T.; Tu, S.; Qin, Y.; Shen, C.; Li, J.; Choi, M.K.; Duan, F.; Zhang, Y. Redundant neural circuits regulate olfactory integration. *PLoS Genet.* **2022**, *18*, 1010029. [CrossRef]
40. Sun, J. D-Limonene: Safety and clinical applications. *Altern. Med. Rev.* **2007**, *12*, 259–264.
41. Lopes, G.R.; Petronilho, S.; Ferreira, A.S.; Pinto, M.; Passos, C.P.; Coelho, E.; Rodrigues, C.; Figueira, C.; Rocha, S.M.; Coimbra, M.A. Insights on Single-Dose Espresso Coffee Capsules' Volatile Profile: From Ground Powder Volatiles to Prediction of Espresso Brew Aroma Properties. *Foods* **2021**, *10*, 2508. [CrossRef]
42. Scandiffio, R.; Geddo, F.; Cottone, E.; Querio, G.; Antoniotti, S.; Gallo, M.P.; Maffei, M.E.; Bovolin, P. Protective Effects of (E)-β-Caryophyllene (BCP) in Chronic Inflammation. *Nutrients* **2020**, *12*, 3273. [CrossRef]
43. Takemoto, Y.; Kishi, C.; Sugiura, Y.; Yoshioka, Y.; Matsumura, S.; Moriyama, T.; Zaima, N. Distribution of inhaled volatile β-caryophyllene and dynamic changes of liver metabolites in mice. *Sci. Rep.* **2021**, *11*, 1728. [CrossRef] [PubMed]

fermentation

Article

Aroma Profiles of *Vitis vinifera* L. cv. Gewürztraminer Must Fermented with Co-Cultures of *Saccharomyces cerevisiae* and Seven *Hanseniaspora* spp.

Jennifer Badura [1], Florian Kiene [1], Silvia Brezina [1], Stefanie Fritsch [1], Heike Semmler [1], Doris Rauhut [1], Isak S. Pretorius [2], Christian von Wallbrunn [1] and Niël van Wyk [1,2,*]

[1]   Department of Microbiology and Biochemistry, Hochschule Geisenheim University, Von-Lade-Strasse 1, 65366 Geisenheim, Germany
[2]   ARC Centre of Excellence in Synthetic Biology, Department of Molecular Sciences, Macquarie University, Sydney, NSW 2113, Australia
*   Correspondence: niel.wyk@hs-gm.de

Abstract: In this study, the aroma-production profiles of seven different *Hanseniaspora* strains, namely *H. guilliermondii*, *H. meyeri*, *H. nectarophila*, *H. occidentalis*, *H. opuntiae*, *H. osmophila* and *H. uvarum* were determined in a simultaneous co-inoculation with the wine yeast *Saccharomyces cerevisiae* Champagne Epernay Geisenheim (Uvaferm CEG). All co-inoculated fermentations with *Hanseniaspora* showed a dramatic increase in ethyl acetate levels except the two (*H. occidentalis* and *H. osmophila*) that belong to the so-called slow-evolving clade, which had no meaningful difference, compared to the *S. cerevisiae* control. Other striking observations were the almost complete depletion of lactic acid in mixed-culture fermentations with *H. osmophila*, the more than 3.7 mg/L production of isoamyl acetate with *H. guilliermondii*, the significantly lower levels of glycerol with *H. occidentalis* and the increase in certain terpenols, such as citronellol with *H. opuntiae*. This work allows for the direct comparison of wines made with different *Hanseniapora* spp. showcasing their oenological potential, including two (*H. meyeri* and *H. nectarophila*) previously unexplored in winemaking experiments.

Keywords: *Hanseniaspora*; mixed-starter culture fermentation; wine; aroma

Citation: Badura, J.; Kiene, F.; Brezina, S.; Fritsch, S.; Semmler, H.; Rauhut, D.; Pretorius, I.S.; von Wallbrunn, C.; van Wyk, N. Aroma Profiles of *Vitis vinifera* L. cv. Gewürztraminer Must Fermented with Co-Cultures of *Saccharomyces cerevisiae* and Seven *Hanseniaspora* spp. *Fermentation* **2023**, *9*, 109. https://doi.org/10.3390/fermentation9020109

Academic Editor: Amparo Gamero

Received: 9 December 2022
Revised: 12 January 2023
Accepted: 18 January 2023
Published: 24 January 2023

1. Introduction

Yeasts belonging to the genus *Hanseniaspora* are among the most commonly isolated in vitivinicultural settings, and their role within grape must fermentations has been the topic of investigation among wine microbiologists for many years [1–3]. They are also common isolates on other fruits and have an influential role on the outcomes of fermentations, ranging from apple cider to coffee and chocolate [4–6]. During grape must fermentations, in general, the population of *Hanseniaspora* spp. drops significantly within the first couple of days due to a number of factors, including the accumulative exposure to an anaerobic environment and, intriguingly, the killer ability of fermenting yeast, such as *S. cerevisiae* [7–9]. Nevertheless, their impact on the final wine product can be meaningful as they compete for nutrients needed for fermenting yeasts to complete the fermentation, and most importantly, are capable of producing a large array of important aroma-active compounds. *Hanseniaspora* spp. are well-known for their production of high-levels of acetate esters, particularly ethyl acetate, which would often exceed concentrations deemed to be pleasant (~150 mg/L) and imparting a solvent or nail polish remover aroma [10]. Although *Hanseniaspora* has often been cited as being a spoilage yeast within winemaking [11,12], researchers are re-evaluating the overall impact that some species of *Hanseniaspora* can have as co-partners with *S. cerevisiae* in mixed-culture fermentations, with many reporting on the beneficial effects in adding to the aroma complexity of a wine [2]. *Hanseniaspora*-initiated fermentations have also been shown to reduce final ethanol levels [13], increase overall glycerol concentration [14], modulate acid composition [15], as well as causing a change in anthocyanin content and therefore the colour of the wine [16,17]. One species in particular, namely *H. vineae*, has shown repeatedly to provide several oenological benefits, which led to the development of a commercially available starter culture by the name of Fermivin VINEAE supplied by Oenobrands [18].

In mixed-culture fermentations, the most popular inoculation modality is a so-called sequential inoculation where the non-*Saccharomyces* yeast (NSY) starter culture is inoculated at the onset, allowing for NSY to exert an effect without the influence of *S. cerevisiae* [19]. One to three days later or once the sugar consumption has reached a certain level, the *Saccharomyces* culture is normally added. Alternatively, both starter cultures can be added simultaneously at the onset of fermentations. Studies where both modalities were conducted show the dramatic difference in the final outcome of the wine, which emphasizes the complexity and unpredictability of the interaction between the different yeasts [20–23].

The recent influx of whole-genome sequencing of representatives of each species within the *Hanseniaspora* genus, along with accompanying comparative studies, have revealed interesting data regarding the genomic make-up of this genus. Phylogenetic analyses have separated *Hanseniaspora* into two major clades, namely the fast-evolving lineage (FEL) and the slow-evolving lineage (SEL) [24,25]. Genes involved with cell cycle and genome integrity, thought to be conserved within ascomycetes, are not present within the genus. These genomic features of *Hanseniaspora* could assist in explaining the biology of the genus that becomes rapidly abundant as the sugar content in fruits increases during ripening [26]. Moreover, *H. vineae* was recently shown to undergo a rapid loss in cell viability during the stationary phase warranting more in-depth research in this genus, in particular how the two different lineages differ from each other [27].

In this study, we report on the co-fermentation of seven different *Hanseniaspora* spp. (two belonging to the SEL and five to the FEL) and how the respective species affected the aroma profile of a Gewürztraminer wine.

2. Materials and Methods

2.1. Yeast Strains Used in the Study

The *Hanseniaspora* strains *H. guilliermondii* NRRL-Y 1625, *H. meyeri* NRRL-Y 27,513 and *H. osmophila* NRRL Y-1613 were obtained from the Agriculture Resource Service Culture Collection (NRRL) (Peoria, IL, USA). The strain *H. uvarum* DSM2768 was a gift from Professor Jürgen Heinisch (University Osnabrück). The strains *H. nectarophila* GYBC-283, *H. occidentalis* GYBC-211 and *H. opuntiae* GYBC-284 were obtained from the Geisenheim yeast breeding culture collection. The *S. cerevisiae* strain Uvaferm CEG (Eaton, Nettersheim, Germany) was used as the fermenting yeast.

2.2. Microvinification

Pasteurised Gewürztraminer (GT) grape must (harvested from the vineyards belonging to the Geisenheim University in the Rheingau wine region of Germany) were used in the study. The GT must had a total sugar concentration of 247.8 g/L (123.3 g/L glucose and 124.5 g/L fructose). The main wine acids, tartaric and malic acid, were measured to be 3.7 g/L and 2.1 g/L, respectively. The must also had a citric acid content of 0.18 g/L. Its yeast available nitrogen was calculated as 65 mg/L by determining the primary free amino acid content using the spectrophotometrically-based nitrogen by the *o*-phthaldialdehyde method [28] and the free ammonium content using the rapid ammonium kit (Megazyme, Bray, Ireland). Opti-MUM White^{TM} (Lallemand, Montreal, QC, Canada), at a concentration of 20 g/hL, was added as a supplement. Fermentations were conducted in 250 mL Schott Duran flasks filled with 150 mL of GT must. All yeasts were precultured in YPD medium (20 g/L glucose, 20 g/L peptone, 10 g/L yeast extract) and washed once with phosphate-buffered saline. The *Hanseniaspora* cultures were inoculated at a concentration of ~1 × 10^7 cells/mL, as determined by haemocytometer, whereas the *S. cerevisiae* strain was inoculated at a concentration of 1 × 10^6 cells/mL. Airlocks were added to the flasks filled with approximately 3 mL of water. Fermentations were conducted at 22 °C and flasks were weighed daily until no further mass-loss was recorded. Samples were then centrifuged and subsequently prepared for high-performance liquid chromatography (HPLC) and gas-chromatography mass-spectrometry (GC-MS) analyses.

A one-day fermentation was also conducted where 1 μL of citronellol (Thermo Fisher Scientific, Geel, Belgium) was added to pure inoculations of *H. guilliermondii* and *H. uvarum*.

2.3. Analysis of the Must and Wines

Standard operating procedures, as established by the analysis team at the Institute of Microbiology and Biochemistry at Hochschule Geisenheim University, were followed to analyse the grape must and final wines produced in the study.

2.3.1. HPLC

For the quantification of the major organic acids, sugars, ethanol and glycerol of the wines and must, HPLC was implemented with a method previously described [29]. An HPLC Agilent Technologies Series 1100 (Agilent Technologies, Steinheim, Germany) was used built-in with an autosampler, a multi-wavelength (MWD) and refractive index (RID) detector and a binary pump. An HPLC column, 250 mm in length (Allure Organic Acids Column, Restek, Bad Homburg v. d. Höhe, Germany), with an inside diameter of 4.6 mm and a particle size of 5 μm was used for the separation of the different analytes. The MWD was set at a wavelength of 210 nm for the detection of organic acids and the RID was used to detect the sugars, organic acids, glycerol and ethanol. The isocratic eluent was comprised of deionized water with 0.5% ethanol and acidified with 0.0139% concentrated sulphuric acid (95–97%). The flow rate was 0.6 mL/min and column temperatures were 29 °C and 46 °C. Chemstation software (Agilent, Steinheim, Germany) was used to analyse, integrate and determine the concentrations of each analyte.

2.3.2. GC-MS
Aroma Bouquet Analysis

To detect and quantify the so-called aroma bouquet of the final wines comprising of expected higher alcohols; medium-chain fatty acids; and acetate and ethyl esters, a targeted headspace solid-phase micro-extraction gas-chromatography mass spectrometry analysis (HS-SPME-GC-MS) was employed using a protocol, as previously outlined [30]. Sample preparation entailed pipetting 5 mL of wine samples along with adding 1.7 g NaCl to a 20 mL headspace vial. Two internal standards, 1-octanol (600 mg/L) and cumene (52 mg/L), were also added. The GC-MS used was a GC 7890 A, equipped with an MS 5975 B (both Agilent, Santa Clara, CA, USA), an MPS robotic autosampler and a CIS 4 (both Gerstel, Mülheim an der Ruhr, Germany). A 65 μm fibre coated with polydimethylsiloxane crosslinked with divinylbenzene (Supelco, Merck, Darmstadt, Germany) was used to carry out the solid-phase microextraction. Separation of the volatiles was performed with a 60 m × 0.25 mm × 1 μm gas chromatography column (Rxi®-5Sil MS w/5 m Integra-Guard, Restek, Bad Homburg v. d. Höhe, Germany) with helium as a carrier gas. Split mode injection was employed (1:10, initial temperature 30 °C, rate 12 °C/s to 240 °C, hold for 4 min). The initial temperature of the GC run was 40 °C for 4 min, and then increased to 210 °C at 5 °C/min and raised again to 240 °C at 20 °C/min and held for 10.5 min. Mass spectral data were acquired in a range of mass-to-charge ratio (m/z) of 35 to 250 and used to determine the concentration values. A 5-point calibration curve was used for each volatile compound within a wine model solution of 12% ethanol with 3% tartaric acid at pH 3.

Terpenes

Free terpenes and C_{13}-norisoprenoids expected in wines were also measured by means of HS-SPME-GC-MS, as detailed previously [29,31]. A GC 6890 and a 5973 N quadrupole MS (Agilent Technologies, Palo Alto, CA, USA) equipped with a Gerstel MPS2 autosampler

(Gerstel, Mülheim an der Ruhr, Germany) was used. Similar to the aroma bouquet sample preparation, 5 mL of wine samples with 1.7 g of NaCl were added to a 20 mL headspace amber vial, along with 10 μL of the internal standard, which contained 30 μg/L 3-octanol, 30 μg/L linalool-d3, 40 μg/L α-terpineol-d3, 10 μg/L β-damascenone-d4, 16 μg/L β-ionone-d3 and 12.5 μg/L of naphthalene-d8 in ethanol solution (Sigma-Aldrich, Merck KGaA, Darmstadt, Germany).

A 100 μm polydimethylsiloxane fibre (Supelco, Merck, Darmstadt, Germany) was used to carry out the headspace solid-phase microextraction. Separation of the volatiles was performed with a 30 m × 0.25 mm × 0.5 μm gas chromatography column (DB-Wax, J & W Scientific, Agilent Technologies, Palo Alto, CA, USA) with helium as a carrier gas. Splitless mode injection was employed (injector temperature: 240 °C). The initial temperature of the GC run was 40 °C for 4 min, and then increased to 190 °C at 5 °C/min and raised again to 240 °C at 10 °C/min and held for 15 min. Mass spectral data were acquired in SIM mode with characteristic ions for each analyte and used to determine the concentration values. Calibration was performed by means of the standard addition in Riesling wine. For both the aroma bouquet and terpene analysis, Masshunter workstation software version B.09.00 (Agilent Technologies, Palo Alto, CA, USA) was used to calculate the concentrations of the aroma compounds.

2.4. Statistical Analysis

All fermentations were conducted in triplicate. Results from the fermentation analyses that are shown in the Tables are the average value of the triplicates followed by the standard deviation of the mean ($\pm$). All statistical analyses were performed using Graphpad Prism version 9.4.1 (GraphPad Software, San Diego, CA, USA). The control experiment in all cases was the fermentation inoculated with only *S. cerevisiae*. Principal component analyses were performed on the fermentation data using RStudio (version 2022.07.0) along with the packages factoextra (version 1.0.7), ggbiplot (version 0.55) and ggplot2 (version 3.3.6).

3. Results and Discussion

3.1. Fermentation Curves

Co-fermentations using *Hanseniaspora* strains with *S. cerevisiae* were carried out at 22 °C using 10:1 ratios of starting cultures. The progression of these fermentations was followed by daily measurements of CO_2 mass-loss (Figure 1). All fermentations were completed by 13 days, which was also confirmed by the HPLC analysis indicating that all fermentations reached dryness (Table 1). There were some deviations in the mass-loss patterns, with some co-fermentations (*H. occidentalis*, *H. osmophila* and *H. nectarophila*) exhibiting slightly less mass-loss throughout the fermentation, yet no notable fermentation burden was observed with any of the co-inoculations. This is in concurrence with previous *Hanseniaspora*-initiated fermentations, where no meaningful influence on the fermentation rate of *S. cerevisiae* was observed [32]. CEG is known for its slow fermentation performance. This was also the reason why we decided to use this *Saccharomyces* yeast.

3.2. Organic Acids

Organic acids, such as tartaric acid, malic acid, lactic acid, acetic acid and citric acid, play a major role in the aroma profile and the mouthfeel of the wine. Table 1 shows the content of the major organic acids of the final GT wines, as determined via HPLC. All fermentations, including the pure *S. cerevisiae* inoculation, led to a reduction in malic acid, yet in some cases (with *H. occidentalis* and *H. opuntiae*) the malic acid content was indeed higher than the control. This observation, especially with *H. occidentalis*, was surprising as it was recently shown that *H. occidentalis* can consume malic acid within grape must [15]. In that study, *H. occidentalis* consumed malic acid in a sequential-type of inoculation modality without an airlock, suggestive of the importance of oxygen in the consumption of malic acid.

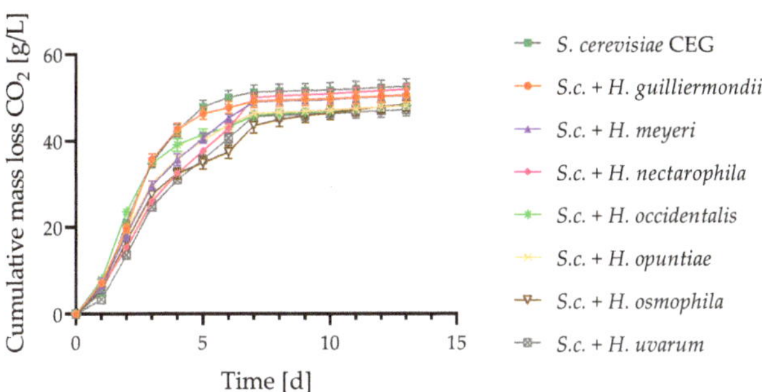

Figure 1. The mass-loss curves (g/L) of the co-fermentations of *Hanseniaspora* spp. conducted in GT must.

Table 1. The major organic acids, ethanol, glycerol and total sugar levels (all g/L) of the GT wines co-fermented with different *Hanseniaspora* spp., as determined with HPLC. Values are the means of three replicate fermentations followed by its standard deviations (±). Two-tailed unpaired *t*-tests with Welch's correction were conducted to compare values to the pure *S. cerevisiae* inoculum control. ↑ indicates significantly more than the control ($p < 0.05$), ↓ indicates significantly less than the control ($p < 0.05$). nd: not detected.

	S. cerevisiae	*H. guilliermondii*	*H. meyeri*	*H. nectarophila*	*H. occidentalis*	*H. opuntiae*	*H. osmophila*	*H. uvarum*
Total sugars	nd	nd	nd	nd	nd	nd	nd	nd
Tartaric acid	2.60 ± 0.12	2.40 ± 0.09	2.53 ± 0.03	2.59 ± 0.01	2.52 ± 0.07	2.45 ± 0.09	2.75 ± 0.02	2.56 ± 0.05
Malic acid	1.48 ± 0.06	1.35 ± 0.02	1.48 ± 0.02	1.57 ± 0.01	1.71 ± 0.05 ↑	1.65 ± 0.03 ↑	1.56 ± 0.02	1.55 ± 0.01
Lactic acid	0.25 ± 0.02	0.23 ± 0.01	0.23 ± 0.01	0.20 ± 0.01 ↓	0.22 ± 0.00	0.22 ± 0.02	0.04 ± 0.00 ↓	0.22 ± 0.01
Acetic acid	0.84 ± 0.04	0.75 ± 0.01	0.75 ± 0.03 ↓	1.10 ± 0.01 ↑	0.56 ± 0.01 ↓	0.94 ± 0.02 ↑	1.23 ± 0.01 ↑	0.68 ± 0.02 ↓
Citric acid	0.14 ± 0.01	0.12 ± 0.00 ↓	0.12 ± 0.00 ↓	0.13 ± 0.01	0.14 ± 0.00	0.11 ± 0.01 ↓	0.14 ± 0.00	0.10 ± 0.01 ↓
Ethanol	117.10 ± 3.20	114.67 ± 1.39	115.75 ± 0.91	117.35 ± 1.34	116.65 ± 2.74	116.37 ± 1.62	118.56 ± 0.59	116.31 ± 0.61
Glycerol	9.39 ± 0.35	9.28 ± 0.15	10.17 ± 0.05	10.05 ± 0.08	7.80 ± 0.12 ↓	10.10 ± 0.22 ↑	8.25 ± 0.08 ↓	9.98 ± 0.02

Regarding acetic acid, *Hanseniaspora*-inoculated fermentations resulted in both a reduction and an increase, depending on the species: for *H. nectarophila*, *H. opuntiae* and *H. osmophila,* co-fermentations led to a significant increase, whereas co-fermentations with *H. occidentalis* and *H. uvarum* led to a reduction. Acetic acid is the main contributor of volatile acidity in wine and along with its activated thioester acetyl-coenzyme (acetyl-CoA) are key participants within the central metabolism of cells. They directly take part in acetate ester formation as acetyl-CoA condenses with ethanol and other higher alcohols, which could explain the reduction in acetic acid in *H. uvarum* co-fermentations, which made the highest levels of ethyl acetate (Table 2). The divergent acetic acid levels with different *Hanseniaspora* spp. additions are consistent with the literature where no clear pattern emerges to what effect it has on the volatile acidity in wine [15,33].

Strikingly, lactic acid (the minor acid) was almost completely consumed within the co-fermentation with *H. osmophila*. It is unclear why this occurred as this has, to our knowledge, not been reported before.

3.3. Ethanol and Glycerol

A microbially-facilitated strategy to reduce ethanol levels in wine is by implementing NSY in co-culturing set-ups [34]. The strategy is based on the idea that the NSY would consume a portion of the initial sugars leaving less sugars to be fermented to ethanol by *Saccharomyces*. With the experimental set-up presented here, no significant reduction of ethanol was observed in any of the *Hanseniaspora* co-inoculums. Even though there have been reports of a reduction in ethanol in final wines with *Hanseniaspora* additions [13,20,35,36], fermentations, especially with simultaneous inoculation modalities did not observe ethanol reductions [37,38].

Table 2. Major wine aroma components of the GT wines, as determined with HS-SPME-GC-MS. Values are means of three replicate fermentations followed by the standard deviations ($\pm$). Two-tailed unpaired t-test with Welch's correction was conducted to compare values with the pure *S. cerevisiae* inoculum control. $\uparrow$ indicates significantly more than the control ($p < 0.05$), $\downarrow$ indicates significantly less than the control ($p < 0.05$). nq: not quantifiable.

	S. cerevisiae	*H. guilliermondii*	*H. meyeri*	*H. nectarophila*	*H. occidentalis*	*H. opuntiae*	*H. osmophila*	*H. uvarum*
Acetate esters (µg/L, except for ethyl acetate (mg/L))								
Ethyl acetate	115.80 ± 6.73	318.93 ± 28.10 ↑	332.86 ± 97.78	312.98 ± 55.29 ↑	108.19 ± 12.18	291.07 ± 21.83 ↑	151.60 ± 34.74	558.07 ± 52.66 ↑
Isoamyl acetate	1089 ± 104	3788 ± 614 ↑	2731 ± 803	2120 ± 351 ↑	646 ± 19 ↓	723 ± 51 ↓	640 ± 197 ↓	2523 ± 383 ↑
2-Phenylethyl acetate	203 ± 16	1185 ± 24 ↑	488 ± 59 ↑	269 ± 22 ↑	225 ± 2	683 ± 45 ↑	1236 ± 48 ↑	466 ± 13 ↑
2-Methylbutyl acetate	108 ± 16	416 ± 63 ↑	390 ± 134	336 ± 127	62 ± 17↓	147 ± 16 ↑	63 ± 10 ↓	339 ± 118
Hexyl acetate	3 ± 2	24 ± 5 ↑	9 ± 5	13 ± 4 ↑	nq	nq	3 ± 1	15 ± 2 ↑
Ethyl esters (µg/L)								
Ethyl propionate	217 ± 17	238 ± 24	274 ± 70	253 ± 53	482 ± 34 ↑	265 ± 19 ↑	122 ± 29 ↓	425 ± 67 ↑
Ethyl butyrate	222 ± 21	312 ± 37 ↑	248 ± 55	241 ± 39	182 ± 7	177 ± 14 ↓	227 ± 46	272 ± 30
Ethyl hexanoate	76 ± 21	92 ± 40	51 ± 59	66 ± 47	nq	nq	35 ± 32	77 ± 11
Ethyl octanoate	494 ± 99	459 ± 184	266 ± 268	541 ± 305	nq	nq	317 ± 140	473 ± 33
Ethyl decanoate	468 ± 44	525 ± 85	425 ± 88	666 ± 318	172 ± 48 ↓	106 ± 44 ↓	462 ± 138	576 ± 205
Medium-chain fatty acids (mg/L)								
Hexanoic acid	5.63 ± 0.06	5.56 ± 0.04	5.44 ± 0.05 ↓	5.53 ± 0.05	5.37 ± 0.02 ↓	5.30 ± 0.01 ↓	5.47 ± 0.02 ↓	5.57 ± 0.01
Octanoic acid	3.17 ± 0.07	3.09 ± 0.04	2.97 ± 0.03 ↓	3.02 ± 0.03 ↓	2.88 ± 0.00 ↓	2.86 ± 0.01 ↓	2.98 ± 0.03 ↓	3.02 ± 0.00
Higher alcohols (mg/L)								
Isobutanol	141.80 ± 12.75	149.54 ± 18.24	161.05 ± 41.06	150.83 ± 26.32	121.33 ± 16.12	190.04 ± 23.72	116.23 ± 13.69	140.45 ± 0.83
Isoamyl alcohol	668.82 ± 40.95	539.02 ± 54.58 ↓	581.12 ± 112.97	493.20 ± 92.87	510.68 ± 64.69 ↓	544.60 ± 66.44	429.67 ± 37.26 ↓	540.64 ± 7.27 ↓
2-Methyl butanol	125.16 ± 3.01	97.85 ± 8.83 ↓	101.78 ± 21.19	104.83 ± 19.05	53.08 ± 6.59 ↓	105.55 ± 9.04	99.63 ± 2.63 ↓	102.37 ± 1.21 ↓
1-Hexanol	1411.90 ± 71.14	1099.56 ± 46.86↓	1348.98 ± 34.27	1257.44 ± 161.69	1029.03 ± 40.98 ↓	1431.11 ± 46.72	1312.47 ± 25.20	1168.21 ± 26.82 ↓
2-Phenyl ethanol	44.34 ± 0.43	28.34 ± 0.45 ↓	41.82 ± 5.42	27.75 ± 4.54 ↓	59.40 ± 4.51 ↑	48.52 ± 2.11	36.31 ± 2.32 ↓	37.70 ± 1.17 ↓

As with acetic acid, the co-fermentations produced both significantly more (*H. opuntiae*) or less glycerol (the two members of the SEL, *H. occidentalis* and *H. osmophila*) than the control. Glycerol levels are often elevated with NSY additions [39] and often coincide with an increase in acetic acid [40]. This is often explained within the context of cofactor maintenance, as the enzymes directly responsible for glycerol and acetic acid production, glycerol-3-phosphate dehydrogenase and acetaldehyde dehydrogenase, respectively, require and produce NADH. Curiously, in this experiment, the co-fermentation with *H. occidentalis* led to an unexpected reduction in both acetic acid and glycerol, a finding which was not observed when the same yeast was co-fermented in a sequential-type inoculation [15].

3.4. Aroma Analysis

Since *Hanseniaspora* is known for acetate ester formation and produces large amounts of ethyl acetate, a solvent-like odour, too much of which can quickly spoil the wine, the volatile aroma compounds (VOC) were measured. Table 2 shows the concentrations of many of the expected esters, medium-chain fatty acids, and higher alcohols in the final wines co-cultured with *Hanseniaspora* spp. With regards to the ethyl esters and medium-chain fatty acids, little modulation was observed with the *Hanseniaspora* co-cultured fermentations, when compared to the *S. cerevisiae* control, apart from the concentration of ethyl propionate (an aroma compound imparting a pineapple-like odour). The ethyl propionate concentrations in wines fermented with *H. occidentalis* and *H. uvarum* were measured to be more than double than that of the control.

As expected, a large level of variability was measured looking at acetate esters. For all of the five acetate esters, the concentrations were significantly more in *H. guilliermondii* co-fermentations, whereas four of the five acetate esters were higher with *H. nectarophila* and *H. uvarum*. An increase in all of these esters within a wine will generally be considered as an oenological benefit as they contribute to a fruity or flowery aroma, except for ethyl acetate. Ethyl acetate measured in the five FEL members were approximately three times more than the control with *H. uvarum* co-fermentations achieving levels of more than 500 mg/L. All of these levels far exceed what is considered to be pleasant (<100 mg/L). With the ethyl acetate levels obtained from the two members of the SEL, no significant difference with the control was observed. Yet for *H. osmophila*, more than six times more 2-phenethyl acetate, a key rose-like aroma component, was recorded. Furthermore, of note are the quantitatively high levels of the beneficial acetate esters produced by *H. guilliermondii*.

The majority of the higher alcohols were either unchanged or comparatively lower than the control. This is directly connected with their conversion to their corresponding acetate esters. Only co-fermentations with *H. occidentalis*, however, showed to have higher amounts of one of the higher alcohol than the controls, namely 2-phenyl ethanol.

3.5. Terpenes

Typical of GT must is that it has a high terpene potential. Terpenes are often bound to sugar moieties in grape must and require deglycosylation for their release, in order to be perceived. This release can be facilitated enzymatically via the action of β-glucosidases or non-specific β-glucanases and it has been shown that NSY, including *Hanseniaspora* spp., have superior terpene-releasing abilities than *Saccharomyces* [38,41–43]. The conversion of terpenes can also occur, which can be catalyzed by several enzymes, including dehydrogenases, oxygenases and reductases [44]. Table 3 shows the terpene content of the GT wines co-fermented with the different *Hanseniaspora* spp. We observed in certain cases significant modulation of the terpenes within the co-fermentations. With *H. opuntiae*, in particular, many of the terpenes were significantly affected like citronellol and β-myrcene, which is suggestive of the release of terpenes from sugar moieties. Curiously, with some co-fermentations (*H. guilliermondii*, *H nectarophila* and *H. uvarum*) citronellol levels were significantly lower than those of the control. A short fermentation of must spiked with higher levels of citronellol in wines co-fermented with *H. guilliermondii* and *H. uvarum* was conducted to see if other peaks related to citronellol could be observed. A peak corresponding to the acetate ester of citronellol (i.e., citronellyl acetate) was detected in the GC-MS chromatograms (Figure 2). This strongly suggests that these *Hanseniaspora* strains converted citronellol (which is a terpenol with a primary alcohol functional group) to its acetate ester, presumably via the same action as the esterification of other alcohols.

3.6. PCA

To analyse the overall outcome of VOC production in the different co-fermentations, a principal component analysis was conducted, including aroma compounds, organic acids, ethanol and glycerol (Figure 3). Large variations, especially in the ethyl ester content, of co-fermentations with *H. nectarophila* and *H. meyeri* replicates caused the overlaps of the data points with other groups. The PCA indicated that the two members of the SEL separated from the other groups regarding their wine profiles. It is also noteworthy that the bulk of the acetate esters production was associated with members of the FEL, such as *H. guilliermondii* and *H. uvarum*. The PCA also shows the noteworthy acid modulation displayed by the fermentation with *H. osmophila* with its high production of acetic acid coinciding with the possible consumption of lactic acid.

Table 3. Terpene and norisoprenoid content (all µg/L) of the GT wines, as measured with HS-SPME-GC-MS. Values are the means of three replicate fermentations followed by the standard deviations ($\pm$). Unpaired *t*-test with Welch's correction was conducted to compare values with the pure *S. cerevisiae* inoculum control. ↑ indicates significantly more (<0.05) than the control (*S. cerevisiae*) ↓ indicates significantly less (<0.05) than the control (*S. cerevisiae*).

	S. cerevisiae	*H. guilliermondii*	*H. meyeri*	*H. nectarophila*	*H. occidentalis*	*H. opuntiae*	*H. osmophila*	*H. uvarum*
β-myrcene	6.71 ± 0.31	9.82 ± 1.26 ↑	7.87 ± 1.05	9.37 ± 0.58 ↑	7.95 ± 1.74	9.13 ± 0.17 ↑	8.87 ± 0.67 ↑	9.06 ± 0.24 ↑
limonene	0.96 ± 0.01	0.99 ± 0.02	0.97 ± 0.02	0.99 ± 0.01 ↑	0.96 ± 0.03	1.01 ± 0.01 ↑	0.98 ± 0.01	1.02 ± 0.01 ↑
cis-rose oxide	0.42 ± 0.03	0.24 ± 0.03 ↓	0.35 ± 0.07	0.38 ± 0.04	0.19 ± 0.07 ↓	0.36 ± 0.03 ↓	0.24 ± 0.04 ↓	0.45 ± 0.02
trans-rose oxide	0.15 ± 0.01	0.09 ± 0.01 ↓	0.14 ± 0.02	0.15 ± 0.01	0.09 ± 0.02 ↓	0.13 ± 0.01 ↓	0.11 ± 0.01 ↓	0.17 ± 0.01
cis-linalool oxide	19.74 ± 1.03	19.68 ± 0.13	19.11 ± 0.19	18.37 ± 0.24	18.13 ± 0.54	18.84 ± 1.43	18.54 ± 0.65	19.12 ± 0.68
nerol oxide	1.29 ± 0.04	1.35 ± 0.07	1.27 ± 0.11	1.32 ± 0.09	1.25 ± 0.18	1.45 ± 0.05 ↑	1.29 ± 0.16	1.38 ± 0.07
trans-linalool oxide	6.20 ± 0.67	5.75 ± 0.18	5.43 ± 0.34	5.77 ± 0.21	5.45 ± 0.20	5.72 ± 0.25	5.63 ± 0.52	5.52 ± 0.03
vitispirane	0.56 ± 0.01	0.54 ± 0.00 ↓	0.55 ± 0.01	0.54 ± 0.01 ↓	0.54 ± 0.01 ↓	0.55 ± 0.01	0.52 ± 0.01 ↓	0.57 ± 0.00
linalool	72.43 ± 3.71	72.93 ± 1.80	71.43 ± 0.69	72.13 ± 3.01	71.18 ± 2.51	73.91 ± 3.70	72.04 ± 5.49	72.50 ± 0.45
hotrienol	36.61 ± 6.24	29.52 ± 1.85	31.48 ± 4.25	28.38 ± 3.47	39.98 ± 3.95	43.04 ± 3.24	35.04 ± 0.35	31.78 ± 1.97
α-terpineol	55.46 ± 9.81	47.13 ± 2.74	50.75 ± 6.32	46.91 ± 5.59	68.66 ± 7.75	72.42 ± 4.46	69.19 ± 2.49	49.01 ± 3.90
citronellol	26.97 ± 4.40	9.52 ± 0.56 ↓	18.96 ± 4.08	13.83 ± 1.22 ↓	27.37 ± 4.97	38.08 ± 1.52 ↑	20.31 ± 2.57	13.85 ± 1.88 ↓
β-damascenone	0.46 ± 0.01	0.56 ± 0.04	0.53 ± 0.05	0.61 ± 0.04 ↑	0.64 ± 0.14	0.69 ± 0.03 ↑	0.70 ± 0.10	0.55 ± 0.06

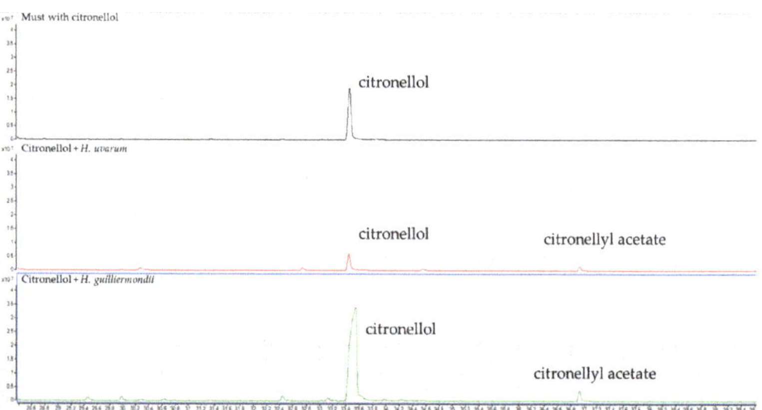

Figure 2. GC-MS chromatograms of the one-day must fermentations with *H. guilliermondii* and *H. uvarum* supplemented with higher levels of citronellol.

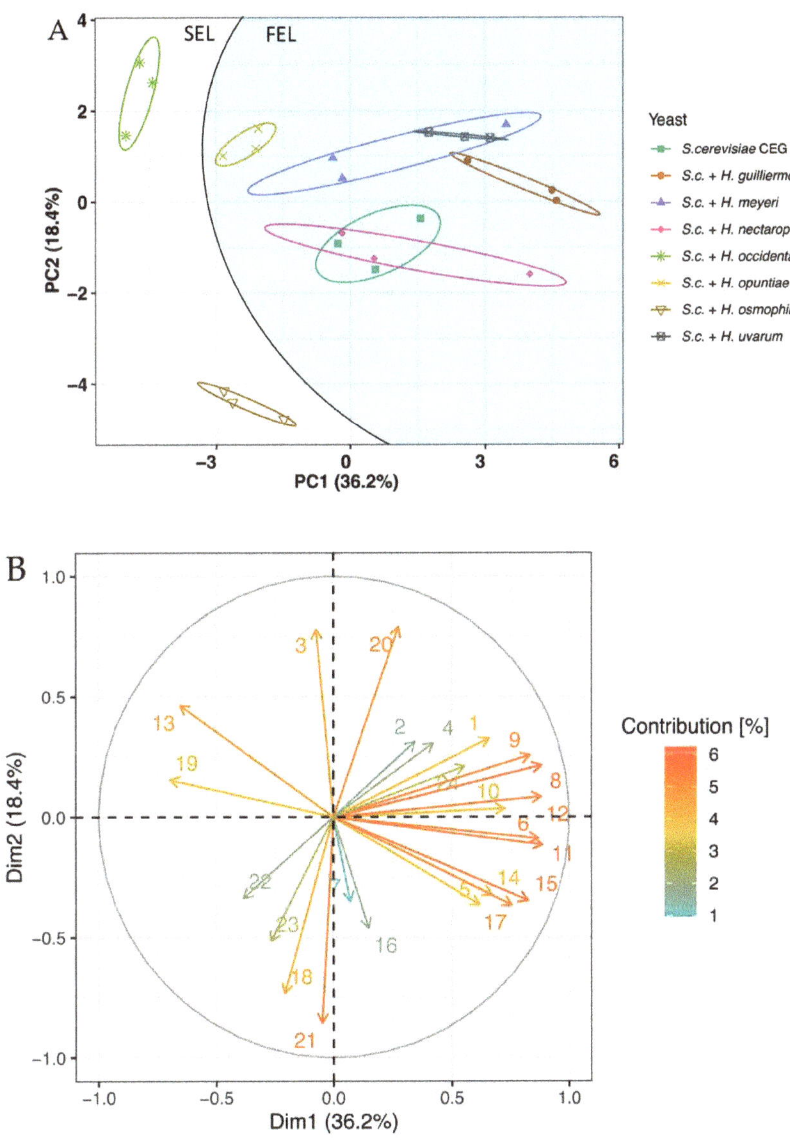

Figure 3. PCA of the final wines co-fermented with *Hanseniaspora* spp. All of the measured wine parameters, excluding the terpene data, were used to compose the PCA. Score plot (**A**) in the first two PCs: fermentation replicates are shown with the same shape and colour. The blue-and-white separation indicates the division of the two lineages (FEL and SEL). (**B**) Corresponding loading map. Each arrow relates to the tip of a vector starting from the origin. The closer the variable is to the circle, the better it is explained by the components. Contribution values (%) are shown in a gradient scale of colours with corresponding values. Numbers in the loading map correspond to: (1) ethyl acetate, (2) isobutanol, (3) ethyl propionate, (4) isoamyl alcohol, (5) 2-methyl-1-butanol, (6) ethyl butanoate, (7) hexan-1-ol, (8) isoamyl acetate, (9) 2-methyl-1-butyl acetate, (10) hexanoic acid, (11) ethyl hexanoate, (12) hexyl acetate, (13) 2-phenylethanol, (14) octanoic acid, (15) ethyl octanoate, (16) 2-phenylethyl acetate, (17) ethyl decanoate, (18) tartaric acid, (19) malic acid, (20) lactic acid, (21) acetic acid, (22) citric acid, (23) ethanol, (24) glycerol.

4. Conclusions

Even though non-*Saccharomyces* yeasts are increasingly used in wine fermentations, reproducibility is a concern. This is not only evident in our study, but also emerges from the literature on mixed-culture fermentations. The interaction of different yeast species is complex and strongly dependent on the initial conditions in the must. In conclusion, the different *Hanseniaspora* species contributed in various ways to the aroma profile of the wine. The SEL yeasts (*H. occidentalis* and *H. osmophila*) showed little change in ethyl acetate formation, which is typical for *Hanseniaspora*, in contrast to the control fermentation with *S. cerevisiae*. In addition, some of the *Hanseniaspora* spp. contributed significantly to the complex aroma profile through the depletion of lactic acid, an increase in acetate esters or terpenols, or through the conversion of citronellol to citronellyl acetate. Finally, this study shows for the first time the use of *H. nectarophila* and *H. meyerii* in winemaking and the characterization of the resulting wine. In particular, the wines fermented with *H. nectarophila* showed increased amounts of acetate esters, compared to the control wine.

Author Contributions: Conceptualization, J.B. and N.v.W.; methodology, J.B. and N.v.W.; validation, J.B.; formal analysis, H.S., S.B. and S.F.; investigation, J.B. and N.v.W.; resources, C.v.W.; data curation, J.B.; writing—original draft preparation, N.v.W.; visualization, J.B. and F.K., supervision, D.R., I.S.P. and C.v.W.; project administration, C.v.W.; funding acquisition, C.v.W. All authors have read and agreed to the published version of the manuscript.

Funding: The authors would like to thank the Hessen State Ministry of Higher Education, Research and the Arts for the financial support within the Hessen initiative for scientific and economic excellence (LOEWE) in the framework of AROMAplus (https://www.hs-geisenheim.de/aromaplus/, (accessed on 25 October 2022)). The authors thank Geisenheim University and Macquarie University for the co-funding of this project and the research fellowship of N.v.W. I.S.P. is a team member of the Macquarie-led national Centre of Excellence in Synthetic Biology funded by the Australian Government through its agency, the Australian Research Council.

Acknowledgments: I.S.P. is a team member of the Macquarie-led national Centre of Excellence in Synthetic Biology funded by the Australian Government thorough its agency, the Australian Research Council. The authors would like to thank the Hessen State Ministry of Higher Education, Research and the Arts for the financial support within the Hessen initiative for scientific and economic excellence (LOEWE) in the framework of AROMAplus (https://www.hs-geisenheim.de/aromaplus/, (accessed on 25 October 2022)). Jürgen Heinisch is thanked for providing the yeast strain DSM2768. Florian Michling is thanked for providing the yeast strain NRRL Y-1613, NRRL-Y 1625, and NRRL-Y 27513. Georgia Tylaki and Kerstin Zimmer are thanked for technical assistance.

Conflicts of Interest: The authors declare no conflict of interest.

References

1. Van Wyk, N.; Badura, J.; Von Wallbrunn, C.; Pretorius, I.S. Exploring future applications of the apiculate yeast Hanseniaspora. *Crit. Rev. Biotechnol.* **2023**, *in press*. [CrossRef]
2. Martin, V.; Valera, M.; Medina, K.; Boido, E.; Carrau, F. Oenological impact of the Hanseniaspora/Kloeckera yeast genus on wines—A review. *Fermentation* **2018**, *4*, 76. [CrossRef]
3. Díaz-Montaño, D.M.; de Jesús Ramírez Córdova, J. The fermentative and aromatic ability of Kloeckera and Hanseniaspora yeasts. In *Yeast Biotechnology: Diversity and Applications*; Satyanarayana, T., Kunze, G., Eds.; Springer: Dordrecht, The Netherlands, 2009; pp. 282–301, ISBN 9781402082917.
4. Wei, J.; Zhang, Y.; Qiu, Y.; Guo, H.; Ju, H.; Wang, Y.; Yuan, Y.; Yue, T. Chemical composition, sensorial properties, and aroma-active compounds of ciders fermented with Hanseniaspora osmophila and Torulaspora quercuum in co- and sequential fermentations. *Food Chem.* **2020**, *306*, 125623. [CrossRef] [PubMed]
5. Elhalis, H.; Cox, J.; Frank, D.; Zhao, J. The crucial role of yeasts in the wet fermentation of coffee beans and quality. *Int. J. Food Microbiol.* **2020**, *333*, 108796. [CrossRef]
6. Ho, V.T.T.; Zhao, J.; Fleet, G. Yeasts are essential for cocoa bean fermentation. *Int. J. Food Microbiol.* **2014**, *174*, 72–87. [CrossRef] [PubMed]
7. Wang, C.; Mas, A.; Esteve-Zarzoso, B. Interaction between Hanseniaspora uvarum and Saccharomyces cerevisiae during alcoholic fermentation. *Int. J. Food Microbiol.* **2015**, *206*, 67–74. [CrossRef] [PubMed]
8. Hierro, N.; Esteve-Zarzoso, B.; Mas, A.; Guillamón, J.M. Monitoring of Saccharomyces and Hanseniaspora populations during alcoholic fermentation by real-time quantitative PCR. *FEMS Yeast Res.* **2007**, *7*, 1340–1349. [CrossRef] [PubMed]

9.   Moreira, N.; Mendes, F.; Guedes de Pinho, P.; Hogg, T.; Vasconcelos, I. Heavy sulphur compounds, higher alcohols and esters production profile of Hanseniaspora uvarum and Hanseniaspora guilliermondii grown as pure and mixed cultures in grape must. *Int. J. Food Microbiol.* **2008**, *124*, 231–238. [CrossRef]

10.   Rojas, V.; Gil, J.V.; Piñaga, F.; Manzanares, P. Acetate ester formation in wine by mixed cultures in laboratory fermentations. *Int. J. Food Microbiol.* **2003**, *86*, 181–188. [CrossRef]

11.   Loureiro, V.; Malfeito-Ferreira, M. Spoilage yeasts in the wine industry. *Int. J. Food Microbiol.* **2003**, *86*, 23–50. [CrossRef]

12.   Malfeito-Ferreira, M. Yeasts and wine off-flavours: A technological perspective. *Ann. Microbiol.* **2011**, *61*, 95–102. [CrossRef]

13.   Rossouw, D.; Bauer, F.F. Exploring the phenotypic space of non-Saccharomyces wine yeast biodiversity. *Food Microbiol.* **2016**, *55*, 32–46. [CrossRef] [PubMed]

14.   Medina, K.; Boido, E.; Fariña, L.; Gioia, O.; Gomez, M.E.; Barquet, M.; Gaggero, C.; Dellacassa, E.; Carrau, F. Increased flavour diversity of Chardonnay wines by spontaneous fermentation and co-fermentation with Hanseniaspora vineae. *Food Chem.* **2013**, *141*, 2513–2521. [CrossRef]

15.   Van Wyk, N.; Scansani, S.; Beisert, B.; Brezina, S.; Fritsch, S.; Semmler, H.; Pretorius, I.S.; Rauhut, D.; von Wallbrunn, C. The use of Hanseniaspora occidentalis in a sequential must inoculation to reduce the malic acid content of wine. *Appl. Sci.* **2022**, *12*, 6919. [CrossRef]

16.   Manzanares, P.; Rojas, V.; Genovés, S.; Vallés, S. A preliminary search for anthocyanin-β-D-glucosidase activity in non-Saccharomyces wine yeasts. *Int. J. Food Sci. Technol.* **2000**, *35*, 95–103. [CrossRef]

17.   Manuel, J.; Escott, C.; Carrau, F.; Herbert-Pucheta, J.E.; Vaquero, C.; Gonz, C.; Morata, A. Improving aroma complexity with Hanseniaspora spp.: Terpenes, acetate esters, and safranal. *Fermentation* **2022**, *8*, 654. [CrossRef]

18.   Carrau, F.; Henschke, P.A. Hanseniaspora vineae and the concept of friendly yeasts to increase autochthonous wine flavor diversity. *Front. Microbiol.* **2021**, *12*, 702093. [CrossRef]

19.   Van Wyk, N.; Grossmann, M.; Wendland, J.; von Wallbrunn, C.; Pretorius, I.S. The whiff of wine yeast innovation: Strategies for enhancing aroma production by yeast during wine fermentation. *J. Agric. Food Chem.* **2019**, *67*, 13496–13505. [CrossRef]

20.   Tristezza, M.; Tufariello, M.; Capozzi, V.; Spano, G.; Mita, G.; Grieco, F. The oenological potential of Hanseniaspora uvarum in simultaneous and sequential co-fermentation with Saccharomyces cerevisiae for industrial wine production. *Front. Microbiol.* **2016**, *7*, 670. [CrossRef]

21.   Chen, L.; Li, D. Effects of simultaneous and sequential cofermentation of Wickerhamomyces anomalus and Saccharomyces cerevisiae on physicochemical and flavor properties of rice wine. *Food Sci. Nutr.* **2021**, *9*, 71–86. [CrossRef] [PubMed]

22.   Gobbi, M.; Comitini, F.; Domizio, P.; Romani, C.; Lencioni, L.; Mannazzu, I.; Ciani, M. Lachancea thermotolerans and Saccharomyces cerevisiae in simultaneous and sequential co-fermentation: A strategy to enhance acidity and improve the overall quality of wine. *Food Microbiol.* **2013**, *33*, 271–281. [CrossRef] [PubMed]

23.   Viana, F.; Belloch, C.; Vallés, S.; Manzanares, P. Monitoring a mixed starter of Hanseniaspora vineae-Saccharomyces cerevisiae in natural must: Impact on 2-phenylethyl acetate production. *Int. J. Food Microbiol.* **2011**, *151*, 235–240. [CrossRef]

24.   Steenwyk, J.L.; Opulente, D.A.; Kominek, J.; Shen, X.; Zhou, X.; Labella, A.L.; Bradley, N.P.; Eichman, B.F.; Libkind, D.; Devirgilio, J.; et al. Extensive loss of cell-cycle and DNA repair genes in an ancient lineage of bipolar budding yeasts. *PLoS Biol.* **2019**, *1*, e3000255. [CrossRef] [PubMed]

25.   Phillips, M.A.; Steenwyk, J.L.; Shen, X.-X.; Rokas, A. Examination of gene loss in the DNA mismatch repair pathway and its mutational consequences in a fungal phylum. *Genome Biol. Evol.* **2021**, *13*, evab219. [CrossRef] [PubMed]

26.   Čadež, N.; Bellora, N.; Ulloa, R.; Tome, M.; Petković, H.; Groenewald, M.; Hittinger, C.T.; Libkind, D. Hanseniaspora smithiae sp. nov., a novel apiculate yeast species from Patagonian forests that lacks the typical genomic domestication signatures for fermentative environments. *Front. Microbiol.* **2021**, *12*, 679894. [CrossRef]

27.   Schwarz, L.V.; Valera, M.J.; Delamare, A.P.L.; Carrau, F.; Echeverrigaray, S. A peculiar cell cycle arrest at g2/m stage during the stationary phase of growth in the wine yeas Hanseniaspora vineae. *Curr. Res. Microb. Sci.* **2022**, *3*, 100129. [CrossRef]

28.   Dukes, B.C.; Butzke, C.E. Rapid determination of primary amino acids in grape juice using an o-phthaldialdehyde/N-acetyl-L-cysteine spectrophotometric assay. *Am. J. Enol. Vitic.* **1998**, *49*, 125–134. [CrossRef]

29.   Scansani, S.; van Wyk, N.; Nader, K.B.; Beisert, B.; Brezina, S.; Fritsch, S.; Semmler, H.; Pasch, L.; Pretorius, I.S.; von Wallbrunn, C.; et al. The film-forming Pichia spp. in a winemaker's toolbox: A simple isolation procedure and their performance in a mixed-culture fermentation of Vitis vinifera L. cv. Gewürztraminer must. *Int. J. Food Microbiol.* **2022**, *365*, 109549. [CrossRef]

30.   Jung, R.; Kumar, K.; Patz, C.; Rauhut, D.; Tarasov, A.; Schuessler, C. Influence of transport temperature profiles on wine quality. *Food Packag. Shelf Life* **2021**, *29*, 100706. [CrossRef]

31.   Brandt, M. The Influence of Abiotic Factors on the Composition of Berries, Juice and Wine in *Vitis vinifera* L. cv. Riesling. Ph.D. Thesis, Hochschule Geisenheim University, Geisenheim, Germany, 2021.

32.   González-Robles, I.W.; Estarrón-Espinosa, M.; Díaz-Montaño, D.M. Fermentative capabilities and volatile compounds produced by Kloeckera/Hanseniaspora and Saccharomyces yeast strains in pure and mixed cultures during Agave tequilana juice fermentation. *Antonie Van Leeuwenhoek* **2015**, *108*, 525–536. [CrossRef]

33.   Del Fresno, J.M.; Loira, I.; Escott, C.; Carrau, F.; Gonz, C.; Cuerda, R.; Morata, A. Application of Hanseniaspora vineae yeast in the production of Rosé wines from a blend of Tempranillo and Albillo grapes. *Fermentation* **2021**, *7*, 141. [CrossRef]

34.   Contreras, A.; Hidalgo, C.; Henschke, P.A.; Chambers, P.J.; Curtin, C.; Varela, C. Evaluation of non-Saccharomyces yeasts for the reduction of alcohol content in wine. *Appl. Environ. Microbiol.* **2014**, *80*, 1670–1678. [CrossRef] [PubMed]

35.	Mestre, M.V.; Maturano, Y.P.; Gallardo, C.; Combina, M.; Mercado, L.; Toro, M.E.; Carrau, F.; Vazquez, F.; Dellacassa, E. Impact on sensory and aromatic profile of low ethanol Malbec wines fermented by sequential culture of Hanseniaspora uvarum and Saccharomyces cerevisiae native yeasts. *Fermentation* **2019**, *5*, 65. [CrossRef]

36.	Zhang, B.; Shen, J.Y.; Duan, C.Q.; Yan, G.L. Use of indigenous Hanseniaspora vineae and Metschnikowia pulcherrima co-fermentation with Saccharomyces cerevisiae to improve the aroma diversity of Vidal blanc icewine. *Front. Microbiol.* **2018**, *9*, 2303. [CrossRef]

37.	Capozzi, V.; Berbegal, C.; Tufariello, M.; Grieco, F.; Spano, G.; Grieco, F. Impact of co-inoculation of Saccharomyces cerevisiae, Hanseniaspora uvarum and Oenococcus oeni autochthonous strains in controlled multi starter grape must fermentations. *LWT Food Sci. Technol.* **2019**, *109*, 241–249. [CrossRef]

38.	Hu, K.; Jin, G.J.; Xu, Y.H.; Tao, Y.S. Wine aroma response to different participation of selected Hanseniaspora uvarum in mixed fermentation with Saccharomyces cerevisiae. *Food Res. Int.* **2018**, *108*, 119–127. [CrossRef] [PubMed]

39.	Goold, H.D.; Kroukamp, H.; Williams, T.C.; Paulsen, I.T.; Varela, C.; Pretorius, I.S. Yeast's balancing act between ethanol and glycerol production in low-alcohol wines. *Microb. Biotechnol.* **2017**, *10*, 264–278. [CrossRef]

40.	van Wyk, N.; Kroukamp, H.; Espinosa, M.I.; von Wallbrunn, C.; Wendland, J.; Pretorius, I.S. Blending wine yeast phenotypes with the aid of CRISPR DNA editing technologies. *Int. J. Food Microbiol.* **2020**, *324*, 108615. [CrossRef]

41.	van Wyk, N.; Pretorius, I.S.; von Wallbrunn, C. Assessing the oenological potential of Nakazawaea ishiwadae, Candida railenensis and Debaryomyces hansenii strains in mixed-culture grape must fermentation with Saccharomyces cerevisiae. *Fermentation* **2020**, *6*, 49. [CrossRef]

42.	Garcia, A.; Carcel, C.; Dulau, L.; Samson, A.; Aguera, E.; Agosin, E.; Günata, Z. Influence of a mixed culture with Debaryomyces vanriji and Saccharomyces cerevisiae on the volatiles of a Muscat wine. *J. Food Sci.* **2002**, *67*, 1138–1143. [CrossRef]

43.	Fernández-González, M.; Di Stefano, R.; Briones, A. Hydrolysis and transformation of terpene glycosides from muscat must by different yeast species. *Food Microbiol.* **2003**, *20*, 35–41. [CrossRef]

44.	Vaudano, E.; Moruno, E.G.; Di Stefano, R. Modulation of geraniol metabolism during alcohol fermentation. *J. Inst. Brew.* **2004**, *110*, 213–219. [CrossRef]

 *fermentation*

Article

Monitoring the Aroma Compound Profiles in the Microbial Fermentation of Seaweeds and Their Effects on Sensory Perception

Yueh-Hao Ronny Hung [1,†], Chien-Yu Peng [1,†], Mei-Ying Huang [2], Wen-Jung Lu [1], Hsuan-Ju Lin [1], Chih-Ling Hsu [1], Ming-Chih Fang [1,*] and Hong-Ting Victor Lin [1,3,*]

[1] Department of Food Science, National Taiwan Ocean University, No. 2, Pei-Ning Road, Keelung 202301, Taiwan
[2] Division of Aquaculture, Fisheries Research Institute, Council of Agriculture, No. 199, Hou-Ih Road, Keelung 202301, Taiwan
[3] Center of Excellence for the Oceans, National Taiwan Ocean University, No. 2, Pei-Ning Road, Keelung 202008, Taiwan
* Correspondence: mcfang@mail.ntou.edu.tw (M.-C.F.); hl358@ntou.edu.tw (H.-T.V.L.); Tel.: +886-2-24622192 (ext. 5111) (M.-C.F.); +886-2-24622192 (ext. 5121) (H.-T.V.L.); Fax: +886-2-24634203 (H.-T.V.L.)
† These authors contributed equally to this work.

Abstract: Seaweeds have a variety of biological activities, and their aromatic characteristics could play an important role in consumer acceptance. Here, changes in aroma compounds were monitored during microbial fermentation, and those most likely to affect sensory perception were identified. *Ulva* sp. and *Laminaria* sp. were fermented and generally recognized as safe microorganisms, and the profile of volatile compounds in the fermented seaweeds was investigated using headspace solid-phase microextraction with gas chromatography–mass spectrometry. Volatile compounds, including ketones, aldehydes, alcohols, and acids, were identified during seaweed fermentation. Compared with lactic acid bacteria fermentation, *Bacillus subtilis* fermentation could enhance the total ketone amount in seaweeds. *Saccharomyces cerevisiae* fermentation could also enhance the alcohol content in seaweeds. Principal component analysis of volatile compounds revealed that fermenting seaweeds with *B. subtilis* or *S. cerevisiae* could reduce aldehyde contents and boost ketone and alcohol contents, respectively, as expected. The odor of the fermented seaweeds was described by using GC–olfactometry, and *B. subtilis* and *S. cerevisiae* fermentations could enhance pleasant odors and reduce unpleasant odors. These results can support the capability of fermentation to improve the aromatic profile of seaweeds.

Keywords: aroma compounds; seaweeds; odor modification; HS-SPME; GC–MS; GC–O

Citation: Hung, Y.-H.R.; Peng, C.-Y.; Huang, M.-Y.; Lu, W.-J.; Lin, H.-J.; Hsu, C.-L.; Fang, M.-C.; Lin, H.-T.V. Monitoring the Aroma Compound Profiles in the Microbial Fermentation of Seaweeds and Their Effects on Sensory Perception. *Fermentation* **2023**, *9*, 135. https://doi.org/10.3390/fermentation9020135

Academic Editor: Niel Van Wyk

Received: 26 December 2022
Revised: 20 January 2023
Accepted: 29 January 2023
Published: 31 January 2023

1. Introduction

Food aroma compounds are volatile molecules that can be released during eating to reach the olfactory receptors. The aroma compounds in seaweeds serve as sex pheromones and chemical defense compounds against herbivores and pathogens [1]. Seaweeds and their aroma compounds have significant application potential in processing food products as seasonings because of their unique and strong aroma and flavors, such as marine, green, and umami aromas [2]. The aroma compounds in seaweeds can differ among species, including the most abundant ones, such as hydrocarbons, ketones, aldehydes, alcohols, esters, halogenated compounds, carboxylic acids, furans, phenols, sulfur compounds, pyrazines, pyridines, and amines [3,4].

Seaweeds have a variety of biological activities, such as antiviral, anti-oxidant, anti-inflammation, and anti-cancer activities, because of their availability, diversity, and productivity [5], and they have been used as food and medicine for centuries. In Asian countries, seaweeds are frequently consumed fresh and dried; however, seaweeds are not normally ingested in the unprocessed form in Western societies, where seaweeds remain of minor importance in spite of their nutritional benefits [6]. For example, the ingestion of red seaweed *Bangia fuscopurpurea* potentially reduces the risk of cardiovascular and chronic metabolic diseases, but its fishy malodor may limit consumers' acceptance [7]. Food sensory properties can play an important role in consumers' preferences and acceptance, of which the aroma compound is of prime importance [8]. Consequently, the identification of seaweed volatile compounds has drawn increasing attention for enhancing their application potential in food [3,9].

Microbial fermentation is often used to improve/enhance the sensory quality of food, such as cereals, meats, vegetables, and dairy foods, by removing undesirable off-flavors and/or generating new aroma compounds [10]. Recently, the application of microbial fermentation in removing undesirable odors in seaweeds has drawn increasing attention. Seo et al. [11] indicated that the inoculation of the fungus *Aspergillus oryzae* could decrease isovaleric acid and allyl isothiocyanate and reduce the peculiar smell of seaweed kelp extracts. In addition, the co-fermentation of *Bacillus subtilis* (BS) and *Lactobacillus plantarum* of the microalga *Spirulina* could reduce off-odors and produce creamy flavor compounds, such as acetoin and ethyl lactate [12]. However, research on sensory profiles of seaweeds during fermentation remains limited, not to mention that there are thousands of seaweed species.

Therefore, the aroma compound profiles of seaweeds should be investigated during microbial fermentation to promote the development of new products and to widen the application of seaweeds in food or beverage sectors. This study aimed to qualitatively and quantitatively characterize the volatiles and odor-active compounds in the green seaweed *Ulva* sp. and brown seaweed *Laminaria* sp. fermented with various microorganisms using headspace solid-phase microextraction (HS-SPME) coupled with gas chromatography–mass spectrometry (GC–MS) and GC–olfactometry (GC–O). Consequently, the changes in aroma compounds were monitored during microbial fermentation, and those most likely to impact sensory perception were identified.

2. Materials and Methods

2.1. Seaweeds

Dried green seaweed *Ulva* sp. was provided by Taiwan Yes (Taiwan Fertilizer Co., Ltd.) in Hualien County, Taiwan, and dried brown seaweed *Laminaria* sp. was purchased in Penghu County, Taiwan. The dried seaweed powders were prepared according to Lu et al. [13]. In summary, the seaweeds were washed, air-dried (40 °C), ground, and sieved through 0.25 mm pores and stored in a freezer until use.

2.2. Fermentation Strains

Five microorganisms were used in this study. *Bacillus subtilis* BCRC 10255, *Saccharomyces cerevisiae* BCRC 21685, *Lactobacillus delbrueckii* subsp. *bulgaricus* BCRC 10696 and *Lactobacillus casei* BCRC 10697 were purchased from Bioresource Collection and Research Center (BCRC) in Hsinchu City, Taiwan. *Bacillus subtilis* was cultured in Luria-Bertani medium (LB) at 30 °C at 150 rpm, *Saccharomyces cerevisiae* was cultured in Yeast Extract Peptone Dextrose medium (YPD) (Formedium, Norfolk, UK) at 24 °C at 150 rpm, and three lactic acid bacteria were cultured at Lactobacilli MRS medium (BD Difco, Franklin Lakes, NJ, USA) at 37 °C. The microbial cells were cultured to OD 0.6–0.8 (during the exponential phase) prior to the inoculation for seaweed fermentations.

2.3. Fermentation on Seaweeds

The seaweed suspension preparation and microbial fermentation were performed according to Hung et al. [14] with modifications. Seaweed powder (*Ulva* or *Laminaria* sp.) was mixed with distilled water in a 500 mL flask to make seaweed suspension (5%, *w/v*). The seaweed suspension was sterilized by autoclaving (121 °C/20 min). The microbial cultures were inoculated into the sterile seaweed suspension with a 0.1% (*v*/*v*) inoculation for fermentations. The fermentation for *Bacillus subtilis*, *Saccharomyces cerevisiae*, and the lactic acid bacteria was conducted at 30 °C /150 rpm, 24 °C/150 rpm, and 37 °C/static, respectively. The samples were taken at 0, 12, 24, 48, and 72 h during fermentation for the analyses of microbial counts by using aerobic plate counts. The volatile compounds were extracted by using headspace-solid phase microextraction after fermentation and analyzed by using gas chromatography-mass spectrometry. The control groups were the seaweed suspensions (*Ulva* or *Laminaria* sp.) without any microbial inoculation.

2.4. Headspace-Solid Phase Microextraction (HS-SPME)

The HS-SPME analysis was performed according to López-Pérez, Picon, and Nuñez [9] with some modifications. The volatile compounds were extracted by the 50/30 μm divinyl-benzene/carboxen/polydimethylsiloxane (DVB/CAR/PDMS) fiber method (Supelco Inc., Bellefonte, PA, USA), and has been used previously for extracting volatile compounds from seaweeds [9,15]. First, 4 mL of fermentation broth was mixed with 4 mL distilled water, and 1.5 g sodium chloride (NaCl) was added to improve extractive efficiency in HS-SPME. Zhang et al. [16] indicate that adding NaCl can improve ionic strength in the solution and further impact the extractive efficiency in HS-SPME. Five μL of 100 ppm ethyl cinnamate (Sigma-Aldrich, St. Louis, MO, USA) was added as the internal standard. The prepared sample was placed in a 60 °C water bath for 20 min with stirring for balancing, and then the SPME fiber was exposed to the headspace of the sample at the same temperature for 30 min. Finally, the SPME fiber was inserted into the GC injector for desorption at 250 °C for 5 min under non-splitting mode.

2.5. Gas Chromatography–Mass Spectrometry (GC–MS) and Gas Chromatography Olfactometry (GC–O)

The GC–MS analysis was performed on an Agilent Technologies-6890N GC coupled with Agilent 5973I Mass Selective Detector (Agilent Technologies, Santa Clara, CA, USA). The GC–MS system was equipped with a DB-WAX capillary column (30 m $\times$ 0.25 mm $\times$ 0.15 μm) (Agilent Technologies, Santa Clara, CA, USA). Helium was used as a carrier gas at a constant flow rate of 1.0 mL/min. The oven temperature was initially set at 40 °C for 2 min, then heated to 160 °C by 6 °C/min, and finally to 225 °C at 10 °C/min, maintained for 10 min. The mass spectrometry was operated in the electron impact (EI) mode at 70 eV and screened from 33 to 450 m/z. Library WILEY 275L was used to identify the volatile compounds for fermentation samples. The concentration of the identified volatile compounds was calculated by the relative peak area between the internal standard and the analyte [17].

The equation is followed by

$$C_2 \ (\mu g/L) = C_1 \ (\mu g/L) \times (A_2/A_1) \tag{1}$$

$C_1 \ (A_1)$: concentration (peak area) of the internal standard, $C_2 \ (A_2)$: concentration (peak area) of the identified volatile compound

The GC–O system shared the same equipment and analysis condition on the GC with the GC–MS. An olfactory detection port was equipped (OPD-3, Gerstel, Linthicum, MD, USA) on the GC. Once the analyte was separated by the GC, the column fluent was split by 1:1 to the mass spectrometry and the sniffing port individually. The odor descriptions of volatile compounds in the GC–O were followed by Ning et al. [18] with modifications. Three trained panelists evaluated the odor intensity by indicating strong (S), medium (M), and weak (W), and the sniffed retention indices and the notes of the odor were recorded as well.

2.6. Statistical Analysis

All experiments and analyses were conducted in triplicate. Data were expressed as mean $\pm$ SD and statistically analyzed by using IBM SPSS Statistics 23.0 (IBM, Armonk,

NY, USA). The statistically significant differences between the samples were determined by one-way analysis of variance (ANOVA), and Scheffe's test was used to indicate the significant differences where $p < 0.05$. Principle component analysis (PCA) and hierarchical clustering heatmaps were carried out by using MetaboAnalyst 5.0. These two analyses were applied to understand the changes in volatile compounds between fermented and unfermented seaweeds.

3. Results and Discussion

3.1. Seaweed Fermentation by the GRAS Microorganisms

In this study, five GRAS microorganisms were selected, i.e., *B. subtilis* (BS), *Saccharomyces cerevisiae* (SC), and *Lb. acidophilus* (LA), *Lb. delbrueckii* subsp. *bulgaricus* (LB), and *Lb. casei* (LC) for seaweed fermentations. The cell growth of these five strains in *Ulva* and *Laminaria* suspensions is shown in Figure S1. In *Ulva* and *Laminaria* suspensions, the cell count of BS and SC increased at 0–24 h, and the growth of lactic acid bacteria was observed during 0–12 h. In addition, the yeast maintained their viable cell number at 5–6 log CFU/mL in both seaweed suspensions during fermentation, and the bacteria maintained their viable cell number at 7–9 log CFU/mL, indicating that seaweed suspension could serve as the sole biomass for microbial fermentations.

3.2. Monitoring the Volatile Compound Profiles in Seaweed Fermentation

3.2.1. Ulva sp. Fermentation

A total of 51 volatile compounds were identified in the unfermented and fermented *Ulva* sp. suspension (Table 1) using HS-SPME–GC–MS. The profile of volatile compounds varies among the *Ulva* sp. suspensions fermented with five microorganisms. Twelve ketones were detected in the unfermented *Ulva* sp., and 20, 13, 11, 11, and 10 ketones were determined in the *Ulva* sp. suspension fermented with BS, SC, LA, LB, and LC, respectively. In the unfermented *Ulva* sp., aliphatic ketones with long chains ($\geq$C7), such as 2,2,6-Trimethylcyclohexanone (floral), 6-Methyl-5-hepten-2-one (also known as sulcatone; citral and musty), and β-Ionone (violet, floral) [19], were identified. β-ionone is a potent odorant in seafood, and sulcatone has been identified in various foods as a metabolite of lycopene [20]. These ketone compounds could be detected in the commonly dehydrated *Ulva lactuca* [9]. The total amount of ketones was significantly increased in the *Ulva* suspension after BS fermentation, particularly short-chain aliphatic ketones ($C \leq 7$), such as pentanone (fruity and pungent) and heptanone (cheesy, fruity, and spicy). In general, aliphatic ketones with shorter chains have a strong aroma, and they can be generated through the lipoxygenation of fatty acids during fermentation [21]. On the contrary, lactic acid bacteria fermentations could reduce the total amount of ketones

Table 1. Changes of the aromatic profile in the green seaweed *Ulva* sp. fermented by various microorganisms.

Compounds	Concentration (µg/L)					
	Unfermented	BS	SC	LA	LB	LC
Ketones						
2-Butanone	nd	56.33 ± 2.05 [a]	11.18 ± 1.80 [b]	10.53 ± 4.07 [b]	2.67 ± 0.34 [b]	7.29 ± 0.20 [b]
2-Pentanone	nd	164.64 ± 6.60 [a]	81.24 ± 3.89 [b]	nd	nd	nd
3-Hexanone	nd	7.93 ± 0.14	14.78 ± 0.01	nd	nd	nd
2-Hexanone	nd	33.03 ± 0.12	nd	nd	nd	nd
5-Methyl-2-hexanone	nd	23.45 ± 1.86	nd	nd	nd	nd
2-Heptanone	nd	436.59 ± 4.50	nd	nd	nd	nd
6-Methyl-2-heptanone	nd	12.85 ± 1.70	nd	nd	nd	nd
2,2,6-Trimethylcyclohexanone	39.66 ± 1.64 [a]	30.28 ± 1.32 [b]	nd	nd	6.03 ± 1.92 [c]	nd

Table 1. *Cont.*

Compounds	Concentration (µg/L)					
	Unfermented	BS	SC	LA	LB	LC
6-Methyl-5-hepten-2-one	23.07 ± 3.21 [c]	47.32 ± 4.38 [b]	nd	114.31 ± 7.44 [a]	62.84 ± 2.52 [b]	47.43 ± 2.14 [b]
3,5,5-Trimethyl-2-cyclohexen-1-one	31.33 ± 2.60 [a]	4.57 ± 0.08 [b]	5.92 ± 0.13 [b]	7.55 ± 1.15 [b]	5.33 ±0.27 [b]	5.68 ± 0.11 [b]
3,5-Octadien-2-one	47.78 ± 1.12 [cd]	34.12 ± 0.37 [d]	77.71 ± 3.92 [b]	147.08 ± 5.40 [a]	48.77 ± 3.48 [cd]	61.86 ± 10.29 [bc]
6-Methyl-3,5-heptadien-2-one	31.68 ± 3.58 [d]	50.78 ± 5.93 [b]	33.91 ± 1.21 [cd]	69.67 ± 3.29 [a]	46.88 ± 0.57 [bc]	60.75 ± 1.35 [ab]
4-Ketoisophorone	19.68 ± 1.48 [b]	1.08 ± 0.22 [c]	56.07 ± 3.39 [a]	11.94 ± 2.77 [b]	nd	nd
Propiophenone	61.68 ± 1.12 [a]	33.20 ± 0.50 [b]	61.16 ± 3.09 [a]	nd	nd	nd
5-Ethyl-2(5H)-furanone	15.33 ± 0.33 [ab]	7.34 ± 0.34 [c]	18.09 ± 0.32 [a]	12.96 ± 1.12 [b]	5.81 ± 0.28 [c]	6.42 ± 1.22 [c]
trans-β-Ionone	nd	36.35 ± 0.32	nd	nd	nd	nd
β-Ionone	211.59 ± 3.93 [ab]	149.52 ± 2.60 [d]	201.70 ± 6.00 [bc]	183.55 ± 1.56 [c]	147.39 ± 3.51 [d]	223.88 ± 8.74 [a]
5,6-Epoxide-β-Ionone	210.31 ± 0.64 [a]	114.60 ± 2.73 [cd]	189.46 ± 3.09 [a]	149.01 ± 11.20 [b]	104.40 ± 3.81 [d]	143.16 ± 11.36 [bc]
Total ketones	738.05 ± 38.28 [b]	1243.98 ± 19.14 [a]	751.21 ± 16.28 [b]	706.59 ± 25.97 [b]	430.11 ± 7.18 [d]	556.47 ± 26.58 [c]
Aldehydes						
3-Methylbutanal	99.47 ± 1.69 [a]	nd	nd	34.47 ± 3.47 [b]	28.33 ± 2.90 [b]	nd
Hexanal	234.47 ± 4.27 [a]	nd	58.38 ± 1.21 [b]	44.98 ± 6.08 [b]	20.55 ± 1.80 [c]	56.88 ± 1.82 [b]
(E)-2-Pentenal	61.75 ± 1.52 [a]	nd	29.38 ± 0.06 [b]	23.20 ± 1.52 [bc]	11.59 ± 1.54 [d]	22.48 ± 0.66 [c]
2-Methyl-2-pentenal	8.61 ± 1.11 [a]	nd	nd	nd	7.14 ± 0.91 [a]	nd
Heptanal	194.50 ± 12.07 [a]	nd	nd	108.34 ± 10.39 [b]	52.29 ± 4.77 [c]	nd
(E)-2-Hexenal	63.60 ± 5.26 [a]	nd	nd	45.49 ± 3.77 [b]	18.09 ± 0.11 [c]	20.88 ± 2.13 [c]
Furfural	49.72 ± 2.77 [a]	nd	nd	51.72 ± 3.24 [a]	26.02 ± 1.51 [b]	22.69 ±1.37 [b]
5-Methylfurfural	560.93 ± 25.70 [a]	nd	501.50 ± 12.66 [b]	405.88 ± 7.23 [c]	308.09 ± 11.48 [d]	321.86 ± 11.53 [d]
2,4-Heptadienal	25.02 ± 1.78 [a]	nd	16.98 ± 0.59 [b]	nd	nd	nd
Safranal	43.37 ± 4.61 [b]	nd	93.64 ± 4.49 [a]	44.26 ± 0.76 [b]	6.38 ± 0.46 [c]	1.81 ± 0.57 [c]
Total aldehydes	1341.44 ± 30.80 [a]	nd	699.87 ± 17.50 [b]	758.34 ± 28.53 [b]	478.48 ± 8.92 [c]	446.59 ± 12.27 [c]
Alcohols						
Ethanol	nd	nd	7.78 ± 0.73	nd	nd	nd
1-Butanol	nd	5.01 ± 0.93	nd	nd	nd	nd
1-Penten-3-ol	61.13 ± 3.21 [a]	32.03 ± 5.10 [b]	40.44 ± 3.71 [b]	27.75 ± 0.56 [bc]	12.90 ± 0.77 [c]	nd
1-Pentanol	nd	21.92 ± 1.34 [b]	61.16 ± 0.34 [a]	nd	nd	nd
1-Hexanol	nd	nd	289.18 ± 12.93	nd	nd	nd
1-Heptanol	nd	nd	116.55 ± 5.13	nd	nd	nd
1-Octanol	nd	nd	20.19 ± 0.76	nd	nd	nd
2,6-Dimethylcyclohexanol	109.89 ± 4.73 [a]	nd	106.64 ± 0.26 [a]	nd	nd	nd
5-Methyl-2-furanmethanol	nd	28.48 ± 1.22	nd	nd	nd	nd
Total alcohols	171.02 ± 5.90 [b]	87.44 ± 5.47 [c]	641.94 ± 18.53 [a]	27.75 ± 0.46 [d]	12.90 ± 0.63 [d]	nd
Acids						
Hexanoic acid	nd	nd	nd	7.87 ± 0.57 [a]	4.93 ± 0.39 [b]	5.86 ± 0.13 [b]
Esters						
Methyl salicylate	163.25 ± 1.61 [c]	240.20 ± 12.70 [ab]	190.91 ± 4.58 [c]	219.98 ± 1.52 [b]	70.57 ± 6.66 [d]	257.76 ± 4.48 [a]
Benzenes and benzene derivatives						
Benzaldehyde	321.98 ± 4.94 [a]	25.18 ± 0.49 [c]	30.14 ± 3.05 [c]	299.63 ± 14.09 [a]	242.48 ± 9.25 [b]	238.97 ± 10.15 [b]
Benzyl alcohol	nd	14.81 ± 0.44 [b]	23.34 ± 0.69 [a]	nd	nd	nd
Benzoic acid	nd	nd	nd	nd	8.21 ± 0.54 [b]	16.93 ± 1.21 [a]
Phenol	nd	20.90 ± 0.28	nd	nd	nd	nd
2,4-Di-tert-butylphenol	7.65 ± 1.25 [a]	7.47 ± 0.37 [a]	5.56 ± 0.57 [ab]	5.67 ± 0.14 [ab]	3.40 ± 0.25 [b]	4.56 ± 0.44 [b]
Total benzenes and benzene derivatives	329.64 ± 5.87 [a]	68.56 ± 0.64 [c]	59.03 ± 2.17 [c]	305.30 ± 11.51 [a]	254.09 ± 9.09 [b]	260.46 ± 9.71 [b]

Table 1. *Cont.*

Compounds	Concentration (µg/L)					
	Unfermented	BS	SC	LA	LB	LC
Hydrocarbons						
Toluene	173.85 ± 1.65 [a]	39.45 ± 1.94 [c]	62.61 ± 2.18 [b]	15.35 ± 0.40 [e]	26.97 ± 1.89 [d]	nd
Furans						
2-Ethylfuran	299.42 ± 23.69 [a]	nd	38.19 ± 2.18 [b]	61.68 ± 0.39 [b]	14.32 ± 1.12 [b]	16.55 ± 0.66 [b]
Pyrazines						
2,5-Dimethylpyrazine	nd	159.03 ± 5.90	nd	nd	nd	nd
Sulfur-containing						
Dimethyl sulfide	nd	26.71 ± 3.56	nd	nd	nd	nd
Isobutyl isothiocyanate	nd	3.16 ± 0.52	nd	nd	nd	nd
Miscellaneous						
Methylethylmaleimide	19.43 ± 0.23 [ab]	6.32 ± 0.29 [e]	16.14 ± 1.59 [bc]	20.84 ± 0.46 [a]	15.10 ± 1.38 [cd]	12.03 ± 1.05 [d]
Dihydroactinidiolide	56.51 ± 3.48 [ab]	58.12 ± 0.63 [a]	47.21 ± 5.19 [ab]	57.08 ± 1.09 [a]	43.17 ± 1.64 [b]	48.28 ± 2.49 [ab]
Total Miscellaneous	75.94 ± 4.45 [a]	64.44 ± 0.38 [b]	63.35 ± 5.28 [b]	77.92 ± 1.81 [a]	58.27± 1.67 [b]	60.31 ± 1.22 [b]

All measurements were performed in triplicate and expressed as mean ± SD (n = 3). Different letters within the same row represent significant differences; nd: non-detected.

Ten aldehydes were detected in the unfermented *Ulva* sp., including linear, branched, or unsaturated aldehydes (Table 1). Most aldehydes have short chains (C ≤ 7), such as furfural, hexanal, and heptanal. According to Peinado et al. [22], the notes of hexanal and heptanal were described as fishy odors. Aldehydes such as hexanals and heptanals were reduced after all microbial fermentations, in which BS fermentation was of the most significance. Moreover, the amount of safranal, a compound that contributes a spicy saffron-like note with herbaceous, tobacco facets, and floral undertones, was doubled after SC fermentation.

As for alcohols, the aroma compound 1-penten-3-ol (green) [20] was identified in the unfermented *Ulva* sp., and it was decreased in each microbial fermentation. The total alcohol content decreased after microbial fermentations except for SC (Table 1). Five additional alcohols were generated in the *Ulva* sp. suspension after SC fermentation, such as 1-pentanol (fermented, oily, sweet) and 1-hexanol (fruity, floral aromatic) [20,23]. In the metabolism of brewing yeast, the Ehrlich pathway is a metabolic route to produce high alcohol content from amino acids, involving transamination, decarboxylation, and reduction [24]. This finding might explain the alcohol production during the SC fermentation of *Ulva* sp.

Benzaldehyde (bitter almond and burnt sugar) was identified in the unfermented *Ulva* sp., and it was converted into benzyl alcohol (floral and rose) [20] by alcohol dehydrogenase after BS or SC fermentation [25]. Our results were similar to those reported by Wang et al. [26]; that is, benzaldehyde and benzyl alcohol were decreased and increased, respectively, through SC fermentation. On the contrary, benzaldehyde was oxidized into benzoic acid (sweet and pleasant odor) [27] during lactic acid fermentation, which was also observed in lactic acid bacteria-fermented dairy products [28]. In addition, the number of other aroma compounds, such as furan 2-ethyfuran (coca), was reduced after microbial fermentations, but BS fermentation increased the amount of 2,5-dimethylpyrazine (roasted, nutty) and dimethyl sulfide (boiled cabbage). The overall changes in volatile compounds for the unfermented and fermented *Ulva* sp. were summarized in the heat map (Figure 1). Compared with the unfermented *Ulva* sp., BS or SC fermentation could reduce aldehyde contents, which might mitigate potential unpleasant odors from seaweed. BS fermentation improved ketone production, whereas SC increased alcohol production during fermentation. By contrast, volatile acids were found in the *Ulva* sp. suspension fermented with lactic acid bacteria (LA, LB, and LC).

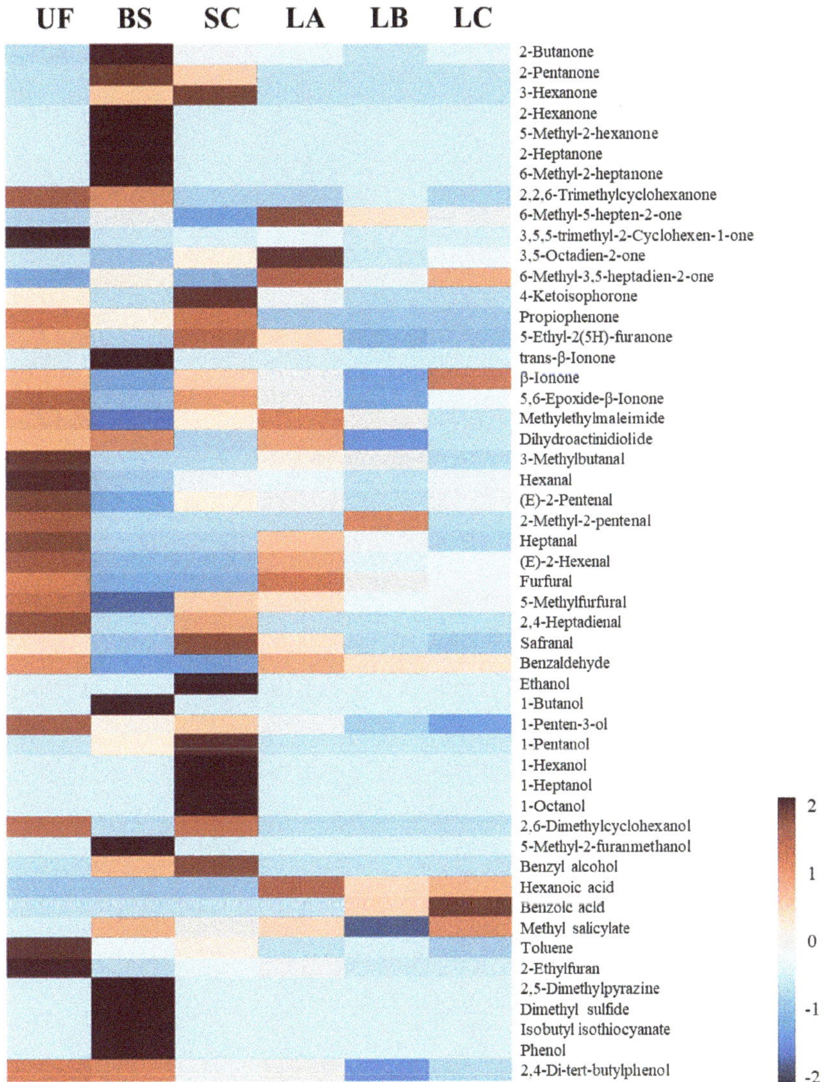

Figure 1. Heat map of the volatile compounds in the green seaweed *Ulva* sp. fermented by various microorganisms. Colors indicate the Z-score distances to the mean levels of the observations. A Z-score greater than 1.645 (red color) or lower than −1.645 (blue color) is significant at the 0.05 level. UF: unfermented; BS: *B. subtilis*; SC: *S. cerevisiae*; LA: *Lb. acidophilus*; LB: *Lb. delbrueckii* subsp. *bulgaricus*; LC: *Lb. casei*.

3.2.2. *Laminaria* sp. Fermentation

A total of 55 volatile compounds were identified in the unfermented and fermented *Laminaria* sp. (Table 2). In addition, a total of 31 volatile compounds were identified in the unfermented *Laminaria* sp., and 36, 40, 41, 40, and 40 volatile compounds were found in the *Laminaria* sp. suspension fermented with BS, SC, LA, LB, and LC, respectively.

Table 2. Changes of the aromatic profile in the brown seaweed *Laminaria* sp. fermented by various microorganisms.

Compounds	Concentration (µg/L)					
	Unfermented	BS	SC	LA	LB	LC
Ketones						
2,3-Butanedione	nd	29.03 ± 8.54	nd	nd	nd	nd
1-Penten-3-one	18.97 ± 2.08 [a]	8.65 ± 0.78 [b]	3.32 ± 0.57 [b]	7.03 ± 1.33 [b]	20.77 ± 2.45 [a]	22.18 ± 0.84 [a]
3-Octanone	nd	113.96 ± 9.33 [b]	215.13 ± 7.16 [a]	33.38 ± 0.88 [d]	53.68 ± 1.83 [c]	46.85 ± 2.08 [cd]
3-Hydroxy-2-butanone	nd	672.20 ± 35.20	nd	nd	nd	nd
2-Octanone	nd	nd	40.54 ± 1.44	nd	nd	nd
1-Octen-3-one	254.73 ± 4.42 [bc]	207.96 ± 2.92 [d]	87.66 ± 0.70 [e]	228.68 ± 4.92 [cd]	277.10 ± 11.00 [ab]	297.45 ± 16.51 [a]
3-Octen-2-one	nd	nd	11.49 ± 0.77	nd	nd	nd
3,5-Octadien-2-one	9.86 ± 0.22 [d]	10.12 ± 1.75 [d]	21.77 ± 1.03 [a]	11.99 ± 0.97 [cd]	13.29 ± 0.85 [bc]	15.54 ± 0.66 [b]
β-Ionone	3.82 ± 0.58 [c]	3.71 ± 0.21 [c]	7.85 ± 0.79 [b]	11.78 ± 1.25 [a]	13.47 ± 0.76 [a]	12.77 ± 0.25 [a]
Total ketones	278.38 ± 7.98 [c]	1048.63 ± 37.34 [a]	387.76 ± 8.23 [b]	285.83 ± 6.33 [c]	378.31 ± 9.95 [b]	394.79 ± 23.44 [b]
Aldehydes						
3-Methylbutanal	2.56 ± 0.05 [b]	nd	nd	6.42 ± 0.74 [a]	1.90 ± 0.08 [b]	1.42 ± 0.36 [b]
2-Butenal	55.74 ± 0.89 [ab]	28.78 ± 1.89 [cd]	24.42 ± 3.09 [d]	41.96 ± 4.29 [bc]	67.56 ± 3.39 [a]	64.68 ± 7.23 [a]
Hexanal	265.90 ± 15.47 [a]	95.22 ± 3.50 [cd]	60.55 ± 2.27 [d]	112.22 ± 10.99 [c]	191.26 ± 9.72 [b]	162.73 ± 13.50 [b]
(E)-2-Pentenal	18.89 ± 1.92 [ab]	10.38 ± 0.97 [c]	10.02 ± 0.35 [c]	13.64 ± 0.71 [bc]	21.36 ± 1.84 [a]	20.71 ± 2.57 [a]
Heptanal	26.92 ± 2.84 [a]	11.99 ± 1.57 [b]	15.39 ± 1.12 [b]	11.45 ± 0.74 [b]	14.69 ± 2.82 [b]	13.66 ± 1.70 [b]
(E)-2-Hexenal	29.48 ± 0.83 [a]	13.27 ± 0.80 [c]	10.22 ± 0.19 [c]	18.70 ± 2.31 [b]	28.88 ± 0.91 [a]	28.26 ± 1.01 [a]
(Z)-4-Heptenal	10.94 ± 2.58 [a]	4.42 ± 0.21 [b]	4.40 ± 0.06 [b]	9.92 ± 0.32 [ab]	11.78 ± 1.72 [a]	15.23 ± 1.24 [a]
Octanal	40.10 ± 1.26 [b]	nd	nd	35.97 ± 1.29 [b]	60.53 ± 3.28 [a]	61.45 ± 1.74 [a]
(E)-2-Heptenal	1179.00 ± 45.48 [a]	538.71 ± 31.97 [d]	429.48 ± 7.48 [d]	761.00 ± 32.68 [c]	931.45 ± 7.23 [b]	1053.81 ± 37.85 [ab]
Nonanal	20.36 ± 0.86 [ab]	6.56 ± 1.58 [d]	21.16 ± 1.58 [a]	11.82 ± 0.59 [c]	16.55 ± 0.46 [abc]	16.22 ± 1.46 [bc]
Furfural	65.36 ± 1.74 [a]	nd	nd	26.26 ± 1.93 [b]	22.71 ±2.21 [b]	23.45 ± 2.80 [b]
(E)-2-Octenal	nd	nd	nd	95.66 ± 7.85 [b]	163.08 ± 17.07 [a]	106.47 ± 3.97 [b]
2,4-Heptadienal	76.42 ± 8.21 [a]	39.33 ± 2.57 [cd]	34.32 ± 3.78 [d]	49.91 ± 1.59 [bcd]	62.51 ± 6.82 [ab]	57.43 ± 0.86 [bc]
5-Methylfurfural	8.22 ± 0.61 [b]	nd	nd	6.92 ± 0.79 [b]	11.52 ± 0.40 [a]	10.95 ± 1.11 [a]
(E,Z)-2,6-Nonadienal	5.68 ± 0.48 [c]	nd	6.60 ± 1.48 [c]	24.14 ± 0.87 [b]	27.55 ± 0.56 [a]	21.88 ± 0.49 [b]
(E)-2-Decenal	76.05 ± 5.16 [d]	107.63 ± 4.87 [d]	155.49 ± 4.71 [c]	264.40 ± 16.70 [b]	358.26 ± 14.80 [a]	362.13 ± 14.09 [a]
2,4-Nonadienal	15.89 ± 0.73 [c]	5.15 ± 0.80 [d]	nd	31.85 ± 0.92 [a]	24.02 ± 1.59 [b]	28.67 ±1.00 [a]
2,4-Decadienal	49.62 ± 2.56 [b]	15.08 ± 1.12 [c]	24.91 ± 3.88 [bc]	106.58 ± 2.20 [a]	121.19 ± 9.81 [a]	118.74 ± 16.88 [a]
Total aldehydes	1947.12 ± 53.11 [b]	876.51 ± 33.6 [d]	796.96 ± 12.63 [d]	1628.84 ± 47.68 [c]	2136.79 ± 50.62 [a]	2167.90 ± 48.24 [a]
Alcohols						
Ethanol	nd	nd	12.63 ± 0.80	nd	nd	nd
Cyclopentanol	nd	nd	11.70 ± 0..58	nd	nd	nd
3-Methylbutanol	nd	nd	23.46 ± 0.96	nd	nd	nd
1-Pentanol	nd	7.30 ± 0.62 [b]	16.89 ± 1.66 [a]	4.64 ± 0.54 [b]	7.90 ± 0.35 [b]	7.53 ± 0.51 [b]
1-Hexanol	nd	127.85 ± 3.58 [c]	259.36 ± 2.34 [a]	128.01 ± 8.90 [c]	201.03 ± 10.89 [b]	211.71 ± 11.57 [b]
1-Octen-3-ol	529.91 ± 24.37 [a]	538.68 ± 16.09 [a]	253.74 ± 3.03 [b]	544.97 ±18.97 [a]	514.24 ±31.25 [a]	593.38 ± 3.70 [a]
1-Heptanol	nd	nd	910.33 ± 23.18	nd	nd	nd
2,3-Butanediol	nd	456.59 ± 4.97	nd	nd	nd	nd
1-Octanol	90.09 ± 1.90 [d]	201.19 ± 14.91 [c]	455.22 ± 4.54 [a]	196.11 ± 2.55 [c]	262.85 ± 2.90 [b]	290.12 ± 11.55 [b]
(E)-2-Octen-1-ol	191.70 ± 4.63 [b]	202.29 ± 31.14 [b]	192.55 ± 10.89 [b]	192.66 ± 4.30 [b]	274.78 ± 19.36 [a]	288.80 ± 15.10 [a]
2,7-Octadien-1-ol	51.53 ± 1.56 [d]	64.25 ± 1.84 [c]	57.60 ± 2.07 [cd]	76.57 ± 4.31 [b]	111.36 ± 1.79 [a]	104.31 ± 2.80 [a]
1-Decanol	nd	nd	23.09 ± 1.34	nd	nd	nd
Total alcohols	863.23 ± 30.80 [e]	1598.15 ± 39.04 [b]	2216.58 ± 24.71 [a]	1142.96 ± 14.19 [d]	1372.16 ± 62.92 [c]	1495.85 ± 42.46 [bc]

Table 2. *Cont.*

Compounds	Concentration (µg/L)					
	Unfermented	BS	SC	LA	LB	LC
Acids						
Acetic acid	nd	17.36 ± 0.43	nd	nd	nd	nd
2-Methylbutanoic acid	nd	49.21 ± 44.15	nd	nd	nd	nd
Hexanoic acid	nd	nd	nd	10.91 ± 0.48 [b]	15.98 ± 1.17 [a]	15.26 ± 0.78 [a]
Octanoic acid	nd	12.20 ± 0.55 [a]	6.31 ± 0.67 [b]	10.29 ± 0.92 [a]	10.23 ± 0.86 [a]	12.45 ± 0.76 [a]
Nonanoic acid	nd	11.07 ± 0.93 [a]	nd	14.96 ± 1.84 [a]	13.94 ± 3.46 [a]	14.33 ± 0.76 [a]
Decanoic acid	nd	nd	nd	4.20 ± 0.61 [a]	4.94 ± 0.44 [a]	5.67 ± 0.60 [a]
Dodecanoic acid	nd	nd	nd	5.32 ± 0.48	nd	nd
Tetradecanoic acid	18.80 ± 1.95 [d]	24.98 ± 1.55 [d]	48.72 ± 4.91 [bc]	56.89 ± 2.48 [ab]	62.49 ± 0.94 [a]	40.36 ± 1.68 [c]
Hexadecanoic acid	nd	12.89 ± 1.71 [c]	24.13 ± 3.76 [b]	41.40 ± 0.76 [a]	35.01 ± 3.87 [ab]	31.15 ± 3.54 [ab]
Total acids	18.80 ± 1.95 [d]	127.71 ± 1.58 [ab]	79.17 ± 8.81 [c]	143.97 ± 2.80 [a]	142.58 ± 4.55 [a]	119.21 ± 5.36 [b]
Esters						
Methyl salicylate	46.54 ± 6.17 [d]	59.44 ± 4.59 [d]	131.18 ± 7.47 [c]	296.03 ± 8.07 [a]	219.04 ± 6.85 [b]	118.80 ± 2.05 [c]
Benzenes and benzene derivatives						
Benzaldehyde	37.53 ± 1.77 [b]	nd	10.57 ± 0.60 [c]	43.14 ± 1.38 [ab]	41.65 ± 1.80 [b]	47.99 ± 2.40 [a]
Benzyl alcohol	nd	nd	8.49 ± 045	nd	nd	nd
Phenethyl alcohol	nd	nd	39.24 ± 2.20	nd	nd	nd
2,4-Di-tert-butylphenol	3.10 ± 0.26 [c]	4.21 ± 0.07 [bc]	3.96 ± 0.09 [c]	5.42 ± 0.10 [ab]	5.55 ± 0.27 [a]	5.31 ± 0.59 [ab]
Total benzenes and benzene derivatives	40.63 ± 1.58 [c]	4.21 ± 0.07 [d]	62.27 ± 2.75 [a]	48.56 ± 1.31 [b]	47.20 ± 1.60 [bc]	53.31 ± 2.70 [b]
Miscellaneous						
Methylethylmaleimide	3.71 ± 0.60 [b]	2.68 ± 0.37 [b]	6.32 ± 0.24 [a]	3.84 ± 0.58 [b]	4.03 ± 0.56 [b]	4.20 ± 0.36 [ab]
Dihydroactinidiolide	8.22 ± 0.64 [c]	9.68 ± 0.48 [bc]	9.00 ± 0.15 [bc]	12.19 ± 0.42 [a]	10.86 ± 0.34 [ab]	10.77 ± 0.74 [ab]
Total Miscellaneous	11.94 ± 0.39 [b]	12.36 ± 0.30 [b]	15.32 ± 0.42 [a]	16.03 ± 0.81 [a]	14.89 ± 0.96 [a]	14.97 ± 1.25 [a]

All measurements were performed in triplicate and expressed as mean ± SD (n = 3). Different letters within the same row represent significant differences; nd: non-detected.

Total ketones were increased after fermentation with BS, SC, LB, or LC (Table 2), with BS showing the most increases. 1-Octen-3-one (metallic and mushroom) [29] and β-ionone were identified in the fermented and unfermented groups. Notably, 3-octanone, described as herb, butter, and resin notes [30], was generated in the fermented *Laminaria* sp. suspension, and 3-hydroxy-2-butanone (sweet), also known as acetoin, was generated in the BS-fermented group. β-ionone, which was responsible for the rose aroma [31] and sweet note in green tea [32], was enhanced during SC, LA, LB, or LC fermentation.

Linear, branched, or volatile unsaturated aldehydes, which were identified in brown seaweeds, may contribute to seaweed-like or seafood-like odors [11,22]. Our data indicated that the total aldehyde content was decreased in the *Laminaria* sp. suspension fermented with BS, SC, and LA (Table 2). Pan et al. [33] have reported that yeast fermentation can improve the odor of tiger puffer skin gelatin and reduce its volatile aldehydes. Heptanal (dry fish notes), hexanal (fishy notes) [22,34], and furfural (burnt note) were reduced in the fermented *Laminaria* sp. suspension, which is consistent with the observation in the fermented *Ulva* sp. (Table 1), indicating the potential application of fermentation in removing off-odors from seaweeds.

The major alcohol compound detected in the unfermented *Laminaria* sp. is 1-octen-3-ol (529.91 µg/L, Table 2), which is described as fishy and grassy notes [34]. Based on the report of López-Pérez, Picon, and Nuñez [9], 1-octen-3-ol has a high odor activity value and low perception threshold, and it is considered the major alcohol compound in brown seaweeds

(*Himanthalia elongata* and *Laminaria ochroleuca*). SC fermentation could significantly reduce 1-Octen-3-ol. In addition, a high content of alcohols was produced via fermentation, such as 1-pentanol (fermented, oily, sweet) and 1-hexanol (fruity, floral aromatic), which is consistent with our observations in the fermented *Ulva* sp. (Table 1). The EMP pathway and Ehrlich pathway are two possible metabolic routes for converting carbohydrates and amino acids into branched alcohols during yeast fermentation, respectively [34]. Thus, more alcoholic compounds were identified in the *Laminaria* sp. fermented with SC compared with the unfermented group.

With regard to volatile acids, tetradecanoic acid was identified in the unfermented *Laminaria* sp. and acid varieties, and total acid contents increased after microbial fermentation (Table 2). Saturated fatty acids, such as hexanoic acid (fatty, cheesy), octanoic acid (fatty, waxy) [20], and decanoic acid, were generated during microbial fermentation. In addition, a branched carboxylic acid (2-methylbutanoic acid) was identified only in *Laminaria* fermented with BS. Based on the report of Leejeerajumnean et al. [35], 2-methylbutanoic acid was identified in soybean products fermented with *Bacillus* sp.

Therefore, changes in volatile compounds in the unfermented and fermented *Laminaria* sp. were demonstrated in the heat map (Figure 2). Compared with the unfermented *Laminaria* sp., BS and SC seemed to significantly decrease aldehyde content during fermentation. In addition, BS fermentation enhanced ketone production, and SC fermentation boosted alcohol formation in fermented *Laminaria* sp. suspensions, which were also observed in *Ulva* sp. fermentation (Figure 1; Table 1). On the contrary, lactic acid bacteria (LA, LB, and LC) might slightly reduce aldehyde production when fermenting *Laminaria* sp., compared with BS and SC. Based on changes in the profile of volatile compounds among fermentations, BS and SC have shown application potential in mitigating off-odors from seaweeds (*Ulva* sp. and *Laminaria* sp.) and modifying their aromatic perceptions after fermentation. Thus, principal component analysis (PCA) and odor analysis of fermented seaweeds were performed.

3.3. Principal Component Analysis (PCA) of the Volatile Compounds in the Fermented Seaweeds

3.3.1. Ulva sp.

The PCA of chemical groups of the volatile compounds in the unfermented and fermented *Ulva* sp. is shown in Figure 3. The PCA can explain 84.1% of the total variances in the dataset (59.5% and 24.6% by principal component 1 [PC1] and principal component 2 [PC2], respectively). Based on the score plot (Figure 3A), the BS is in the first quadrant of the plot, whereas the unfermented *Ulva* sp. is in the second quadrant. This result shows the difference in the composition of volatile compounds between the BS and the unfermented *Ulva* sp. Moreover, in the loading plot (Figure 3B), the ketone family is observed in the positive axis of the PC1, whereas the aldehyde family is observed in the negative axis of the PC1. This negative correlation between the ketone and aldehyde family indicates the difference in volatile compounds between the BS and the unfermented *Ulva*. The aforementioned statement is supported by Table 1, i.e., ketones are increased, and aldehydes are reduced after BS fermentation. On the contrary, alcohol formation is associated with SC fermentation, which demonstrates the highest alcohol content among other fermentations.

In evaluating the specific compound affecting the difference among fermentations, the PCA of a single volatile compound is shown in Figure 4. The PC1 and PC2 can explain 57.1% and 20.5% of the total variances in the dataset, respectively (total: 77.6%). Based on the score plot (Figure 4A), the BS and unfermented group are in a different quadrant of PC1, which correlates to Figure 3A. In addition, 2-heptanone is identified in the positive axis of PC1 in the loading plot (Figure 4B), which indicates that this compound likely affects the difference between the BS and the unfermented *Ulva*.

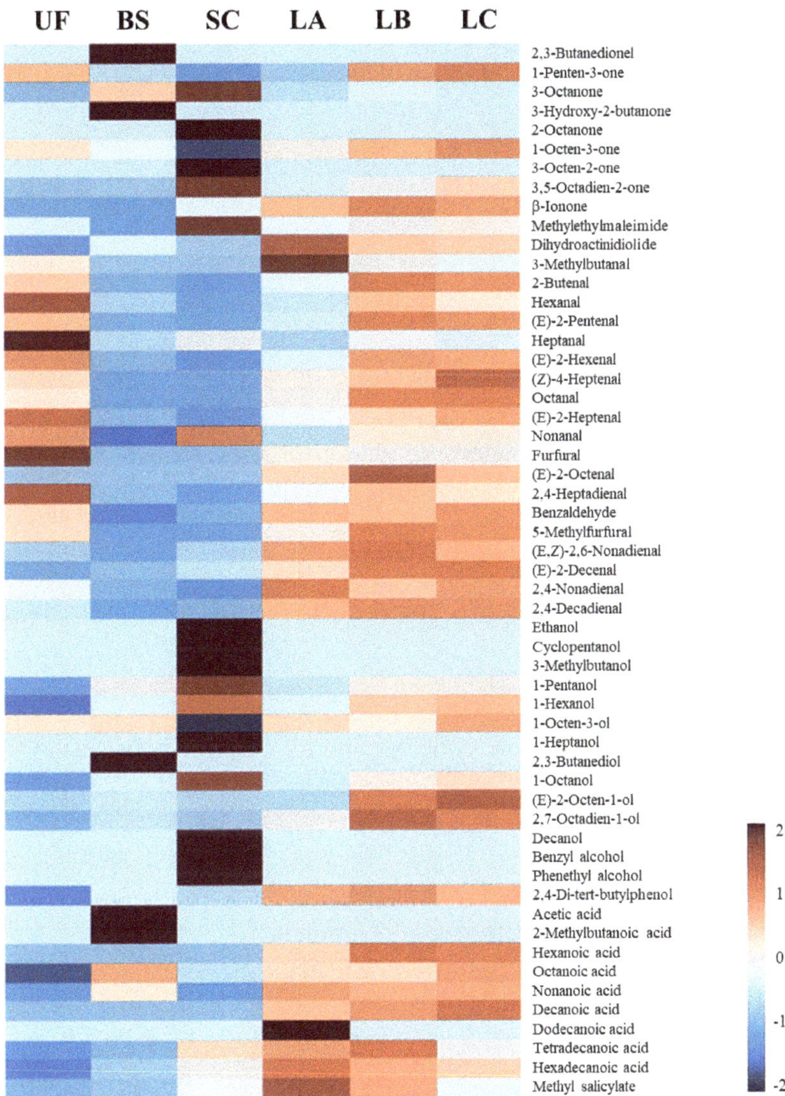

Figure 2. Heat map of the volatile compounds in the brown seaweed *Laminaria* sp. fermented by various microorganisms. Colors indicate the Z-score distances to the mean levels of the observations. A Z-score greater than 1.645 (red color) or lower than −1.645 (blue color) is significant at the 0.05 level. UF: unfermented; BS: *B. subtilis*; SC: *S. cerevisiae*; LA: *Lb. acidophilus*; LB: *Lb. delbrueckii* subsp. *bulgaricus*; LC: *Lb. casei*.

Moreover, 2-heptanone (436.59 μg/L) was formed after fermentation with BS, and its content was dominant at around 33% of total ketones (1308.42 μg/L) in the fermented *Ulva* (Table 1). The difference in volatile compounds between the SC and the unfermented group based on PC2 is shown in Figure 4A. Based on the loading plot (Figure 4B), 1-hexanol is in the positive axis of PC2, which indicates its contribution to the difference in volatile compounds between the SC and the unfermented *Ulva*. This result might correspond to the formation of 1-hexanol (289.18 μg/L) after SC fermentation (Table 1). Based on our GC–O analysis, the odor intensity of 1-hexanol, described as floral and sweet, is enhanced by SC fermentation. Therefore, 2-heptanone might contribute to the difference between

Ulva fermented with BS and unfermented *Ulva*. Furthermore, 1-hexanol plays an important role in the fermentation of *Ulva* with SC compared with the unfermented *Ulva* (Figure 4). These results are consistent with the PCA of chemical groups shown in Figure 3.

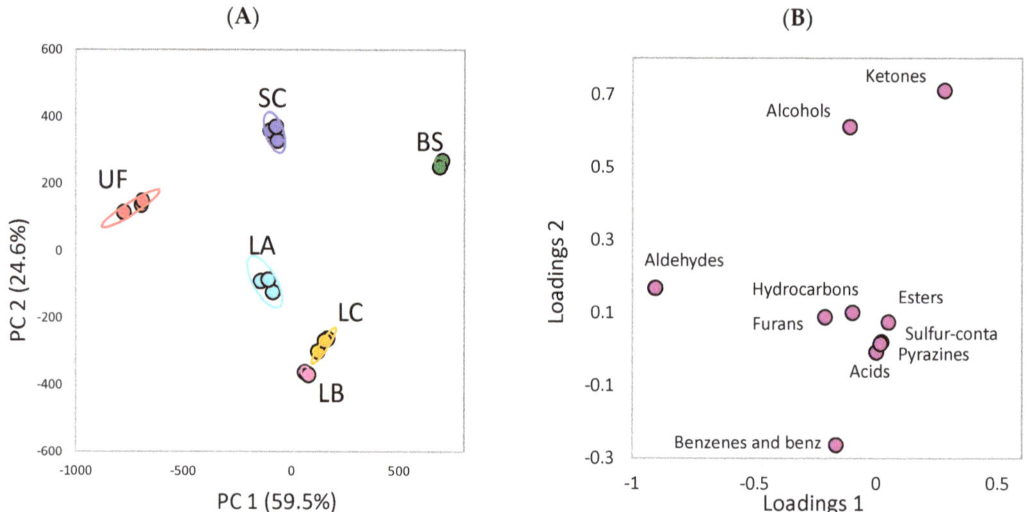

Figure 3. Principal component analysis plots of chemical groups of the volatile compounds in the green seaweed *Ulva* sp. fermented by various microorganisms: (**A**) score plot (**B**) loading plot. UF: unfermented; BS: *B. subtilis*; SC: *S. cerevisiae*; LA: *Lb. acidophilus*; LB: *Lb. delbrueckii* subsp. *bulgaricus*; LC: *Lb. casei*. Sulfur-conta, sulfur containing compounds; benz, benzene derivatives.

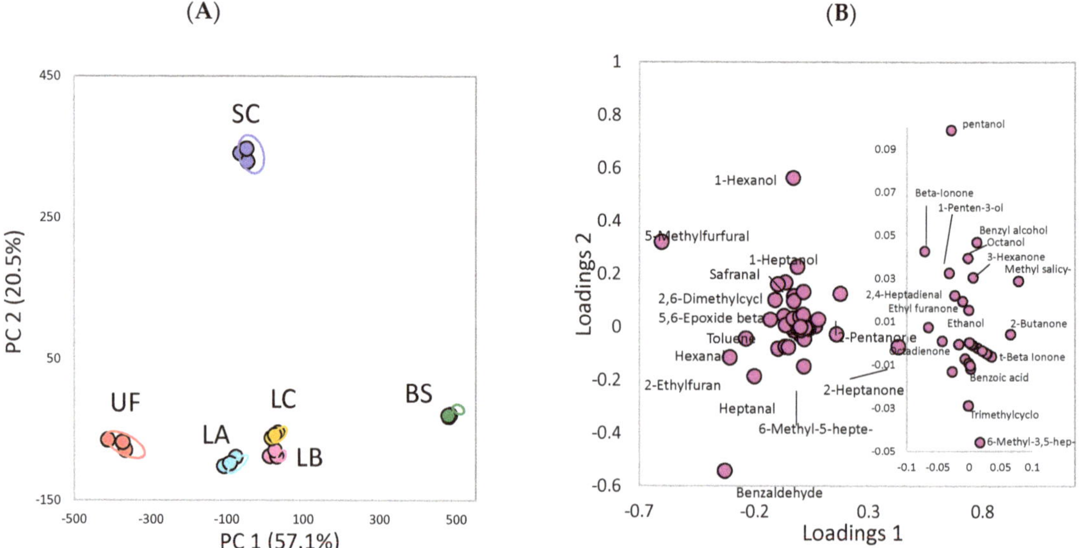

Figure 4. Principal component analysis plots of the volatile compounds in the green seaweed *Ulva* sp. fermented by various microorganisms: (**A**) score plot (**B**) loading plot. UF: unfermented; BS: *B. subtilis*; SC: *S. cerevisiae*; LA: *Lb. acidophilus*; LB: *Lb. delbrueckii* subsp. *bulgaricus*; LC: *Lb. casei*.

3.3.2. *Laminaria* sp.

The PCA of chemical groups of volatile compounds in the unfermented and fermented *Laminaria* sp. is shown in Figure 5. The results showed that these variables could explain 91.7% of the total variances among the samples. BS and SC are located in the positive axis of PC1, whereas the unfermented group is in the negative axis of PC1 (Figure 5A). Therefore, fermenting *Laminaria* sp. by BS or SC can change the composition of chemical groups in volatile compounds. In addition, the chemical groups of volatile compounds are distinct in BS and SC because these two groups are in the positive and negative axes of PC2, respectively. Based on the loading plot (Figure 5B), ketones in BS and alcohols in SC primarily cause the difference. As shown in Table 2, after fermentation, BS acquired higher ketone content than SC, whereas SC obtained higher alcohol content than BS.

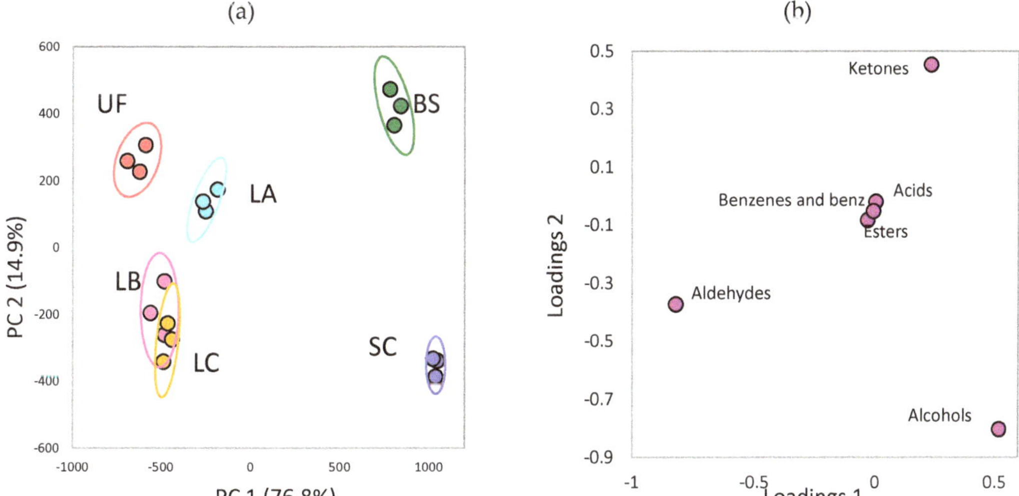

Figure 5. Principal component analysis plots of chemical groups of the volatile compounds in the brown seaweed *Laminaria* sp. fermented by various microorganisms: (**A**) score plot (**B**) loading plot. UF: unfermented; BS: *B. subtilis*; SC: *S. cerevisiae*; LA: *Lb. acidophilus*; LB: *Lb. delbrueckii* subsp. *bulgaricus*; LC: *Lb. casei*. Sulfur-conta, sulfur containing compounds; benz, benzene derivatives.

Another PCA was conducted to evaluate the correlation between a single volatile compound and the type of fermentation (Figure 6). PC1 and PC2 explained 55.6% and 29.9%, respectively, of the total variance in the dataset (Total: 85.5%). Consistent with Figure 5, the volatile compound compositions of BS and SC are differentiated from the unfermented *Laminaria* sp. (Figure 6A). Moreover, based on the loading plot (Figure 6B), acetoin (i.e., 3-hydroxy-2-butanone) and 2,3-butanediol are positively correlated with BS, and 1-heptanol and 1-octanol are positively correlated with SC, thereby causing the difference in volatile compounds between the two fermented *Laminaria*. This finding is consistent with that shown in Figure 5, i.e., the difference between BS and SC is associated with the ketone and alcohol family.

As shown in Table 2, acetoin accounts for the highest content (63.4%) in the ketone family from BS fermentation, followed by 1-heptanol (41.1%) and 1-octanol (20.5%) in the alcohol family from SC fermentation. On the contrary, in the third quadrant of the score plot (Figure 6A), the unfermented group and fermentation with lactic acid bacteria (LA, LB, and LC) are in the negative axis of PC1 and PC2. This finding indicates that lactic acid bacteria may not significantly modify the composition of volatile compounds as compared with the unfermented group. Based on the loading plot (Figure 6B), (E)-2-heptenal shows a higher correlation with the unfermented group and fermentation with lactic acid bacteria. Furthermore, (E)-2-heptenal was significantly reduced from 1179.00 to 538.71 and 429.48 µg/L by fermenting with BS and SC, respectively (Table 2). In addition, (E)-2-heptenal contents were significantly lower in BS and SC fermentation than in lactic acid bacteria fermentation (Table 2). This finding may support the result that the groups of lactic acid bacteria and BS/SC are positioned on different quadrants in PCA.

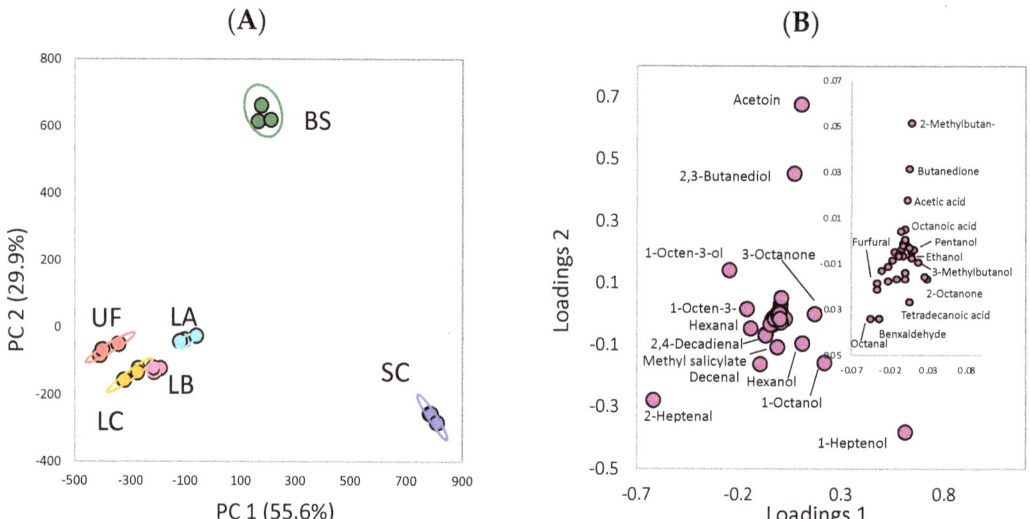

Figure 6. Principal component analysis plots of the volatile compounds in the brown seaweed *Laminaria* sp. fermented by various microorganisms: (**A**) score plot (**B**) loading plot. UF: unfermented; BS: *B. subtilis*; SC: *S. cerevisiae*; LA: *Lb. acidophilus*; LB: *Lb. delbrueckii* subsp. *bulgaricus*; LC: *Lb. casei*.

Therefore, according to the PCA (Figures 3–6), fermenting *Ulva* sp. or *Laminaria* sp. with BS or SC can modify the composition of volatile compounds compared with unfermented seaweeds. These results can be supported by Tables 1 and 2. Volatile and non-volatile compounds can contribute to the sensory characteristics of foods, which may further affect consumer preference [36–38]. Hence, GC–O was applied to evaluate the odor description and intensity of the fermented seaweeds.

3.4. Sensory Evaluation of the Intensity of the Odor-Active Compounds in the Fermented Seaweeds

GC–O is considered as a major technique for analyzing the effect of volatile compounds on sensory characteristics, which has been applied to seaweeds [3]. The results of GC–O analysis for the unfermented and fermented *Ulva* sp. are summarized in Table 3.

A total of 22 odor-active compounds were detected, including six ketones, four aldehydes, four alcohols, one ester, one pyrazine, and six unidentified compounds. Among the pleasant odors, floral, sweet, and fruity were described. After fermentation with BS and SC, aromatic notes were enhanced, compared with the unfermented *Ulva*. For example, benzyl alcohol could contribute to floral and fruity aromas after BS and SC fermentation. Benzyl alcohol is described as flower, apple peel, and sweet aromas in Chinese bayberry [39], and it has been reported in many fruits [40]. In addition, 1-hexanol and 1-heptanol could provide extra floral/sweet and sweet aromas in the SC group, and the floral aroma of 6-methyl-5-hepten-2-one was enhanced in the BS group (Table 3). Notably, after fermentation with BS, 2,5-dimethylpyrazine is formed (Table 1), and it contributes to the nutty and roasted aromas in the fermented *Ulva* sp. (Table 3).

This odor description is consistent with the report of Guo et al. [41], indicating that 2,5-dimethylpyrazine is the critical contributor to roasted peanut flavor in Oolong tea. According to Leejeerajumnean, Duckham, Owens, and Ames [35], 2,5-dimethylpyrazine was found in several *Bacillus*-fermented soybean products. With regard to strong odor intensity, β-ionone (violet, floral) is the only compound that is described as having a strong and pleasant odor in all groups. Based on the report of Buttery et al. [42], β-ionone has a lower odor threshold (0.007 ppb) than 6-methyl-5-hepten-2-one (50 ppb) in water. This finding might indicate that the odor intensity of β-ionone is strong and greater than 6-methyl-5-hepten-2-one. In strengthening the pleasant odor of *Ulva* sp. by fermenting with BS and SC, both fermentations can mitigate the unpleasant odors (Table 3). For example, the odor intensity of hexanal (fishy, grassy) was reduced in the BS- and SC-fermented *Ulva* sp. suspensions. Described as caramelly, fried, and toasty, 5-Methylfurfural was not detected in the BS-fermented *Ulva* sp. suspension by GC–O. These reductions are consistent with that shown in Table 1 in this study. BS and SC have demonstrated their application potential in improving pleasant odors and lightening unpleasant odors of *Ulva* by fermentation.

Table 3. Odor description and intensity of the odor-active compounds in the green seaweed *Ulva* sp. fermentations.

Compounds	Odor Description [a]	Odor Intensity [b]		
		UF	BS	SC
Pleasant odors				
6-Methyl-5-hepten-2-one	Floral	W	M	
3,5-Octadien-2-one	Fruity	W	W	W
Propiophenone	Floral	M	W	M
trans-β-Ionone	Floral		W	
β-Ionone	Violet, floral	S	S	S
5,6-Epoxide-β-Ionone	Floral, sweet	M	M	M
Benzaldehyde	Floral, sweet	M	W	W
1-Hexanol	Floral, sweet			M
1-Heptanol	Sweet			W
1-Octanol	Waxy, citrus			M
Benzyl alcohol	Floral, fruity		M	M
2,5-Dimethylpyrazine	Nutty, roasted		M	
Unknown 1	Floral		M	M
Unpleasant odors				
Hexanal	Fishy, grassy	M		W
5-Methylfurfural	Caramellic, fried, toasty	S		S
Unknown 2	Fishy	M	M	M
Unknown 3	Earthy, musty	M	M	M
Unknown 4	Fatty, oily	W	W	W
Neutral odors				
Safranal	Herbal	W		W
Methyl salicylate	Minty, wintergreen	M	M	M
Unknown 5	Tea	M	W	M
Unknown 6	Plastic	W		W

UF: unfermented; BS: *Bacillus subtilis*; SC: *Saccharomyces cerevisiae*. [a] Odor description perceived at the olfactory detection port. [b] The perceived odor intensity was evaluated and recorded using the degree of strong (S), medium (M), and weak (W).

The results of the GC–O analysis for the unfermented and fermented *Laminaria* sp. are outlined in Table 4. A total of 36 odor-active compounds were detected, including 7 ketones, 11 aldehydes, 8 alcohols, 1 acid, 1 ester, and eight unidentified compounds. Only three compounds, namely, β-ionone (violet, floral), benzaldehyde (fruity, sweet), and 1-octanol (waxy, citrus), were detected as pleasant odors in the unfermented *Laminaria* sp. suspension. In addition, the pleasant odors could be strengthened in the *Laminaria* suspension fermented with BS and SC. For example, 2,3-butanedione, described as yogurt and creamy, has shown a medium odor intensity in BS, whereas 1-heptanol, described as sweet, has shown a strong

Funding: This research was funded by the Center of Excellence for the Oceans, National Taiwan Ocean University from the Featured Areas Research Center Program within the framework of the Higher Education Sprout Project by the Ministry of Education (MOE) in Taiwan (NTOU-RD-AA-2021-1-02018).

Institutional Review Board Statement: Not applicable.

Informed Consent Statement: Not applicable.

Data Availability Statement: The data presented in this study are available on request from the corresponding author.

Acknowledgments: We thank Chorng-Lan Pan from National Taiwan Ocean University for providing the lactic acid bacteria strains.

Conflicts of Interest: The authors declare no conflict of interest.

References

1. Amsler, C.D.; Fairhead, V.A. Defensive and sensory chemical ecology of brown algae. *Adv. Bot. Res.* **2005**, *43*, 1–91. [CrossRef]
2. Mouritsen, O.G.; Rhatigan, P.; Pérez-Lloréns, J.L. The rise of seaweed gastronomy: Phycogastronomy. *Bot. Mar.* **2019**, *62*, 195–209. [CrossRef]
3. Vilar, E.G.; O'Sullivan, M.G.; Kerry, J.P.; Kilcawley, K.N. Volatile compounds of six species of edible seaweed: A review. *Algal Res.* **2020**, *45*, 101740. [CrossRef]
4. Gressler, V.; Stein, É.M.; Dörr, F.; Fujii, M.T.; Colepicolo, P.; Pinto, E. Sesquiterpenes from the essential oil of Laurencia dendroidea (Ceramiales, Rhodophyta): Isolation, biological activities and distribution among seaweeds. *Rev. Bras. De Farmacogn.* **2011**, *21*, 248–254. [CrossRef]
5. Smit, A.J. Medicinal and pharmaceutical uses of seaweed natural products: A review. *J. Appl. Phycol.* **2004**, *16*, 245–262. [CrossRef]
6. Qin, Y. *Bioactive Seaweeds for Food Applications: Natural Ingredients for Healthy Diets*; Academic Press: Cambridge, MA, USA, 2018.
7. Du, X.; Xu, Y.; Jiang, Z.; Zhu, Y.; Li, Z.; Ni, H.; Chen, F. Removal of the fishy malodor from *Bangia fusco-purpurea* via fermentation of *Saccharomyces cerevisiae*, *Acetobacter pasteurianus*, and *Lactobacillus plantarum*. *J. Food Biochem.* **2021**, *45*, e13728. [CrossRef]
8. Francezon, N.; Tremblay, A.; Mouget, J.-L.; Pasetto, P.; Beaulieu, L. Algae as a source of natural flavors in innovative foods. *J. Agric. Food Chem.* **2021**, *69*, 11753–11772. [CrossRef]
9. López-Pérez, O.; Picon, A.; Nuñez, M. Volatile compounds and odour characteristics of seven species of dehydrated edible seaweeds. *Food Res. Int.* **2017**, *99*, 1002–1010. [CrossRef]
10. Hernández, T.; Estrella, I.; Pérez-Gordo, M.; Alegría, E.G.; Tenorio, C.; Ruiz-Larrrea, F.; Moreno-Arribas, M. Contribution of malolactic fermentation by *Oenococcus oeni* and *Lactobacillus plantarum* to the changes in the nonanthocyanin polyphenolic composition of red wine. *J. Agric. Food Chem.* **2007**, *55*, 5260–5266. [CrossRef]
11. Seo, Y.-S.; Bae, H.-N.; Eom, S.-H.; Lim, K.-S.; Yun, I.-H.; Chung, Y.-H.; Jeon, J.-M.; Kim, H.-W.; Lee, M.-S.; Lee, Y.-B. Removal of off-flavors from sea tangle (*Laminaria japonica*) extract by fermentation with *Aspergillus oryzae*. *Bioresour. Technol.* **2012**, *121*, 475–479. [CrossRef]
12. Bao, J.; Zhang, X.; Zheng, J.-H.; Ren, D.-F.; Lu, J. Mixed fermentation of *Spirulina platensis* with *Lactobacillus plantarum* and *Bacillus subtilis* by random-centroid optimization. *Food Chem.* **2018**, *264*, 64–72. [CrossRef] [PubMed]
13. Lu, W.-J.; Lin, H.-J.; Hsu, P.-H.; Lai, M.; Chiu, J.-Y.; Lin, H.-T.V. Brown and red seaweeds serve as potential efflux pump inhibitors for drug-resistant *Escherichia coli*. *Evid.-Based Complement Altern. Med.* **2019**, *2019*, 1836982. [CrossRef] [PubMed]
14. Hung, Y.-H.R.; Chen, G.-W.; Pan, C.-L.; Lin, H.-T.V. Production of ulvan oligosaccharides with antioxidant and angiotensin-converting enzyme-inhibitory activities by microbial enzymatic hydrolysis. *Fermentation* **2021**, *7*, 160. [CrossRef]
15. Vilar, E.G.; O'Sullivan, M.G.; Kerry, J.P.; Kilcawley, K.N. A chemometric approach to characterize the aroma of selected brown and red edible seaweeds/extracts. *J. Sci. Food Agric.* **2021**, *101*, 1228–1238. [CrossRef]
16. Zhang, Z.; Li, T.; Wang, D.; Zhang, L.; Chen, G. Study on the volatile profile characteristics of oyster *Crassostrea gigas* during storage by a combination sampling method coupled with GC/MS. *Food Chem.* **2009**, *115*, 1150–1157. [CrossRef]
17. Hosoglu, M.I. Aroma characterization of five microalgae species using solid-phase microextraction and gas chromatography–mass spectrometry/olfactometry. *Food Chem.* **2018**, *240*, 1210–1218. [CrossRef] [PubMed]
18. Li, N.; Zheng, F.-p.; Chen, H.-t.; Liu, S.-y.; Gu, C.; Song, Z.-y.; Sun, B.-g. Identification of volatile components in Chinese Sinkiang fermented camel milk using SAFE, SDE, and HS-SPME-GC/MS. *Food Chem.* **2011**, *129*, 1242–1252. [CrossRef]
19. Paparella, A.; Shaltiel-Harpaza, L.; Ibdah, M. β-Ionone: Its occurrence and biological function and metabolic engineering. *Plants* **2021**, *10*, 754. [CrossRef]
20. Neta, M.; Narain, N. Volatile components in seaweeds. *Examines Mar. Biol. Oceanogr.* **2018**, *2*, 195–201.
21. Park, M.K.; Kim, Y.-S. Comparative metabolic expressions of fermented soybeans according to different microbial starters. *Food Chem.* **2020**, *305*, 125461. [CrossRef]
22. Peinado, I.; Girón, J.; Koutsidis, G.; Ames, J. Chemical composition, antioxidant activity and sensory evaluation of five different species of brown edible seaweeds. *Food Res. Int.* **2014**, *66*, 36–44. [CrossRef]

23. Kiritsakis, A. Flavor components of olive oil—A review. *J. Am. Oil Chem. Soc.* **1998**, *75*, 673–681. [CrossRef]
24. Hazelwood, L.A.; Daran, J.-M.; Van Maris, A.J.; Pronk, J.T.; Dickinson, J.R. The Ehrlich pathway for fusel alcohol production: A century of research on *Saccharomyces cerevisiae* metabolism. *Appl. Environ. Microbiol.* **2008**, *74*, 2259–2266. [CrossRef] [PubMed]
25. Long, A.; James, P.; Ward, O. Aromatic aldehydes as substrates for yeast and yeast alcohol dehydrogenase. *Biotechnol. Bioeng.* **1989**, *33*, 657–660. [CrossRef]
26. Wang, R.; Sun, J.; Lassabliere, B.; Yu, B.; Liu, S.Q. Biotransformation of green tea (*Camellia sinensis*) by wine yeast *Saccharomyces cerevisiae*. *J. Food Sci.* **2020**, *85*, 306–315. [CrossRef]
27. Moreira, R.; Trugo, L.; Pietroluongo, M.; De Maria, C. Flavor composition of cashew (*Anacardium occidentale*) and marmeleiro (*Croton species*) honeys. *J. Agric. Food Chem.* **2002**, *50*, 7616–7621. [CrossRef]
28. Sieber, R.; Bütikofer, U.; Bosset, J. Benzoic acid as a natural compound in cultured dairy products and cheese. *Int. Dairy J.* **1995**, *5*, 227–246. [CrossRef]
29. Lubran, M.B.; Lawless, H.T.; Lavin, E.; Acree, T.E. Identification of metallic-smelling 1-octen-3-one and 1-nonen-3-one from solutions of ferrous sulfate. *J. Agric. Food Chem.* **2005**, *53*, 8325–8327. [CrossRef]
30. Synos, K.; Reynolds, A.; Bowen, A. Effect of yeast strain on aroma compounds in Cabernet franc icewines. *LWT-Food Sci. Technol.* **2015**, *64*, 227–235. [CrossRef]
31. Brunke, E.J.; Hammerschmidt, F.J.; Schmaus, G. Scent of roses–recent results. *Flavour Fragr. J.* **1992**, *7*, 195–198. [CrossRef]
32. Hattori, S.; Takagaki, H.; Fujimori, T. Identification of volatile compounds which enhance odor notes in Japanese green tea using the OASIS (Original aroma simultaneously input to the sniffing port) method. *Food Sci. Technol. Res.* **2005**, *11*, 171–174. [CrossRef]
33. Pan, J.; Jia, H.; Shang, M.; Li, Q.; Xu, C.; Wang, Y.; Wu, H.; Dong, X. Effects of deodorization by powdered activated carbon, β-cyclodextrin and yeast on odor and functional properties of tiger puffer (*Takifugu rubripes*) skin gelatin. *Int. J. Biol. Macromol.* **2018**, *118*, 116–123. [CrossRef] [PubMed]
34. Giri, A.; Osako, K.; Ohshima, T. Identification and characterisation of headspace volatiles of fish miso, a Japanese fish meat based fermented paste, with special emphasis on effect of fish species and meat washing. *Food Chem.* **2010**, *120*, 621–631. [CrossRef]
35. Leejeerajumnean, A.; Duckham, S.C.; Owens, J.D.; Ames, J.M. Volatile compounds in *Bacillus*-fermented soybeans. *J. Sci. Food Agric.* **2001**, *81*, 525–529. [CrossRef]
36. King, E.S.; Kievit, R.L.; Curtin, C.; Swiegers, J.H.; Pretorius, I.S.; Bastian, S.E.; Francis, I.L. The effect of multiple yeasts co-inoculations on Sauvignon Blanc wine aroma composition, sensory properties and consumer preference. *Food Chem.* **2010**, *122*, 618–626. [CrossRef]
37. Kim, Y.; Lee, K.-G.; Kim, M.K. Volatile and non-volatile compounds in green tea affected in harvesting time and their correlation to consumer preference. *J. Food Sci. Technol.* **2016**, *53*, 3735–3743. [CrossRef] [PubMed]
38. Choi, S.M.; Lee, D.-J.; Kim, J.-Y.; Lim, S.-T. Volatile composition and sensory characteristics of onion powders prepared by convective drying. *Food Chem.* **2017**, *231*, 386–392. [CrossRef] [PubMed]
39. Kang, W.; Li, Y.; Xu, Y.; Jiang, W.; Tao, Y. Characterization of aroma compounds in Chinese bayberry (*Myrica rubra* Sieb. et Zucc.) by gas chromatography mass spectrometry (GC-MS) and olfactometry (GC-O). *J. Food Sci.* **2012**, *77*, C1030–C1035. [CrossRef]
40. Burdock, G.A. *Fenaroli's Handbook of Flavor Ingredients*; CRC Press: Boca Raton, FL, USA, 2016.
41. Guo, X.; Song, C.; Ho, C.-T.; Wan, X. Contribution of L-theanine to the formation of 2,5-dimethylpyrazine, a key roasted peanutty flavor in Oolong tea during manufacturing processes. *Food Chem.* **2018**, *263*, 18–28. [CrossRef]
42. Buttery, R.G.; Teranishi, R.; Ling, L.C.; Turnbaugh, J.G. Quantitative and sensory studies on tomato paste volatiles. *J. Agric. Food Chem.* **1990**, *38*, 336–340. [CrossRef]
43. Mall, V.; Sellami, I.; Schieberle, P. New degradation pathways of the key aroma compound 1-penten-3-one during storage of not-from-concentrate orange juice. *J. Agric. Food Chem.* **2018**, *66*, 11083–11091. [CrossRef] [PubMed]
44. Ritter, S.W.; Gastl, M.I.; Becker, T.M. The modification of volatile and nonvolatile compounds in lupines and faba beans by substrate modulation and lactic acid fermentation to facilitate their use for legume-based beverages—A review. *Compr. Rev. Food Sci. Food Saf.* **2022**, *21*, 4018–4055. [CrossRef] [PubMed]

fermentation

Article

Combined Use of *Schizosaccharomyces pombe* and a *Lachancea thermotolerans* Strain with a High Malic Acid Consumption Ability for Wine Production

Javier Vicente [1], Niina Kelanne [2], Eva Navascués [3,4], Fernando Calderón [3], Antonio Santos [1], Domingo Marquina [1], Baoru Yang [2] and Santiago Benito [3,*]

[1] Unit of Microbiology, Genetics, Physiology and Microbiology Department, Biology Faculty, Complutense University of Madrid, Ciudad Universitaria, S/N, 28040 Madrid, Spain
[2] Food Science, Department of Life Technology, University of Turku, FI-20014 Turku, Finland
[3] Department of Chemistry and Food Technology, Polytechnic University of Madrid, Ciudad Universitaria, S/N, 28040 Madrid, Spain
[4] Pago de Carraovejas, S.L.U., Camino de Carraovejas, S/N, 47300 Peñafiel, Spain
*   Correspondence: santiago.benito@upm.es; Tel.: +34-9133-63710 or +34-9133-63984

Abstract: The development of new fermentative strategies exploiting the potential of different wine-related species is of great interest for new winemaking conditions and consumer preferences. One of the most promising non-conventional approaches to wine fermentation is the combined use of deacidifying and acidifying yeasts. *Lachancea thermotolerans* shows several other properties besides lactic acid production; among them, high malic acid consumption is of great interest in the production of red wines for avoiding undesirable refermentations once bottled. The combination of a *L. thermotolerans* strain that is able to consume malic acid with a *Schizosaccharomyces pombe* strain helps to ensure malic acid elimination during alcoholic fermentation while increasing the final acidity by lactic acid production. To properly assess the influence of this alternative strategy, we developed combined fermentations between specific strains of *L. thermotolerans* and *S. pombe* under sequential inoculation. Both species showed a great performance under the studied conditions, influencing not only the acidity but also the aromatic compound profiles of the resulting wines. The new proposed biotechnological strategy reduced the final concentrations of ethanol, malic acid and succinic acid, while it increased the concentrations of lactic acid and esters.

Keywords: wine; non-*Saccharomyces*; alternative fermentation strategies; biological acidity management; volatile compounds

Citation: Vicente, J.; Kelanne, N.; Navascués, E.; Calderón, F.; Santos, A.; Marquina, D.; Yang, B.; Benito, S. Combined Use of *Schizosaccharomyces pombe* and a *Lachancea thermotolerans* Strain with a High Malic Acid Consumption Ability for Wine Production. *Fermentation* **2023**, *9*, 165. https://doi.org/10.3390/fermentation9020165

Academic Editors: Niel Van Wyk and Ronnie G. Willaert

Received: 21 December 2022
Revised: 4 February 2023
Accepted: 7 February 2023
Published: 11 February 2023

1. Introduction

The combined use of *Lachancea thermotolerans* and *Schizosaccharomyces pombe* has been recently described as an alternative strategy in winemaking for malolactic fermentation. Malolactic fermentation decreases acidity, since malic acid is more acidic than lactic acid. An excessive reduction in total acidity can lead to spoilage, so winemakers sometimes have to re-acidify wines by adding tartaric acid. The aim of the alternative strategy is to reduce the malic acid content through *S. pombe* metabolism in the red wine while maintaining the total acidity through lactic acid production by *L. thermotolerans* during the alcoholic fermentation [1]. This methodology should be considered for grape juices from warm viticulture areas that possess high sugar concentrations and high pH levels. Under those circumstances, it is difficult to perform classical malolactic fermentation without any deviation.

L. thermotolerans selection procedures have traditionally been focused on the ability to produce L-lactic acid to acidify wine [2–4]. Nevertheless, recent studies have focused on other secondary traits of *L. thermotolerans* that can also improve wine quality [5,6]. Among those secondary objectives, malic acid consumption is one of the most important ones. The first studies that reported malic acid consumption in *L. thermotolerans* stated that most strains consumed 10% to 20% of it [6]. However, novel studies showed that specific *L. thermotolerans* strains may consume up to 50% of malic acid or even more in both single and mixed cultures [4,7]. The deacidification carried out by this non-*Saccharomyces* yeast is of great interest when wines require malic acid stabilization, since this reduction in the content of malic acid may reduce the risk of undesirable malolactic fermentation by lactic acid bacteria. However, any *L. thermotolerans* strain can totally consume all the malic acid present in standard grape musts.

S. pombe is noted for its malic acid consumption. The malic acid content reduction by this species varies from 60% to 100%, depending on the strain [8,9]. Nevertheless, the use of some strains of *S. pombe* is limited by the high production of acetic acid, over 0.8 g/L [8]. High acetic acid production is one of the main collateral effects of *S. pombe* strains, as only 5% of the isolated strains produce acceptable levels. Recent studies have reported characteristics of specific *S. pombe* strains other than malic acid deacidification, such as a high production of polysaccharides, mannoproteins, galactomannoproteins, pyruvic acid, glycerol and stable anthocyanins and a low production of higher alcohols and ethyl carbamate precursors [1,8,9]. Recent studies have started to pay attention to other yeast species besides those in the *Schizosaccharomyces* genus that are able to significantly reduce the initial malic acid concentration, such as *Pichia kudriavzevii* [10,11] and *Hanseniaspora occidentalis* [12]. Those species can reduce the initial malic acid content by about 50%, although their fermentative metabolism is less efficient.

This study proposes the use of an *L. thermotolerans* strain with high lactic acid production and high malic acid consumption and a commercial *S. pombe* strain to ensure proper alcoholic fermentation in order to produce a red wine of the Tempranillo grape variety that does not require classic malolactic fermentation.

2. Materials and Methods

2.1. Microorganisms

The study used the following yeast strains: *Lachancea thermotolerans* L1 (Complutense University of Madrid, Madrid, Spain), *Saccharomyces cerevisiae* AG006 (Agrovín S.L, Alcazar de San Juan, Spain) and *Schizosaccharomyces pombe* Atecrem 12H (Bioenologia, Oderzo, Italy).

2.2. Vinification

All fermentations used the grape juice of *Vitis vinifera* L. cultivar Tempranillo grapes grown at the Cuzcurrita vineyard (Rioja Alta, Spain). The grape juice was taken from a fermentation tank just after the grapes were destemmed, crushed and introduced into the tank, before any inoculation. The must was enriched with 0.15 g/L of di-ammonium phosphate and 0.30 g/L of Actimax Natura (Agrovin, Spain). Then, the must was pasteurized at 105 °C for 1 min in an autoclave: Presoclave 75 (J.P. Selecta, Barcelona, Spain).

All fermentations took place in 250 mL Pyrex™ borosilicate glass reagent bottles, each with a slightly open polypropylene cap and pouring ring, allowing for CO_2 release and preventing microbial contamination. Fermentations were carried out at 25 °C in triplicate.

The initial sum of glucose and fructose concentrations was 251.22 g/L, pH = 3.74, primary amino nitrogen = 172 mg/L, ammonia nitrogen = 26 mg/L, malic acid = 2.03 g/L. The initial lactic and acetic acid concentrations were below 0.1 g/L. Four treatments were used. Table 1 describes the strain combinations used in each treatment.

The preculturing of the strains of *L. thermotolerans* (LT), *S. pombe* (SP) and *S. cerevisiae* (SC) was carried out in 40 mL YMB medium for 24 h, with shaking, at 25 °C and 150 rpm in 100 mL borosilicate bottles. The optical density of the cultures was determined using a spectrophotometer (Genesys 2.0 Spectrophotometer, ThermoFisher, Waltham, MA, USA), and the lowest value was used. The fermentation cultures were inoculated at a concentration of 10^6 cells/mL ($\approx$0.2 O.D.).

Table 1. Must inoculum compositions.

SC	*S. cerevisiae* (10^6 CFU/mL) alone.
LT … SC	*L. thermotolerans* (10^6 CFU/mL) followed by *S. cerevisiae* (10^6 CFU/mL) 5 days later.
LT … SP	*L. thermotolerans* (10^6 CFU/mL) followed by *S. pombe* (10^6 CFU/mL) 5 days later.
SP	*S. pombe* (10^6 CFU/mL) alone.

SC: *S. cerevisiae*, LT: *L. thermotolerans*, SP: *S. pombe*.

Fermentation monitoring was performed by measuring the weight loss every 24 h. Fermentation was considered complete when the weight loss was less than 0.01% for two consecutive days. The initial weight of each fermentation was considered as 100%.

In the sequential fermentations (LT … SC and LT … SP), the yeast of the most fermentative species (*S. cerevisiae* or *S. pombe*) was inoculated 5 days (96 h) after the initial inoculation of *L. thermotolerans*. The alcoholic fermentations took place at 25 °C in a thermostatic chamber.

2.3. Chemical Parameter Measurements

A Y15 Autoanalyzer and its commercial kits (Biosystems, Barcelona, Spain) were used in the determinations of glucose + fructose, L-malic acid, L-lactic acid, acetic acid, succinic acid and glycerol concentrations. The alcohol content was determined using the boiling method of GAB Microebu (http://shop.gabsystem.com (accessed on 20 December 2022)). A Crison pH Meter Basic 20 (Crison, Spain) measured the final pH.

2.4. Volatile Compounds

The volatile compounds of the fermentations were measured according to a previous methodology [13]. The samples were analyzed in triplicate using headspace solid-phase microextraction coupled with gas chromatography–mass spectrometry (HS-SPME-GC-MS). Two milliliters of each sample and 0.2 g of sodium chloride were placed in a 20 mL glass vial, and 10 µL of 4-methyl-2-pentanol solution (802 µg/mL in methanol) was added as an internal standard. The volatile compounds were extracted from the headspace with a 2 cm DVB/CAR/PDMS fiber (50/30 µm, Supelco, Bellefonte, PA, USA) at 45 °C for 30 min after 10 min of incubation. The fiber was conditioned at 250 °C prior to the sample extraction. After the extraction, the SPME fiber was immediately transferred to the injection port of a Trace 1310 gas chromatograph equipped with a TSQ 8000 EVO mass spectrometer (Thermo Fisher Scientific, MA, USA) to be thermally desorbed in splitless mode at 240 °C for 3 min. A DB-WAX polar capillary column (60 m × 0.25 mm i.d. × 0.25 µm film thickness, J&W Scientific, Folsom, CA, USA) was used to separate the volatile compounds of the samples. Helium was used as the carrier gas at a flow rate of 1.6 mL/min. The initial column temperature was set to 50 °C and held for 3 min. Afterwards, the temperature was increased to 220 °C at a rate of 5 °C/min and held at 220 °C for 8 min. Mass spectra were detected in electron impact (EI) mode at 70 eV, with a scan range from m/z 33 to m/z 300. The MS transfer line and the ionization source temperatures were 220 and 240 °C, respectively.

The RIs of the volatiles were calculated via co-injection with an alkane mixture (C7-C21, Sigma-Aldrich, St. Louis, MO, USA). Volatiles were identified by matching the obtained mass spectra with the standard NIST 08 library and by comparing the retention indices (RIs) to those of the compounds reported in the literature and the NIST Webbook (https://webbook.nist.gov/chemistry (accessed on 20 December 2022)). Moreover, the identification of a selected number of volatile compounds was confirmed by comparing the retention indices and mass spectra with those of the authentic reference compounds. Supplementary Table S1 shows the measured RIs and those reported in the literature.

2.5. Color Intensity

A Y350 diode array spectrophotometer (Biosystems, Spain) was used for the analysis. The samples were analyzed in a 1 mm path-length quartz cuvette with a range of

200–1100 nm. Absorbance at 420, 520 and 620 nm was measured. Color intensity was calculated as the sum of absorbance at the three wavelengths.

2.6. Statistical Analyses

All statistical analyses were performed using R software version 4.1.2 (R Development Core Team, Vienna, Austria, 2013). Analysis of variance (ANOVA) and Tukey post hoc tests were applied to compare the different groups and values.

3. Results and Discussion

3.1. Fermentation Kinetics

The fermentations lasted between 18 days (the fermentation conducted by *S. pombe*) and 27 days (the sequential fermentation conducted by the LT and SC strains) (Figure 1). A slowdown took place after 3–4 days of fermentation, and another took place after 9 days. The alcoholic fermentation of the combined fermentation between *L. thermotolerans* and *S. pombe* lasted for 22 days, which is the same time employed by the pure *S. cerevisiae* control. Three previous studies report delays that vary from 4 to 8 days for the new biotechnology compared to the regular *S. cerevisiae* control for alcoholic fermentation, while three other studies reported no differences, as is the case in this study [1]. However, regular *S. cerevisiae* fermentations made in red wines require performing additional malolactic fermentation before bottling to avoid possible refermentation problems. This additional process may require at least 21 additional days to obtain a stable wine from a microbiological point of view.

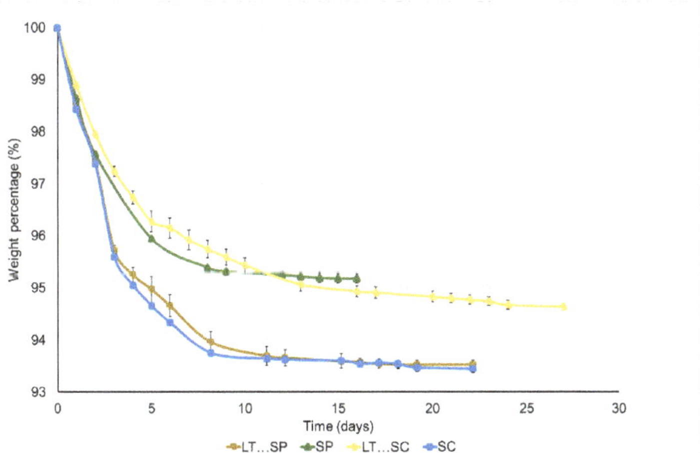

Figure 1. Fermentation kinetics of gravimetrically measured variants by total weight loss during fermentation. *S. cerevisiae* alone (SC); sequential fermentation with *S. cerevisiae* and *L. thermotolerans* (LT . . . SC); sequential fermentation with *S. pombe* and *L. thermotolerans* (LT . . . SP); *S. pombe* alone (SP). Data points represent the averages and standard deviation of the three replicates for each condition. SC fermentations are in blue; SP fermentations are in green; LC . . . SC fermentations are in yellow; LT . . . SP fermentations are in brown.

Previous studies have described that the initial consumption of nutrients by *L. thermotolerans* can compromise the availability of nutrients for the fermentative yeast strain used in the sequential mode to conclude the alcoholic fermentation [1,6,7]. The rapid fermentation by the *Schizosaccharomyces pombe* strain could have taken place due to the lower nutrient demands of this species, despite it normally showing slower kinetics due to its reproduction by bipartition, which requires more time than the budding of most yeasts [14].

3.2. Glucose and Fructose

Most of the fermentations showed final sugar concentrations of glucose and fructose below 2 g/L (Table 2). The only trial that showed a slightly higher concentration of 4.73 g/L was the sequential fermentation with *L. thermotolerans* and *S. cerevisiae*. Previous authors explained this effect by the higher nutrient demands of *L. thermotolerans* that can compromise the performance of *S. cerevisiae* in sequential fermentations. Most manufacturers recommend a second round of nutrient addition concurrently with *S. cerevisiae* inoculation to avoid this problem of a lack of nutrients that can slow down the last stages of alcoholic fermentation [7].

Table 2. Final chemical analysis of fermentations from Tempranillo red grapes: *S. cerevisiae* alone (SC); sequential fermentation with *S. cerevisiae* and *L. thermotolerans* (LT ... SC); sequential fermentation with *S. pombe* and *L. thermotolerans* (LT ... SP); *S. pombe* alone (SP).

	SC	LT ... SC	LT ... SP	SP
L-lactic acid (g/L)	0.11 ± 0.03 [a]	2.26 ± 0.64 [c]	1.52 ± 0.09 [b]	0.16 ± 0.02 [a]
L-malic acid (g/L)	1.41 ± 0.03 [d]	1.11 ± 0.10 [c]	0.11 ± 0.03 [a]	0.48 ± 0.07 [b]
Succinic acid (g/L)	1.47 ± 0.06 [c]	1.41 ± 0.07 [bc]	1.34 ± 0.03 [b]	1.26 ± 0.04 [a]
Acetic acid (g/L)	0.36 ± 0.05 [a]	0.51 ± 0.07 [b]	0.45 ± 0.03 [b]	0.45 ± 0.01 [b]
pH	3.77 ± 0.02 [b]	3.62 ± 0.02 [a]	3.78 ± 0.04 [b]	3.83 ± 0.01 [c]
Ethanol (g/L)	11.63 ± 0.06 [c]	11.21 ± 0.06 [a]	11.46 ± 0.06 [b]	11.72 ± 0.04 [c]
Glucose + Fructose (g/L)	1.24 ± 0.21 [a]	4.73 ± 0.06 [c]	1.85 ± 0.24 [b]	1.56 ± 0.14 [ab]
Glycerol (g/L)	9.25 ± 0.32 [b]	8.96 ± 0.47 [ab]	9.13 ± 0.26 [b]	8.26 ± 0.14 [a]

Results are mean ± SD of three replicates. Different letters indicate statistical significance between groups.

All the employed species employed in this study (*S. cerevisiae*, *S. pombe* and *L. thermotolerans*) are reported to show a preference to consume glucose against fructose [1,6,7,14], so, in demanding situations, significant amounts of fructose may remain in the wine. *S. cerevisiae* [1] and *S. pombe* [14] can completely metabolize glucose and fructose into ethanol for regular dry wines under no severe stress conditions. *L. thermotolerans* is widely reported to not be able to metabolize sugar concentrations over 170 g/L [6,7], which makes it impossible to produce regular wines using only strains of *L. thermotolerans*, although modern studies manage to produce beer or base sparkling wines [7]. In this study, the potential undesirable effect of residual sugars was diminished by combining *L. thermotolerans* with the more fermentative species *S. cerevisiae* and *S. pombe*.

3.3. Ethanol

Fermentations involving *L. thermotolerans* showed significantly lower concentrations of ethanol than fermentations involving *S. pombe* and *S. cerevisiae* (Table 2). The maximum difference was 0.65% (*v/v*) between the sequential fermentation involving *L. thermotolerans* and *S. cerevisiae* and the pure *S. pombe* fermentation. Although there were significant differences from a statistical point of view, the quantitative reduction represented about 1%, which is low compared to other, more effective technologies that are able to significantly reduce the final ethanol content at an industrial scale. *L. thermotolerans* fermentative metabolism is different, since part of the carbon flux is derived through lactic acid and glycerol production, together with the fact that this yeast is not a strongly fermentative one [7]. Previous authors reported differences of up to 3% (*v/v*), whereas others reported no significant differences regarding the ethanol content [1]. Additionally, *S. pombe* produces ethanol from malic acid metabolism and occasionally may increase the final ethanol concentration compared to *S. cerevisiae* controls when the initial concentration of malic acid is high [10].

3.4. L-Lactic Acid

Pure fermentations of *S. pombe* and *S. cerevisiae* did not show a significant production of lactic acid (Table 2). Fermentations involving *L. thermotolerans* produced significant final concentrations of L-lactic acid that clearly influenced the final pH. The combined fermentation with *L. thermotolerans* and *S. cerevisiae* produced 2.26 g/L of lactic acid; this effect reduced the pH in 0.15 units compared to the *S. cerevisiae* control. The combined fermentation with *L. thermotolerans* and *S. pombe* produced 1.52 g/L of lactic acid, and it did not show differences in pH compared to the *S. cerevisiae* control due to the malic acid decrease that compensated the pH reduction by lactic acid. Previous studies reported the opposite effect, as all of them determined that the combined fermentations with *L. thermotolerans* and *S. pombe* produced higher final amounts of lactic acid than the combined fermentations with *L. thermotolerans* and *S. cerevisiae* [1]. These studies have indicated that the slower kinetics of *S. pombe* allow *L. thermotolerans* to survive longer. However, in this study, the combined fermentation using *L. thermotolerans* and *S. pombe* was faster than the combined fermentation using *L. thermotolerans* and *S. cerevisiae* (Figure 1). Recent studies indicate that the interactions between different strains of distinct species vary highly, indicating inhibition or symbiosis, depending on the specific strains [15]. The lactic acid production of *L. thermotolerans* under mixed fermentation conditions is extremely variable, ranging from 0 to 9 g/L; nevertheless, the most common value is around 3 g/L of lactic acid [7].

3.5. Malic Acid

The pure fermentation of *S. cerevisiae* reduced the initial concentration of malic acid by 30.5%. Previous studies reported a strain variability of malic acid production of up to 0.7 g/L and a variability of malic acid degradation of up to 50% [10]. The pure fermentation of *S. pombe* reduced malic acid significantly, by 78%, which clearly influenced the final pH. For *S. pombe*, the malic acid degradation varied from 60% to 100%, depending on the selected strain [16]. Although a 78% reduction is significant from a de-acidification point of view, it is not enough to stabilize a red wine from the malic acid point of view. Most previous studies have reported that combined fermentations with *L. thermotolerans* and *S. pombe* consume 95–100% of malic acid [1]. Similar studies have reported smaller reductions varying from 50% to 75% [1,17,18]. The main use of the commercial *S. pombe* strain employed in the study is to de-acidify wine, but not to stabilize it. The combined fermentation with the *L. thermotolerans* strain, which has a special ability to reduce malic acid, and *S. cerevisiae* reduced the initial concentration of malic acid by 45%. Finally, the combined fermentation with the malic-acid-metabolizing *L. thermotolerans* strain and *S. pombe* reduced the initial concentration of malic acid by almost 100%, achieving malic acid stabilization. The results show that the combined use of more than one malic-acid-consuming microorganism increases the stability of wines from a malic acid point of view.

3.6. Acetic Acid

S. cerevisiae produced the lowest final concentration of acetic acid: 0.36 g/L; the other fermentations produced 0.45 to 0.51 g/L (Table 2). All the final values were moderate and below the faulty threshold of 0.8 g/L [19]. Most studies reported that *Lachancea thermotolerans* species produce less acetic acid than *S. cerevisiae*, although others reported the opposite effect [7]. Most studies report that *S. pombe* produces higher acetic acid concentrations than *S. cerevisiae*, although recent studies reported that specific strains produce similar or lower amounts of acetic acid compared to *S. cerevisiae* [1,14]. Although joint fermentation with *L. thermotolerans* and *S. pombe* produced 0.09 g/L more acetic acid than pure *S. cerevisiae* fermentation, we must consider that this fermentation does not require stabilization from a malic acid point of view. During a regular controlled malolactic fermentation without any deviation, the volatile acidity usually increases in about 0.1 g/L [1,15].

3.7. Succinic Acid

Pure *S. pombe* fermentation produced the lowest concentration of succinic acid (1.26 g/L), and pure *S. cerevisiae* fermentations showed the highest concentration (up to 1.47 g/L) (Table 2). The literature reports the *S. cerevisiae* strain variability that results in succinic acid contents from 0.3 to 1.8 g/L in wine [10]. No previous studies have reported data for *L. thermotolerans* or *S. pombe*.

3.8. Glycerol

S. pombe fermentation resulted in the lowest final concentration of glycerol, whereas the *S. cerevisiae* fermentation resulted in the highest (Table 2). The other trials resulted in intermediate concentrations.

A high glycerol concentration is often related to soft and mouthful sensory properties. The rise of the glycerol concentration in wine is one of the main contributions of some specific non-*Saccharomyces* species to the value of wine [5]. However, this study reports that the pure *S. cerevisiae* control produces higher final glycerol concentrations than its combination with *L. thermotolerans* and the pure *S. pombe* fermentation. This effect can be explained since, although several scientific articles report specific non-*Saccharomyces* as a higher glycerol producer, others report an additional high variability depending on the strain level. Previous studies report a strain variability for *L. thermotolerans* [6,7] and *S. pombe* [14] up to 50% regarding glycerol production. A similar strain variability has been previously reported for *S. cerevisiae*. The *S. cerevisiae* strain employed in this study was selected by the manufacturer, including the glycerol production as a selection parameter, while the employed *L. thermotolerans* strain was selected to produce high lactic acid concentrations, and the *S. pombe* strain was selected to reduce acidity.

The combination between *L. thermotolerans* and *S. pombe* did not show statistical differences compared to the *S. cerevisiae* control. Previous studies report changes in different parameters depending on the interactions between different strains and species that can promote specific metabolic routes or inhibit others. This phenomenon remains widely unknown, but most researchers recommend testing the interactions between different strains before using them at an industry scale to avoid undesirable effects [10].

3.9. Volatile Compounds

Pure fermentations of *S. cerevisiae* and *S. pombe* did not produce any ethyl lactate or isoamyl lactate, and fermentations involving *L. thermotolerans* showed significant final values (almost 90% higher) (Table 3). These results are related to the L-lactic acid production (around 95% higher) observed in *L. thermotolerans* fermentations, which favored the esterification. These lactic acid esters may increase the fruity profiles of the final wines by increasing the fruity and fatty odor series. Fermentations involving *L. thermotolerans* produced less ethyl hexanoate than the others.

The pure fermentations of *S. pombe* were the only ones that did not produce any detectable 3-methyl butanal. The pure *S. cerevisiae* fermentations resulted in the highest values. Additionally, the pure *S. pombe* fermentations produced the highest concentration of 2-nonanol (up to five times higher), and the fermentations involving *S. pombe* were the only ones that generated 1-(1-ethoxyethoxy)pentane. The production of compounds of these types increases the herbaceous and malt aromas in the final wines.

The pure fermentations of *S. pombe* had lower final levels of 3-methyl-1-butanol, ethyl phenylacetate and 2-phenylethyl acetate than pure *S. cerevisiae* fermentation, which can be related to reductions in several undesirable aromas of wines, mainly those related to chemical or synthetic products (Table 3). Previous studies reported that *S. pombe* is a lesser producer of alcohols and esters than *S. cerevisiae* [20,21]. This phenomenon is of interest in order to avoid varietal aroma masking [19]. Additionally, the pure *S. pombe* fermentation had the lowest concentrations of butyrolactone (50%) and phenylethyl alcohol (20%).

Fermentations involving *L. thermotolerans* reduced the final hexanol, dodecanal, ethyl octanoate, ethyl decanoate and hexanoic acid levels compared to the *S. cerevisiae* control by between 20% and 50%, reducing the fruity and floral profiles of the wines, which are usually related with lighter and fresher wines. Additionally, fermentations involving *L. thermotolerans* reduced the concentration of 3-(methylthio)-1-propanol in combined fermentations with *S. cerevisiae*; the opposite effect took place for combined fermentations with *S. pombe* (three times higher).

Table 3. Final volatile compound profiles of fermentations from Tempranillo red grapes.

Compound (Area Units)	SC	SP	LT … SC	LT … SP
Ethyl acetate	1.42 ± 0.02 [a]	1.59 ± 0.2 [a]	0.77 ± 0.41 [b]	1.43 ± 0.14 [a]
3-Methyl butanal *	2.70 ± 0.62 [a]	0.00 ± 0.00 [a]	0.87 ± 0.70 [a]	1.88 ± 2.45 [a]
2-Methylpropyl acetate	0.03 ± 0.00 [a]	0.02 ± 0.00 [a]	0.01 ± 0.01 [b]	0.03 ± 0.00 [a]
Ethyl butanoate	0.05 ± 0.01 [a]	0.05 ± 0.01 [a]	0.01 ± 0.01 [b]	0.053 ± 0.01 [a]
Toluene	0.15 ± 0.14 [a]	0.11 ± 0.10 [a]	0.13 ± 0.11 [a]	0.10 ± 0.09 [a]
Ethyl 3-methylbutanoate *	5.56 ± 1.04 [a]	2.93 ± 0.28 [ab]	1.97 ± 1.89 [b]	4.95 ± 1.43 [ab]
1-(1-Ethoxyethoxy)pentane *	0.00 ± 0.00 [b]	3.24 ± 0.99 [a]	0.00 ± 0.00 [b]	3.17 ± 1.89 [a]
2-Methyl-1-propanol	0.89 ± 0.14 [a]	0.69 ± 0.09 [ab]	0.48 ± 0.08 [b]	0.92 ± 0.25 [a]
3-Methyl-1-butyl acetate	0.55 ± 0.07 [a]	0.39 ± 0.08 [a]	0.25 ± 0.20 [a]	0.26 ± 0.07 [a]
Butanol *	2.58 ± 2.65 [b]	3.73 ± 3.35 [b]	31.42 ± 13.54 [a]	7.88 ± 0.95 [b]
3-Methyl-1-butanol	6.65 ± 0.37 [a]	5.88 ± 0.43 [a]	5.06 ± 0.23 [a]	6.25 ± 1.07 [a]
Ethyl hexanoate	0.48 ± 0.10 [a]	0.51 ± 0.09 [a]	0.21 ± 0.08 [a]	0.42 ± 0.26 [a]
Ethyl lactate	0.03 ± 0.01 [c]	0.02 ± 0.00 [c]	0.19 ± 0.04 [a]	0.11 ± 0.01 [b]
Hexanol	0.29 ± 0.01 [a]	0.28 ± 0.02 [a]	0.26 ± 0.00 [a]	0.29 ± 0.02 [a]
cis-3-Hexen-1-ol *	20.35 ± 1.91 [a]	18.19 ± 1.51 [ab]	17.41 ± 1.23 [ab]	13.38 ± 3.91 [b]
Ethyl octanoate	0.11 ± 0.03 [b]	0.15 ± 0.02 [ab]	0.01 ± 0.00 [c]	0.19 ± 0.03 [a]
2-Nonanol *	4.82 ± 2.04 [b]	10.51 ± 1.95 [a]	2.71 ± 0.23 [b]	4.85 ± 1.33 [b]
Ethyl 3-hydroxybutyrate *	3.76 ± 1.12 [a]	4.44 ± 0.34 [a]	1.70 ± 0.08 [b]	2.99 ± 0.85 [ab]
Benzaldehyde	0.03 ± 0.00 [a]	0.02 ± 0.00 [b]	0.02 ± 0.00 [ab]	0.02 ± 0.00 [ab]
Ethyl nonanoate *	7.92 ± 1.35 [a]	7.96 ± 2.38 [a]	1.53 ± 0.62 [b]	8.51 ± 0.86 [a]
Ethyl 2-hydroxy-4-methylpentanoate *	14.75 ± 0.28 [a]	10.69 ± 0.67 [b]	11.09 ± 0.59 [b]	15.78 ± 1.04 [a]
Octanol *	11.93 ± 3.63 [a]	13.07 ± 1.85 [a]	3.66 ± 1.36 [b]	13.54 ± 2.62 [a]
2-Methyl propanoic acid	0.03 ± 0.00 [a]	0.02 ± 0.00 [a]	0.03 ± 0.01 [a]	0.02 ± 0.00 [a]
Isoamyl lactate	0.00 ± 0.00 [b]	0.00 ± 0.00 [b]	0.03 ± 0.00 [a]	0.03 ± 0.00 [a]
Ethyl 2-furoate *	1.32 ± 0.14 [a]	1.01 ± 0.08 [b]	1.03 ± 0.11 [ab]	1.11 ± 0.12 [ab]
Methyl benzoate *	1.95 ± 0.27 [a]	0.98 ± 0.03 [c]	1.73 ± 0.11 [ab]	1.33 + 0.15 [bc]
Butanoic acid *	2.04 ± 0.30 [a]	1.94 ± 0.31 [a]	1.62 ± 0.16 [a]	1.68 ± 0.52 [a]
Ethyl decanoate	0.01 ± 0.00 [a]	0.02 ± 0.00 [a]	0.00 ± 0.00 [b]	0.02 ± 0.04 [a]
Butyrolactone *	7.92 ± 0.11 [a]	4.08 ± 0.99 [b]	6.52 ± 0.71 [a]	8.26 ± 1.12 [a]
4-methylbenzaldehyde *	9.11 ± 1.74 [a]	6.36 ± 1.15 [a]	7.42 ± 1.12 [a]	7.30 ± 1.12 [a]
Acetophenone *	2.23 ± 0.90 [ab]	1.50 ± 0.36 [b]	2.17 ± 0.77 [ab]	3.79 ± 0.80 [a]
2-methyl butanoic acid	0.01 ± 0.00 [a]	0.01 ± 0.00 [a]	0.01 ± 0.00 [a]	0.01 ± 0.00 [a]
Diethyl succinate	0.08 ± 0.01 [b]	0.05 ± 0.01 [b]	0.04 ± 0.01 [b]	0.27 ± 0.11 [a]
Dodecanal *	4.52 ± 0.57 [ab]	4.68 ± 0.39 [a]	3.26 ± 0.15 [b]	4.21 ± 0.70 [ab]
3-(methylthio)-1-propanol *	7.06 ± 1.32 [b]	6.63 ± 1.27 [b]	3.70 ± 0.89 [c]	10.83 ± 0.72 [a]
Ethyl phenylacetate *	1.70 ± 0.13 [ab]	1.22 ± 0.12 [b]	1.76 ± 0.31 [a]	1.84 ± 0.14 [a]
2-phenylethyl acetate	0.02 ± 0.00 [a]	0.01 ± 0.00 [a]	0.02 ± 0.00 [a]	0.01 ± 0.00 [a]
β-damascenone *	4.75 ± 0.92 [a]	3.73 ± 0.58 [a]	4.61 ± 0.30 [a]	4.82 ± 0.68 [a]
Ethyl dodecanoate *	0.80 ± 0.12 [a]	1.12 ± 0.13 [a]	0.00 ± 0.00 [b]	1.26 ± 0.36 [a]
Hexanoic acid	0.07 ± 0.00 [a]	0.07 ± 0.01 [a]	0.04 ± 0.00 [b]	0.07 ± 0.01 [a]
N-(3-Methylbutyl)acetamide	0.06 ± 0.01 [b]	0.05 ± 0.00 [b]	0.05 ± 0.00 [b]	0.09 ± 0.01 [a]
Butanedioic acid, ethyl 3-methylbutyl ester *	2.30 ± 0.16 [b]	1.77 ± 0.19 [b]	1.22 ± 0.21 [b]	10.54 ± 4.61 [a]
Phenylethyl alcohol	1.44 ± 0.08 [a]	1.22 ± 0.11 [a]	1.43 ± 0.25 [a]	1.55 ± 0.13 [a]
Octanoic acid	0.20 ± 0.02 [a]	0.23 ± 0.01 [a]	0.08 ± 0.00 [b]	0.20 ± 0.01 [a]
Nonanoic acid	0.04 ± 0.00 [a]	0.05 ± 0.02 [a]	0.05 ± 0.01 [a]	0.04 ± 0.02 [a]
Decanoic acid *	25.98 ± 2.68 [a]	31.34 ± 2.19 [a]	28.92 ± 2.54 [a]	17.17 ± 1.96 [b]

S. cerevisiae control (SC); *S. pombe* control (SP); sequential fermentation with *S. cerevisiae* and *L. thermotolerans* (LT … SC) and *S. pombe* and *L. thermotolerans* (LT … SP). Compounds highlighted with an asterisk (*) have a 1000-times-lower area. Results are mean ± SD of three replicates. Different letters indicate statistical significance between groups.

The pure fermentations of *S. cerevisiae* produced the highest concentrations of benzaldehyde, diethyl succinate and 2-methyl butanoic acid (up to two times higher), which are usually related with the fruity, floral and roasted profiles of the final fermentation.

The combined fermentations between *S. pombe* and *L. thermotolerans* had the highest final amounts of N-(3-methylbutyl)acetamide and the lowest finals amounts of decanoic and octanoic acid.

3.10. Color Intensity

The color intensity results were minor and did not show any significant differences (Table 4). Previous studies reported eventual slight differences of up to 10% for *L. thermotolerans* fermentations due to the different colorations of anthocyanins at low pHs and the different yeast strain absorptions [7,17]. *S. pombe* fermentations did not show differences compared to the other trials, although previous studies usually reported higher color intensities due to the formation of highly stable anthocyanin compounds [22].

Table 4. Final color intensity analysis of fermentations from Tempranillo red grapes: *S. cerevisiae* alone (SC); sequential fermentation with *S. cerevisiae* and *L. thermotolerans* (LT … SC); sequential fermentation with *S. pombe* and *L. thermotolerans* (LT … SP); *S. pombe* alone (SP).

	SC	**LT … SC**	**LT … SP**	**SP**
420 nm	0.63 ± 0.02 [a]	0.65 ± 0.04 [a]	0.66 ± 0.03 [a]	0.69 ± 0.04 [a]
520 nm	1.22 ± 0.04 [a]	1.30 ± 0.06 [a]	1.26 ± 0.05 [a]	1.29 ± 0.06 [a]
620 nm	0.19 ± 0.01 [a]	0.21 ± 0.02 [a]	0.19 ± 0.02 [a]	0.21 ± 0.03 [a]
CI	2.04 ± 0.07 [a]	2.16 ± 0.12 [a]	2.11 ± 0.10 [a]	2.19 ± 0.13 [a]

Results are mean ± SD of three replicates. Different letters indicate statistical significance between groups.

4. Conclusions

The results show that the combined use of strongly malic-acid-consuming microorganisms increases the stabilization of wines from a malic acid point of view. The consideration of malic acid degradation in the selection of *L. thermotolerans* strains for fermenting red wine may be of great interest in facilitating future stabilization processes, such as malolactic fermentation. The results of the study showed that, when employing the biotechnology that combines *L. thermotolerans* and *S. pombe*, the *L. thermotolerans* strain should possess a great ability to consume malic acid in order to enhance *S. pombe's* malic acid consumption capacity. The proposed strategy reduces the final concentrations of other chemicals, such as ethanol, malic acid and succinic acid, while increasing the concentrations of lactic acid and esters.

Supplementary Materials: The following supporting information can be downloaded at: https://www.mdpi.com/article/10.3390/fermentation9020165/s1, Table S1. Compounds identified using retention indices (RI) measured in the study and those reported in the literature, as well as using confirmation with reference compounds.

Author Contributions: J.V., S.B. and D.M. conceived and designed the experiments; J.V., S.B., A.S. and D.M. performed the experiments; J.V., S.B. and F.C. performed the chemical analyses; N.K. and B.Y. performed the volatile compounds analyses; J.V., S.B., E.N. and F.C. analyzed the data; J.V., S.B. and D.M. wrote the paper; E.N. and S.B. revised the manuscript. All authors have read and agreed to the published version of the manuscript.

Funding: Funding was provided by the Ministry of Science and Innovation under the framework of Project PID2020-119008RB-I00 and by the Spanish Center for the Development of Industrial Technology under the framework of Project LowpH-Wine (IDI-20210391).

Institutional Review Board Statement: Not applicable.

Informed Consent Statement: Not applicable.

Data Availability Statement: Not applicable.

Conflicts of Interest: The authors declare no conflict of interest.

References

1. Benito, S. Combined Use of *Lachancea thermotolerans* and *Schizosaccharomyces pombe* in Winemaking: A Review. *Microorganisms* **2020**, *8*, 655. [CrossRef]
2. Hranilovic, A.; Gambetta, J.M.; Schmidtke, L.; Boss, P.K.; Grbin, P.R.; Masneuf-Pomarede, I.; Bely, M.; Albertin, W.; Jiranek, V. Oenological Traits of *Lachancea thermotolerans* Show Signs of Domestication and Allopatric Differentiation. *Sci. Rep.* **2018**, *8*, 1–13. [CrossRef]
3. Hranilovic, A.; Albertin, W.; Capone, D.L.; Gallo, A.; Grbin, P.R.; Danner, L.; Bastian, S.E.P.; Masneuf-Pomarede, I.; Coulon, J.; Bely, M.; et al. Impact of *Lachancea thermotolerans* on Chemical Composition and Sensory Profiles of Merlot Wines. *Food Chem.* **2021**, *349*, 129015. [CrossRef]
4. Blanco, P.; Rabuñal, E.; Neira, N.; Castrillo, D. Dynamic of *Lachancea thermotolerans* Population in Monoculture and Mixed Fermentations: Impact on Wine Characteristics. *Beverages* **2020**, *6*, 36. [CrossRef]
5. Jolly, N.P.; Augustyn, O.P.H.; Pretorius, I.S. The Role and Use of Non-*Saccharomyces* Yeasts in Wine Production. *S. Afr. J. Enol. Vitic.* **2006**, *27*, 15–38. [CrossRef]
6. Benito, S. The Impacts of *Lachancea thermotolerans* Yeast Strains on Winemaking. *Appl. Microbiol. Biotechnol.* **2018**, *102*, 6775–6790. [CrossRef]
7. Vicente, J.; Navascués, E.; Calderón, F.; Santos, A.; Marquina, D.; Benito, S.; Fracassetti, D.; Rustioni, L. An Integrative View of the Role of *Lachancea thermotolerans* in Wine Technology. *Foods* **2021**, *10*, 2878. [CrossRef]
8. Benito, Á.; Calderón, F.; Benito, S. Combined Use of *S. pombe* and *L. thermotolerans* in Winemaking. Beneficial Effects Determined through the Study of Wines' Analytical Characteristics. *Molecules* **2016**, *21*, 1744. [CrossRef]
9. Benito, Á.; Calderón, F.; Benito, S. Mixed Alcoholic Fermentation of *Schizosaccharomyces pombe* and *Lachancea thermotolerans* and Its Influence on Mannose-Containing Polysaccharides Wine Composition. *AMB Express* **2019**, *9*, 17. [CrossRef]
10. Vicente, J.; Baran, Y.; Navascués, E.; Santos, A.; Calderón, F.; Marquina, D.; Rauhut, D.; Benito, S. Biological Management of Acidity in Wine Industry: A Review. *Int. J. Food Microbiol.* **2022**, *375*, 109726. [CrossRef]
11. del Mónaco, S.M.; Barda, N.B.; Rubio, N.C.; Caballero, A.C. Selection and Characterization of a Patagonian *Pichia kudriavzevii* for Wine Deacidification. *J. Appl. Microbiol.* **2014**, *117*, 451–464. [CrossRef]
12. van Wyk, N.; Scansani, S.; Beisert, B.; Brezina, S.; Fritsch, S.; Semmler, H.; Pretorius, I.S.; Rauhut, D.; von Wallbrunn, C. The Use of *Hanseniaspora occidentalis* in a Sequential Must Inoculation to Reduce the Malic Acid Content of Wine. *Appl. Sci.* **2022**, *12*, 6919. [CrossRef]
13. Liu, S.; Laaksonen, O.; Yang, B. Volatile Composition of Bilberry Wines Fermented with Non-*Saccharomyces* and *Saccharomyces* Yeasts in Pure, Sequential and Simultaneous Inoculations. *Food Microbiol.* **2019**, *80*, 25–39. [CrossRef]
14. Benito, S. The Impacts of *Schizosaccharomyces* on Winemaking. *Appl. Microbiol. Biotechnol.* **2019**, *103*, 4291–4312. [CrossRef] [PubMed]
15. Urbina, Á.; Calderón, F.; Benito, S. The Combined Use of *Lachancea thermotolerans* and *Lactiplantibacillus plantarum* (Former *Lactobacillus plantarum*) in Wine Technology. *Foods* **2021**, *10*, 1356. [CrossRef]
16. Benito, Á.; Jeffares, D.; Palomero, F.; Calderón, F.; Bai, F.Y.; Bähler, J.; Benito, S. Selected *Schizosaccharomyces pombe* Strains Have Characteristics That Are Beneficial for Winemaking. *PLoS ONE* **2016**, *11*, e0151102. [CrossRef]
17. Chen, K.; Escott, C.; Loira, I.; del Fresno, J.M.; Morata, A.; Tesfaye, W.; Calderon, F.; Suárez-Lepe, J.A.; Han, S.; Benito, S. Use of Non-*Saccharomyces* Yeasts and Oenological Tannin in Red Winemaking: Influence on Colour, Aroma and Sensorial Properties of Young Wines. *Food Microbiol.* **2018**, *69*, 51–63. [CrossRef]
18. YuHua, W.; WenJun, S.; Min, L.; Lan, M.; YuMei, J.; Jing, W. Effect of Sequential Fermentation with *Lachancea thermotolerans* and *Schizosaccharomyces pombe* on the Quality of Merlot Dry Red Wine. *Shipin Kexue/Food Sci.* **2019**, *40*, 102–111.
19. Ruiz, J.; Kiene, F.; Belda, I.; Fracassetti, D.; Marquina, D.; Navascués, E.; Calderón, F.; Benito, A.; Rauhut, D.; Santos, A.; et al. Effects on Varietal Aromas during Wine Making: A Review of the Impact of Varietal Aromas on the Flavor of Wine. *Appl. Microbiol. Biotechnol.* **2019**, *103*, 7425–7450. [CrossRef] [PubMed]
20. Gardoni, E.; Benito, S.; Scansani, S.; Brezina, S.; Fritsch, S.; Rauhut, D. Biological Deacidification Strategies for White Wines. *S. Afr. J. Enol. Vitic.* **2021**, *42*, 114–122. [CrossRef]
21. Scansani, S.; Rauhut, D.; Brezina, S.; Semmler, H.; Benito, S. The Impact of Chitosan on the Chemical Composition of Wines Fermented with *Schizosaccharomyces pombe* and Saccharomyces Cerevisiae. *Foods* **2020**, *9*, 1423. [CrossRef] [PubMed]
22. Benito, Á.; Calderón, F.; Benito, S. The Combined Use of *Schizosaccharomyces pombe* and *Lachancea thermotolerans*—Effect on the Anthocyanin Wine Composition. *Molecules* **2017**, *22*, 739. [CrossRef] [PubMed]

 *fermentation*

Review

Key Aromatic Volatile Compounds from Roasted Cocoa Beans, Cocoa Liquor, and Chocolate

Orlando Meneses Quelal [1,*], David Pilamunga Hurtado [1], Andrés Arroyo Benavides [2], Pamela Vidaurre Alanes [1] and Norka Vidaurre Alanes [1]

[1]   Graduate Department, Universidad Politécnica Estatal del Carchi, Tulcán 040102, Ecuador
[2]   School of Agricultural and Environmental Sciences, Pontificia Universidad Católica del Ecuador-Sede Ibarra, Ibarra 100112, Ecuador
*   Correspondence: orlando.meneses@upec.edu.ec

Abstract: The characteristic aromas at each stage of chocolate processing change in quantity and quality depending on the cocoa variety, the chemical composition of the beans, the specific protein storage content, and the polysaccharides and polyphenols determining the type and quantity of the precursors formed during the fermentation and drying process, leading to the formation of specific chocolate aromas in the subsequent roasting and conching processes. Bean aroma is frequently profiled, identified, and semiquantified by headspace solid-phase microextraction combined with gas chromatography-mass spectrometry (HS-SPMEGC-MS) and by gas chromatography olfactometry (GC-O). In general, the flavors generated in chocolate processing include fruity, floral, chocolate, woody, caramel, earthy, and undesirable notes. Each processing stage contributes to or depletes the aroma compounds that may be desirable or undesirable, as discussed in this report.

Keywords: aroma profile; odor threshold; odor activity value; fermentation; cocoa liquor; chocolate

Citation: Quelal, O.M.; Hurtado, D.P.; Benavides, A.A.; Alanes, P.V.; Alanes, N.V. Key Aromatic Volatile Compounds from Roasted Cocoa Beans, Cocoa Liquor, and Chocolate. *Fermentation* **2023**, *9*, 166. https://doi.org/10.3390/fermentation9020166

Academic Editor: Niel Van Wyk

Received: 27 November 2022
Revised: 31 January 2023
Accepted: 7 February 2023
Published: 11 February 2023

1. Introduction

The volatile aromatic components of cocoa beans vary depending on whether the beans are dry, roasted, or processed in cocoa liquor or chocolate. Different concentrations of the most relevant volatile components in a particular variety (criollo, forastero, and trinitario) rely on geographical origin, the chemical composition of the beans, the postharvest techniques used, such as fermentation and drying, and industrial processes like roasting and conching [1]. Regarding the varieties and genotypes, criollo, trinitario, and national cocoa have higher fine aroma concentrations as fruit (fresh and ripe), floral, herbaceous, wood, nuts, and caramel notes [2–5]; forastero cocoa on the other hand (considered bulk grade), possess higher concentrations of the predominant aromas of malt, honey, roasted, caramel, cocoa, and chocolate [6], as well as low acidic and alcoholic notes [7]. Finally, the high yielding CCN51 cocoa genotype (Colección Castro Naranjal) grown in Ecuador (classified as bulk grade), has volatiles that exude a sweet and fruity or fresh character to the beans [5].

The unique aromatic profiles and diversity of aromas are linked to the number of components exhibited in the beans; the greater the number of components, the more complex the overall aroma of a specific cocoa sample. For instance, the aromatic profiles of criollo and trinitario cocoa from Mexico comprise 46 and 47 of the most relevant volatile components, respectively [8]. Likewise, 69 of the most important volatile components represent Ecuador's national cocoa [5], with 67 to 89% being aromatically desirable components, and the remaining fraction (11 to 33%) exudes unpleasant aromas that may originate during fermentation and drying [9–12]. The main cause of unpleasant aromas comes from excessive fermentation that originates compounds that impart smoky and ham aromas, which have significantly high concentrations that affect the final quality of the beans. The suggested limits for some unpalatable volatile components in fermented beans are

2 µg/kg for 3-ethylphenol (smoked) and 3-propylphenol (smoked, phenolic), 12.20 µg/kg for 4-methylphenol (fecal), 3-methylphenol (smoked) and 4-ethylphenol (smoked), and 70 µg/kg for 2-methoxyphenol [13]. If the drying process is performed slowly, there is a loss of volatile acids and water [14]. This is because the porosity of the husk enclosing the cotyledons increases [15], facilitating volatilization and implying that the dried beans are less acidic and have a higher pH, providing their aromatic potential [14]. As the moisture level decreases, the level of cocoa butter accumulates. Cocoa butter is the most abundant component in dried cocoa beans [16], accounting for half of the components. This butter fraction may include fatty acids (myristic, palmitic, palmitoleic, stearic, oleic, linoleic, and arachidic) and triacylglycerols (composed of about 42% of 1-palmitate-2-oleate-3-stearate triacylglycerol, 24% of 1,3-diestearate-2-oleate triacylglycerol, and 22% of 1,3-dipalmitate-2-oleate triacylglycerol), as the major components [17,18] have a serious impact on the perception of the chocolate flavor and aroma. The release of chocolate aromas during tasting is linked to lipophilic volatile compounds, mainly alkylpyrazines, which are considered to be more complex and more substituted with the alkyl group, therefore having higher lipophilicity [19]. The higher degree of saturation and hydrophobicity of cocoa butter may be associated with decreased pyrazine release [17].

Most aromatic volatile compounds are released during the roasting process. The number of volatile compounds varies with roasting intensity; therefore, a greater amount is formed when roasting at higher temperatures [20]. Several studies detail the diversity of these compounds within the temperature ranges of 95 to 160 °C [6,9,21,22]. The most volatile compounds found in roasted beans correspond to esters, followed by acids, alcohols, aldehydes, and ketones, which are characteristic of a given cocoa variety, and depending on the roasting time and temperature, other compounds, such as pyrroles, pyrazines, and Strecker aldehydes, are reduced or generated [6,9,21,22]. Higher roasting temperatures may generate unpleasant compounds with burnt or smoky notes and a dark brown color in processed chocolate. In the next process in which the roasted beans are ground, the cocoa liquor has particles measuring between 22–26.5 µm [9,23], driving the extra release of aroma compounds. Similarly, during conching, the removal of the remaining moisture and volatile unpleasant acetic acid compounds occurs, acquiring acceptable levels [24]. Prolonged conching (6–10 h at 80 °C) results in the loss of both undesirable volatile flavorings and other flavorings that provide desirable characteristic aromas like sweet, fruity, floral, and chocolate [25]. It has also been reported that the aromatic volatile compound profiles of dark chocolates are strongly affected by the brand-related formulation and processing conditions. In some cases, differences within the same brand have even been shown [26].

Chocolate production generally involves a combination of fermentation, drying, roasting, and conching processes [25]; thus, it is crucial to investigate how these processes impact the levels and types of compounds that determine the chocolate aroma and its origin. Therefore, this review provides information on more relevant volatile organic compounds arising from postharvest and bean manufacturing methods to their implication in the final quality of the chocolate. In addition, the active odor components in dried beans, roasted beans, liquor, and chocolate are identified and traced with emphasis on the most desirable compounds.

2. Extraction and Quantification of Volatile Cocoa Compounds

Several studies have been published on the aromatic components contained in cocoa beans, especially the forastero, national, criollo, and trinitario varieties. Forastero cocoa represents about 85% of world production [27]. It is characterized by producing seeds that, after being fermented, generate volatile floral and sweet aromas [28,29]. The national cocoa grown in Ecuador classified as "fine and aroma grade", produces seeds with fruity, green, and woody volatile aromas. This variety represents 63% of world production [28,30]. On the other hand, Criollo cocoa represents only 5% [27] and its fermented beans are associated with floral, fruity, and woody, volatile aromas [28]. Finally, trinitario cocoa is a criollo hybrid variety and has strong basic chocolate characteristics together with fruity and floral aromas [28]. The characterization and comparison of the seeds in the different varieties are complex since the list of representative genotypes in all cocoa-producing countries is extensive [29,31–33], making the standardization of the preharvest, postharvest, and cocoa manufacturing processes derivatives since each variety has a unique flavor and aroma potential [34].

Cocoa aromas and flavors are detailed through several volatile aroma profiles and samples, such as fermented and roasted dry beans, cocoa liquor, and chocolate. Although there are several methods of characterization, the two most common are detailed. On the one hand, the analysis of volatile compounds is carried out by headspace solid-phase microextraction combined with gas chromatography-mass spectrometry (HS-SPMEGC-MS) [4,11,35,36]. The use of divinylbenzene carboxene polydimethylsiloxane (DVB/CAR/PDMS) fiber has been reported to define the organic volatiles in cocoa [8,12,21,22,28,37–39]. Optimal fiber coating is related to the number of peaks in the chromatogram and generated intensity. When using CAR-PDMS, up to 100 peaks have been detected more intensely, while in PDMS-DVB, the number of peaks is less than 75 [40], implying that a higher number of volatiles are extracted from a specific cocoa matrix.

To cite an example, Ref. [28] identified a total of 121 volatile compounds in criollo, national and forastero cocoa, and 62 were positively identified, including nine organic acids, 12 alcohols, 14 aldehydes and ketones, five esters, 12 hydrocarbons, two amines, two furans, one sulfur, and five unspecified compounds. Likewise, Ref. [12] reported a total of 67 volatile components, including acids, alcohols, aldehydes, esters, ketones, pyrazines, furans, furanones, lactones, pyrans, pyrroles, and terpenes. The total volatiles concentration was higher in cocoa liquors than in chocolates.

In contrast, olfactometry, which collects data that are obtained from gas chromatography olfactometry (GC-O), helps to calculate to what extent a certain compound influences the overall aroma depending on its dilution factor (DF), [6,10,41–43; it detects odors and describes them by means of an "aromagram"–a representation of the dilution factor logarithm versus retention time [10,42,43]. In order to illustrate this, [6] reported the dilution factors of roasted and unroasted forastero cocoa beans. In Table 1, 2- and 3-methylbutanoic acid (FD 8192; sweaty), acetic acid (FD 2048; acidic), and 3-hydroxy-4,5-dimethyl-2 (5H)-furanone (FD 1024; spiced) were detected at the highest dilutions, suggesting that these compounds are among those contributing to the aroma of unroasted cocoa beans.

Both techniques focus on cocoa matrix aroma profile characterization and the evaluation of the most relevant compounds. Volatile profiles can be estimated by the odor activity value (OAV) of a compound present in a specific cocoa matrix. The OAV is defined as the ratio between the compound concentration and the odor threshold value (OTV), for which the contribution to the overall aroma of the cocoa matrix is to be evaluated. The OTV found in the literature determined in a fatty medium (sunflower oil) is a reference for estimating the OTV in cocoa beans, for which the fat content is between 53.3 to 55 % in dry fermented beans. If the concentration of a volatile compound is higher than its respective odor threshold and its OAV > 1, this compound contributes to the overall aroma and flavor of the cocoa matrix. [9,12,44–46], determined that the concentration of 3-methylbutanal (provides malty, cocoa and chocolate notes) in cocoa liquor was 721.44 ng/g, much higher than its OUV: 5.4–80 ng/g. This compound clearly contributes to the overall aroma of cocoa liquor, with an OAV from 9.02 to 133.60. On the other hand, [12] reported OAV = 0.55 for linalool (floral notes) in the chocolate processed from cocoa liquor of the national variety; its concentration was 20.2 ng/g and its OTV = 37 ng/g, meaning this compound does not contribute to the final chocolate aroma. The volatile profiles of other matrixes studied are detailed in Tables 2–6.

Table 1. Dilution factors of active odor compounds from roasted and unroasted forastero cocoa beans.

No	Volatile	Odor	Dilution Factors (FD)	
			Unroasted	Roasted
1	4-hydroxy-2,5-dimethyl-3 (2H)-furanone	Like caramel	128	8192
1	4-Hydroxy-2,5-dimethyl-3(2H)-furanone	Caramel-like	128	8192
2	2- and 3-Methylbutanale	Malty	64	4096
3	Phenylacetaldehyde	Honey-like	64	4096
4	2–and 3-Methylbutanoic acid	Rancid	8192	4096
5	2-Acetyl-1-pyrroline	Popcorn-like	32	1024
6	Acetic acid	Acetic	2048	1024
7	2-Methoxyphenol	Smoky	256	512
8	2-Phenylethanol	Flowery	256	512
9	2-Ethyl-3,5-dimethylpyrazine	Earthy	64	256
10	Linalool	Flowery	64	256
11	Methyl propanoic acid	Rancid	256	256
12	2-Methyl-3-(methyldithio)furan	Cooked meat-like	128	256
13	Ethyl methyl propanoate	Fruity	64	128
14	Ethyl 2-methylbutanoate	Fruity	256	128
15	Dimethyl trisulfide	Sulfur-like	64	128
16	2,3,5-Trimethylpyrazine	Earthy	32	128
17	Phenylethyl acetate	Flowery	64	128
18	δ-Decenolactone	Coconut-like	256	128
19	Phenylethyl acetic acid	Sweet	256	128
20	Ethyl 3-methylbutanoate	Fruity	32	64
21	2,3-Diethyl-5-methylpyrazine	Earthy	2	64
22	Butanoic acid	Rancid	128	64
23	δ-Octenolactone	Coconut-like	32	64
24	cis-Isoeugenol	Smoky	64	64
25	3-Hydroxy-4,5-dimethyl-2(5H)-furanone	Spiced	1024	-

Source: Key aroma compounds in fermented Forastero cocoa beans and changes induced by roasting [6].

3. Flavor Precursors in Cocoa Beans

Aromas and flavors are generated during fermentation. This is a postharvest process carried out by various micro-organisms (yeasts, lactic acid bacteria, and acetic acid bacteria) generating flavor and aroma precursors, like the reducing sugars (glucose and fructose) formed by the action of cotyledon invertase and free amino acids release by carboxypeptidase (optimum pH 5.6) and aspartic protease (optimum pH 3.5), for which, later, during the drying and roasting of the beans, is combined in Maillard reactions to produce a high potential of aromas of different chemical classes [5,47–52]. According to the cut test method, the cotyledons of well-fermented and dry beans from a set of 100 have at least 60% brown beans and a smaller percentage of slaty and defective beans [49,50]. For example, Ref. [21] reported that the percentage of forastero brown beans ranged from 75.33 to 84%. Further, the fermentation index (FI) estimates the fermentation quality and provides a fermented bean final state analytical evaluation. According to [53], well-fermented beans have a (FI) ≥ 1. Ref. [21] reported a FI from 1.04 to 1.22 in preconditioned and fermented beans. Likewise, Ref. [54] reported values from 0.37 to 1.05 in beans fermented from 0 to 6 days, respectively. The FI correlates with the amount of reducing sugars, free amino acids, pH, and cocoa bean cotyledon color [55]. Therefore, the evaluation of precursors formed during fermentation is crucial to determine the quality of cocoa since well-fermented beans have a higher amount of precursors [56].

Regarding sugar reduction, a 78% and 93.5% increase at the end of fermentation was reported for trinitario and forastero cocoa beans, respectively [5,21]. Similarly, Ref. [16] reported a 60% to 70% increase for a forastero cocoa hybrid in Ghana. The difference is due to the postharvest process in which the cob is stored in order to reduce the moisture and the amount of pulp adhering to the beans; this method reduces the concentration of fermentable sugars and, therefore, the acidity at the end of fermentation. Additionally, the increment in reducing sugars is related to the method and fermentation duration, where the greatest amount of fructose is produced with respect to glucose [5,21,47]. A fructose-to-glucose ratio of 2:1 is expected when using the heap and tray fermentation method, 4:1 when the pods have been preconditioned and fermented in heap [5,57], and 4:1 when the pods have been preconditioned and fermented in heap [21]. This ratio gives an indication of fermentation quality in terms of the reducing sugars in cocoa varieties [58].

Likewise, the proteolysis of cocoa globulin (similar to vicilin) (storage protein of 566 amino acids) leads to the breakdown and abundance of free amino acids from 300 to 800 di- and tripeptide units, highlighting acidic, hydrophobic, basic, etc. [59,60]. Ref. [57] reported that hydrophobic amino acids comprised 68–73% of the total amount of amino acids, with leucine, phenylalanine, and alanine being the most abundant in dried forastero cocoa beans. Similarly, Ref. [56] indicated that leucine, phenylalanine, valine, alanine, and isoleucine were the predominant hydrophobic amino acids in dried criollo cocoa beans, comprising 56 to 70% of the total amino acids, while [61] quantified 50% of leucine and phenylalanine in fermented and dried forastero cocoa beans. When roasting the beans, this hydrophobic amino acid fraction partially descends through Maillard reactions, increasing the pyrazine levels [52]. In beans, these hydrophobic amino acid fractions partially decrease via Maillard reactions, increasing the pyrazine levels [62]. Therefore, the level of pyrazines can be used to evaluate the efficiency of roasting during this stage because of its pronounced influence on the cocoa's final aroma [56].

The abundance of peptides depends on geographical origin. Refs. [63,64], reported that high levels of oligopeptides and amino acids are found in cocoa bean samples from Central America, Caribbean Islands, Mexico, Santo Domingo, and Peru, together with samples from Papua New Guinea and the criollo variety from Mexico; in contrast, the samples from Flores, Sulawesi, Malaysia, and Ivory Coast showed low peptide concentrations and free amino acids. The difference in peptide concentration in relation to the origin of the cocoa is due to the nitrogen fertilization of the plants and mainly the pH found in the soil [65]. Furthermore, the season in which the pods are harvested particularly affects the pH and nitrogen concentration of the beans. Ref. [66] reported that the pods harvested in summer had beans with a lower pH than those harvested in winter; likewise, the beans harvested in the dry season had a higher concentration of nitrogen than the ripe beans during the rainy season.

3.1. Volatile Compounds Generated from Precursors

Table 2 indicates the number of compounds in the different chemical groups and cocoa matrixes. It is noted that the availability of the compounds increases as the beans are turned into chocolate. Overall, acids, alcohol, aldehydes, aldehydes, ketones, and esters are present throughout the processing chain. The compounds found represent sour, vinegar, rancid, sweaty, fruity, floral, soapy, creamy, green leafy, herbal, and spicy notes. Obviously, unfermented beans do not have pyrazines, furans, furanones, pyrans, pyrones, or pyrroles as they are more common in well-fermented roasted beans, except for the lower concentrations of pyrazines that are found in dry beans, a metabolic product of *B. subtilis* or *Bacillus megaterium*, which are present at the end of cocoa fermentation [5,38,67]. These compounds have cocoa, chocolate, walnut, popcorn, coconut, and candies notes, although they also have fruity notes like furans (furaneol) [45].

Table 2. Number of common volatile compounds in the cocoa bean processing chain.

* Variety	** Matrix	Volatile Compounds													Total Volatiles	Reference
		Acids	Alcohols	Aldehydes	Ketones	Esters	Ethers	Pyrazines	Lactones	Furans	Furanones	Pyrans, Pyrodons and Pyrroles	Terpenes and Terpenoids	Other		
CR	CL	2	4	4	2	7		7					6	2	34	[2]
TR	DB	3	10	5	6	7		4	1				5		41	[4]
NC, CCN51	RB	4	10	7	10	11		9		4	4	3	6	2	70	[5]
FR	DB	6	3	4	1	3		5	2	1	2		1	3	31	[6]
CR, FR, TR	UDB	4	12	4	4	7	2		2	5			9	3	52	[7]
CR	DB	6	7	10	5	11		5		1			4	1	50	[8]
	CH	6	1	9	7	11		8	1	1	2		1	2	49	[10]
	CH	4	4	7	6	3		6		2		1	2		35	[11]
NC, FR	CL	3	11	8	7	9		12	1	1	2	4	8	1	67	[12]
FR	DB	10	9	3	5	11		1							39	[14]
	RB			9	8			10	1	3	2	1		5	39	[21]
FR	CL	3	7	7	10	8		15		4	3	10	6	7	80	[23]
CR, FR, NC	DB	9	12	14		5				2			12	6	60	[28]
	CH	11	14	13	11	8		16	2	5			3	2	85	[37]
FR	DB	11	12	3	5	20		4						3	58	[39]
FR	CH	4	4	4	3	3		5		1	1	1	6	1	33	[44]
FR	CL	2	9	7	5	9		9		2			6	3	52	[57]

* Cocoa varieties: CR: Criollo, FR: Forastero, TR: Trinitario, NC: Nacional, CCN51: Colección Castro Naranjal. ** Cocoa matrix, UDB: unfermented dried beans, DB: dried beans, RB: roasted beans, CL: cocoa liqueur, CH: dark chocolate.

Dried beans have organic acids that volatilize during roasting [39]. It has been reported that 2- and 3-methylbutanoic acid and acetic acid have been found to be most relevant in dried forastero cocoa beans. During roasting, the acid level decreases significantly depending on the roasting temperature. Thus, at 135 °C and 160 °C, a significant increase in acids is observed, while at lower temperatures, the increase is negligible. The 2- and 3-methylbutanoic acids persist at the end of roasting at a temperature of 95 °C and their concentration reaches 17,300 ng/g ([6]. This concentration exceeds the odor threshold of 203 and 11 ng/g, respectively, indicating that these acids greatly affect the final chocolate aroma. Likewise, Ref. [44] reported that only acetic acid and isovaleric acid are of an active odor in bitter chocolate samples, especially from beans roasted at 100 °C to 140 °C and in concentration rises from 171.26 to 381.4 ng/g. Figure 1 illustrates the compound percentage generated from foreign cocoa beans from Africa, America, and Southeast Asia when roasted at 140 °C for 30 min. There is a rise in aldehydes, phenols, pyrazines, furans, pyrans, and pyrroles at the end of roasting. A reduction in undesirable compounds, like acids and alcohols, is clearly evident. However, not all acids change at the end of the process.

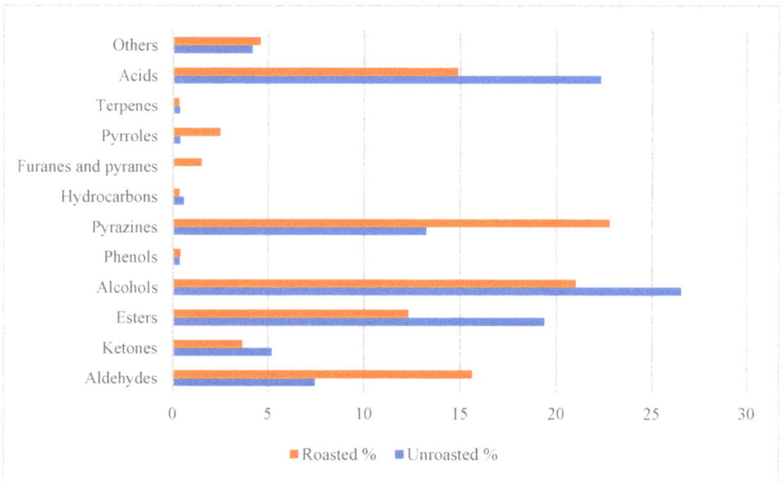

Figure 1. Relative percentage of volatile compounds before and after roasting. Source: taken and adapted from [22].

Cocoa liquor is the product obtained from fermented, dried, roasted, and ground cocoa beans. The concentration of total aromatic volatiles in cocoa liquor is higher than in chocolates. Ref. [12] reported that cocoa liquor had four to seven times more volatiles than the processed national and forastero variety chocolates than cocoa liquor, respectively. These included acids, alcohols, aldehydes, esters, ketones, pyrazines, lactones, and terpenes. Similarly, Ref. [9] highlighted acids, alcohols, esters, terpenes and terpenoids, aldehydes, ketones, pyrazines, furans, furanones, pyrans, pyrones, and pyrroles in the cocoa liquor of the forastero variety. Conversely, in the criollo cocoa variety, Ref. [68] reported that the main components were alcohol, esters, aldehydes, ketones, hydrocarbons, nitrogen and oxygen heterocycles, nitriles, and sulfides. Many of the compounds are similar in each variety, and what differentiates them is the contribution of each of them to the overall aroma of a particular variety. Hence the need to know the aromatic profiles of the varieties.

Finally, processed chocolates have a wide variety of compounds. Table 2 shows that 85 compounds present in chocolate are subdivided into several chemical groups, including acids, alcohols, aldehydes, ketones, esters, and pyrazines as the major compounds. Pyrazines (16 compounds) are most abundant in chocolates and impart their malty, roasted cocoa, nutty, almond, and hazelnut notes. In contrast, Table 2 also indicates the presence of ethers (carbitol, 2-(2-butoxyethoxy) ethanol furfural), which are only present in unroasted trinitario cocoa beans and do not contribute to the processed chocolate [7].

3.2. Specific Volatile Compounds
3.2.1. Fermented and Dried Beans

Table 3 indicates the contribution of the individual compounds to the overall aroma of the beans; their OAV > 1. 20 and 36% out of the compounds impart undesirable aromas. Acetic acid, 2-methylbutanoic acid, 3-methylbutanoic acid, and 2-methylpropanoic acid were common in both studies. They differ by approximately 55 times for acetic acid and 48 times for 3-methylbutanoic acid due to the fermentation method used (wood box vs. pile fermentation), days fermented (5 d vs. 3 d with prefermentation), and the origin of the beans (Costa Rica vs. Ecuador). These compounds are present in the over-fermented beans from leucine metabolism; such acids are undesirable in any cocoa matrix, imparting vinegar, pungent, sweaty, and rancid flavors [5,42]; it was reported in separate and complementary studies that box fermentation offers less aeration compared to the heap method. It was expected that the acetic acid concentration in the heap method would be higher, as aeration drives the emergence of acetic acid bacteria; however, the box method offers a higher concentration. The prefermentation of the beans helped to reduce their moisture and sugar content, thus decreasing the acid content at the end of fermentation. Prefermentation was absent in the box fermentation study. 2-methoxyphenol and 3-ethylphenol are also compounds that impart an unpleasant smoky flavor and odor and have been reported as a crucial marker of this aroma. In this study, these compounds are very significant, presenting concentrations of 221.00 and 7.66 µg/kg and an OAV of 122.8 and 3.5, respectively. Ref. [13] proposed the maximum tolerable concentration based on the rounded threshold value of 2 µg/kg for 3-ethylphenol. Sometimes, the aroma imparted by 2-methoxyphenol goes unnoticed as it is masked by the more odorous compounds in cocoa. Ref. [13] reported a concentration of 70 µg/kg for 2-methoxyphenol, and the reason for this depends on compound concentration and the authentic cocoa odors contributing to the pleasant aroma of cocoa masking the foul odor.

Table 3 shows the difference between the two studies on fruit aromas. The highest amount of these aromas is present in the study of [5], who used beans of the Ecuadorian trinitario variety in their study. In contrast, Ref. [42] proposed a single relevant fruit aroma, as is the case for ethyl phenylacetate. In both studies, the same variety (trinitario) was analyzed; however, there were large differences in the amount of its volatile compounds. The difference may be due to geographical origin, postharvest techniques, the chemical composition of the beans, and the method used for their quantification [22]. Fruity volatile compounds, such as isoamyl acetate, 2-nonanone, 2-heptanol, 2-heptanone, 2-pentyl acetate, and ethyl, in descending order of their OAV, contribute to the overall aroma of the dried beans. These fruity aroma compounds are most relevant in the trinitario cocoa variety from Ecuador, known for aromatic cocoa. The esters synthesized during fermentation are dependent on environmental precursors, as well as on aeration and the presence of alcohol, such as ethanol. For example, the production of isoamyl acetate by yeasts (*Pichia fermentans*) uses isoamyl alcohol as a precursor [69]. According to Table 2, this compound is also odor active (OAV = 4), and its concentration exceeds its OTV; therefore, it remains in the aromatic profile and serves as a precursor for other compounds. 2-pentyl acetate (OAV = 5.47) and ethyl acetate (OAV = 1) can act as the precursors of secondary alcohols, such as 2-pentanol and 2-heptanol [70]. The latter contributes its relevant citrus aroma (OAV = 11.08) to Ecuadorian trinitario cocoa beans.

Floral aromas in both studies include, in descending order, linalool, 2-phenylethanol, 2-phenylacetaldehyde, 2-phenylethyl acetate, acetophenone, and 2-methyl-3-buten-2-ol. Ref. [71] proposed linalool as a grade indicator (fine or basic grade) in some varieties. A linalool/benzaldehyde ratio of greater than 0.3 indicates fine-grade cocoa. Our review showed linalool and benzaldehyde OAVs of 17.38 and 35.15, respectively, with both compounds aromatically active at a 0.3 ratio. Ref. [20] reported a 0.56-to-0.89 linalool/benzaldehyde ratio in roasted Criollo cocoa beans. As linalool is a biosynthesis product, its creation depends on plant varieties, growing, and fermentation conditions. During the roasting process, the linalool content decreases slightly due to volatility, but the relative difference between basic and fine-grade cocoa remains [71].

Table 3. Odor activity values of key aroma compounds in fermented dried beans.

Reference	Variety	Method	Matrix	Volatile	OTV (μg/kg)	OAV	Odor Description
[5]	Trinitario (Sacha Gold)	HS-SPME GC–MS	Dried beans	Isoamyl alcohol	100	4	Banana, fruity, fermented, cognac
				Isoamyl acetate	9.6	91.46	Banana, fruity
				2-heptanone	300–98,000	5.79	Fruity, coconut, floral, cheesy
				2-heptanol	263	11.08	Citrus, fruity
				2-nonanone	100	31.14	Fruity, fresh, sweet
				Ethyl acetate	940–22,000	1	Pineapple, fruity, sweet, grape
				2-pentyl acetate	13–27,000	5.47	Fruity, orange, tropical
				2-Phenylacetaldehyde	22–154	10.82	Floral, honey
				2-phenylethyl acetate	137–233	2.42	Floral, honey
				2-phenylethanol	211	13.12	Floral, honey
				Acetophenone	5629	1	Floral
				Linalool	37	17.38	Floral, pink, sweet, green, citrus
				2-methyl-3-buten-2-ol	480	1	Herbal, earthy
				2-octanol	100	1	Spicy, green, woody, earthy
				2-methylbutanal	2.2–152	85.37	Chocolate
				3-methylbutanal	5.4–80	37.39	Chocolate
				Benzaldehyde	60	35.15	Sweet, bitter almond, cherry, woody
				Trimethylpyrazine	290	1.32	Cocoa, roasted nuts
				Tetramethylpyrazine	38,000	1	Chocolate, cocoa, coffee
				2,3-butanedione	3–10	217.53	Buttery
				Acetic acid	124	154.77	Sour vinegar
				2-methylpropanoic acid	190–755	4.87	Rancid butter
				3-methylbutanoic acid	22	131.78	Rancid sweat
				Ethanol	3×10^4–6×10^5	1	Alcoholic
				2-methyl-1-propanol	1000	1	Wine, ethereal
[42]	Trinitario	GC–OAEDA-AIDS	Dried beans	Acetic acid	124	8467.7	bitter, vinegar
				2-methylbutanoic acid	203	99.0	spicy, sweaty
				3-methylbutanoic acid	11	6590.9	spicy, sweaty
				2-phenylethanol	211	8.5	flowery
				3-methylbutanal	5.4	115.2	of malt
				2-methylbutanal	2.2	320.5	of malt
				Fethyl enilacetate	300	1.0	flowery, fruity
				Ethyl 3-methylbutanoate	0.98	2,8	tasty
				2-phenylethyl acetate	0.137	9489.05	Nuts
				Ethyl 2-methylbutanoate	0.37	59.7	tasty
				4-hydroxy-2,5-dimethyl-3 (2H)-furanone	27	1.0	like caramel
				2-Methoxyphenol	1.8	122.8	Smoked
				3-Ethylphenol	2.2	3.5	phenolic, animalic
				2-ethyl-3,5-dimethylpyrazine	2.2	18.0	earthy
				2-ethyl-3,6-dimethylpyrazine	57	0.01	earthy
				dimethyl trisulfide	0.03	83.3	Cabbage

OTV: Odor threshold values. OAV: Odor activity values.

Schluter et al. (2020) [42] reported a high OAV with a prevalence of chocolate and malt flavors, likely related to the high concentrations of isoleucine, leucine, and phenylalanine during fermentation. They attributed such aromas to Strecker aldehydes, such as 2- and 3-methylbutanal (chocolate, malt flavor) and phenylacetaldehyde (honey flavor), respectively. On the contrary, 2-phenylethyl acetate, trimethylpyrazine, ethyl acetate, acetophenone, 2-methyl-3-buten-2-ol, 2-octanol, tetramethylpyrazine, ethanol, and 2-methyl-1-propanol show a low OAV, which, individually, would not contribute to their respective aromas, but, as a whole, would present a range of odors from wine to floral and honey.

Figure 2 illustrates the key aroma profile in both cocoa samples. The aroma profile of the trinitario Sacha Gold variety (Figure 2b) is broader and has a distinct fine diversity of aromas representing processed chocolate. Figure 2a, on the other hand, presents a limited aroma profile that is not appreciated by the chocolate industry. In both studies, a total of four pyrazines are observed, including 2-ethyl-3,5 and 3,6-dimethylpyrazine, trimethylpyrazine, and tetramethylpyrazine, for which the OAVs are less than 18. During fermentation, both trimethyl and tetramethylpyrazines appeared; however, they represent only 0.6% of the total volatiles [72]. During grain roasting, the number and concentration of pyrazines notably rose: 2,5-dimethylpyrazine (DMP), 2,6-DMP, 2-ethylpyrazine, 2,3-DMP, 2,3,4-trimethylpyrazine (TrMP) and 2,3,5,6-tetramethylpyrazine (TMP) [62], evidenced in the following section.

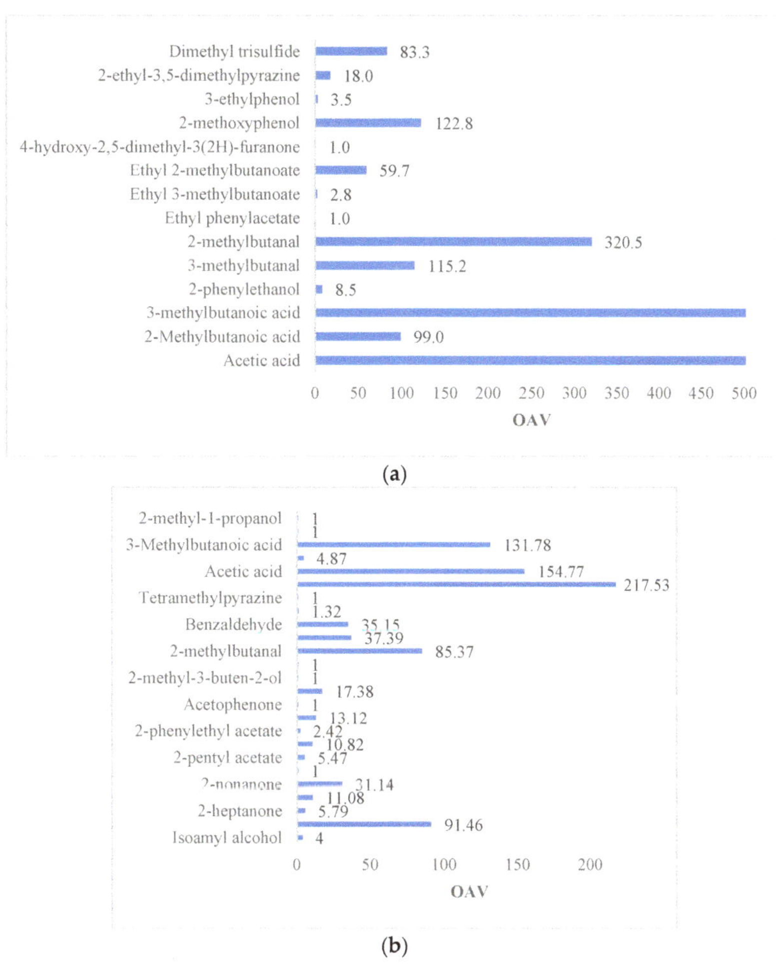

Figure 2. Profile of key volatile compounds present in fermented dry beans determined by (**a**) [42] trinitario cocoa beans and (**b**) [5] trinitario Sacha Gold from Ecuador.

3.2.2. Roasted Beans

Roasting the beans boosts the intensity of some of the volatile compounds present in dried beans, generating the appearance of new compounds. Ref. [6] reported 31 compounds in forastero cocoa beans roasted at 95 °C, 71 compounds in beans (of the same variety) roasted at 140 °C, while [9] reported 78 compounds in forastero cocoa beans roasted at 160 °C. They found 34 compounds that had increased in their concentration during roasting, nine compounds that had increased in concentration up to 140 °C and had then decreased or were undetectable; A total of 26 new compounds were added, and the concentration of nine compounds decreased significantly. After roasting, the distribution of the different classes of compounds changed, increasing in pyrazines (22.79%) and aldehydes (15.62%) and forming new compounds from Maillard reactions, such as pyrroles, furans, and pyrans [22]. Some pyrrole derivatives are formed at moderate roasting temperatures and relatively high humidity [40]. These compounds come from sugar precursor degradation in cocoa and decrease during roasting. Ref. [73] reported that these volatiles are a useful indicator for which their level can be used to monitor the early stages of roasting.

The presence of undesirable compounds, Figure 3, in both studies represents 42% and 27% of the total aroma profile, indicating the presence of acidic, musty/earthy, burnt, and smoky flavors [74]. Its high intensity and overall contribution are reflected in the OAVs ranging from 1.77 for 2-methoxyphenol to 4920 for acetic acid. The concentration of acetic acid is more than half in the unroasted beans when roasted at 130 °C for 30 mins. Roasting at higher temperatures or for a longer period does not significantly affect the acetic acid concentration. On the contrary, the concentration is higher when roasted at 160 °C for 30 min [75]. Other compounds, such as sulfur derivatives, are also present in cocoa beans roasted at temperatures >160 °C. These compounds generate unpleasant smoky, onion, cabbage, and gasoline notes that persist in the final chocolates, albeit in lower concentrations [39,44].

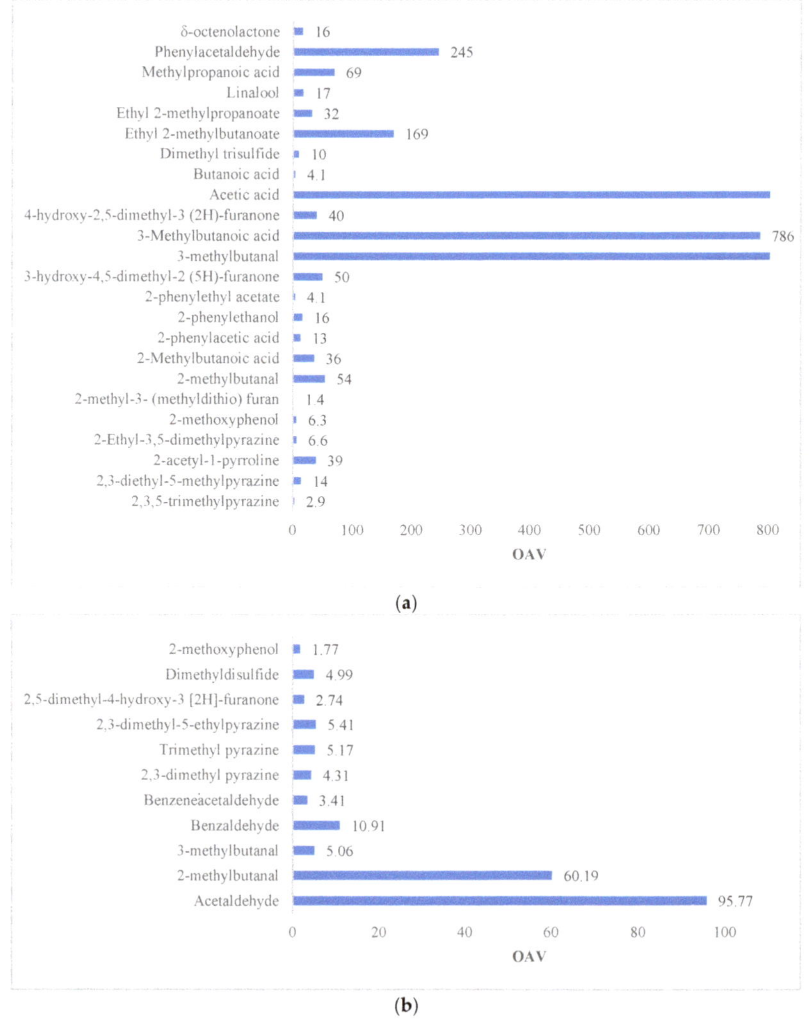

Figure 3. Volatile compounds profile present in the roasted beans determined by (**a**) [6] forastero roasted cocoa beans, and (**b**) [21] forastero roasted cocoa beans.

A percentage greater than 50% of the compounds in Table 4 indicates a contribution of desirable aromas, such as fruity, floral, cocoa, and chocolate, including the Strecker aldehydes (2- and 3-methylbutanal, phenylacetaldehyde) present in dry beans in smaller quantities. 3-methylbutanal increases its contribution during roasting, especially at 95 and 110 °C, while at 125 °C, there is a return to the initial values. This is due to the prevalence of the volatilization phenomenon compared to that of generation [76]. Furthermore, the concentration of benzaldehyde increases during fermentation [5] and during roasting, making it susceptible to losses at temperatures above 125 °C [76]. Throughout the processing of the beans, the concentration of this compound decreases. In dry beans, the concentration is 1.08 µg/g, and its OAV is 18.08, whereas, in roasted beans, it drops to 1.01 µg/g and presents OAV = 16.90 [9,21].

Table 4. Odor activity values of key aroma compounds in roasted beans.

Reference	Variety	Method	Matrix	Volatile	OTV (ng/g)	OAV	Odor Description
[6]	Forastero	SAFE-AEDA	Roasted beans	2. 3. 5-trimethylpyrazine	290	2.9	Earthy
				2. 3-diethyl-5-methylpyrazine	0.5	14	Roasted potato
				2-acetyl-1-pyrroline	0.1	39	Like popcorn
				2-ethyl-3. 5-Dimethylpyrazine	2.2	6.6	Chocolate, sweet
				2-Methoxyphenol	16	6.3	Smoked
				2-methyl-3-(methyldithio) furan	0.4	1.4	Similar to cooked meat
				2-methylbutanal	140	54	Malt
				2-methylbutanoic acid	203	36	Rancid
				2-phenylacetic acid	360	13	Floral smell, nasty geranium
				2-phenylethanol	211	16	Flowery
				2-phenylethyl acetate	233	4.1	Floral, honey
				3-hydroxy-4. 5-dimethyl-2 (5H)-furanone	0.2	50	Spicy
				3-methylbutanal	13	2030	Malt
				3-methylbutanoic acid	22	786	Rancid
				4-hydroxy-2. 5-dimethyl-3 (2H)-furanone	25	40	Like to caramel
				Acetic acid	124	4920	Sour, vinegar
				Butanoic acid	135	4.1	Rancid
				Dimethyl trisulfide	2.5	10	Sulfuric
				Ethyl 2-methylbutanoate	0.26	169	Tasty
				Ethyl 2-methylpropanoate	1.24	32	Tasty
				Linalool	37	17	Flowery
				Methylpropanoic acid	190	69	Rancid
				Phenylacetaldehyde	22	245	Like honey
				δ-octenolactone	4730	16	Like coconut
[21]	Forastero	HS–SPME–GC–MS	Roasted beans	Acetaldehyde	0.22	96.78	Sour, fruity
				2-methylbutanal	2.2–152	53.36	Chocolate
				3-methylbutanal	5.9–80	5.35–79.32	Chocolate
				Benzaldehyde	60	18.08	Sweet almond, bitter, cherry
				Benzeneacetaldehyde	22–154	2.41	Honey, sweet, pink, floral
				2. 3-Dimethyl pyrazine	123	1.45	Caramel, cocoa
				Trimethylpyrazine	290	1.64	Cocoa, toasted walnuts, peanuts
				2. 3-dimethyl-5-ethylpyrazine	60	2.95	Popcorn, burnt cocoa, toasted
				2. 5-dimethyl-4-hydroxy-3 [2H]-furanone	1.6–50	1.15–36.06	-
				Dimethyl disulfide	12	3.68	Sulfur-like, cabbage, onion
				2-Methoxyphenol	10–70	6.81	Smoked, repulsive

Other compounds that turn up after roasting are 4-hydroxy-2,5-dimethyl-3 (2H)-furanone, 2-acetyl-1-pyrroline, 2,3-dimethyl-5-ethylpyrazine, trimethylpyrazine, and 2,3-dimethylpyrazine. 2-acetyl-1-pyrroline, which is typically associated with roasting (Parker, 2015), exudes a popcorn and cracker-like aroma, having a low threshold in oil 0,1 ng/g and a concentration above OTV = 39 ng/g. Therefore, the contribution to the overall aroma of the roasted beans is significant (Table 3). Likewise, the pyrazines generated during roasting at temperatures above 100 °C exude notes of cocoa, baked potato, and chocolate. Ref. [45] reported that simple unsubstituted or monosubstituted pyrazines have a roasted, biscuit aroma and relatively high aroma thresholds, but as the substitution increases, the odor threshold decreases. It is evident from Table 3 that some of the pyrazines have low OTVs: 2,3-diethyl-5-methylpyrazine and 2-ethyl-3,5-dimethylpyrazines, and high OTVs: 2,3,5-trimethylpyrazine and 2,3-dimethyl-5-ethylpyrazine. These compounds exude baked potato, chocolate, earthy, popcorn, and cocoa aromas, respectively. However, their OTVs are inversely related to their OAVs. There are a few fruity and floral aromas, as shown in Table 3, including linalool, 2-phenylethanol, and 2-phenylethyl acetate, in descending order of their aromatic contribution. This deficiency in fruity aromas is common in the forastero variety [28]. Finally, the 59 volatile compounds present in the roasted beans include alkyl pyrazines and aldehydes from reactions between oligopeptides derived from vicilin class globulin (7S) and reducing sugars of the cocoa beans [36]. The author reported eight unidentified volatile compounds likely to expel specific cocoa notes.

3.2.3. Cocoa Liquor

The aroma of cocoa liquor is an important characteristic affecting its quality; therefore, the affecting factors are the origin of the cocoa bean and the postharvest processing (fermentation and drying), roasting, and storage [19]. The origin of the cocoa bean is vital in determining the aroma of all its cocoa products and is related to genetic and environmental factors. In this regard, only a few aromas are characteristic of cocoa liquor from different origins, including 3-methylbutanal (malty), linalool (floral), β-phenylethyl alcohol (pink), benzaldehyde (almond-like), benzeneacetaldehyde (pink), β-phenylethyl acetate (fruity), 3-methylbutyl benzoate (fruity), 2,5-6-dimethylpyrazine (potato-like), ethylpyrazine (popcorn), trimethylpyrazine (roasted), 3-ethyl-2,5-dimethylpyrazine (roasted), tetramethylpyrazine (nutty), 3,5-diethyl-2-methylpyrazine (cocoa), furfural (potato), acetic acid (acid), 3-methylbutanoic acid (stench), and dimethyltrisulfide (onion), which are considered the key active aroma compounds contributing to the overall cocoa liquor odor of the forastero variety. However, the intensity of each aroma among the different cocoa liquors is distinct [77]. Likewise, Ref. [78] reported that the specific aroma of bulk cocoa liquor included sweet, nutty, caramel, and chocolate notes associated with trimethylpyrazine, tetramethylpyrazine, 2,3-butanediol, dodecanoic acid, β-phenylethyl alcohol, 2-acetylpyrrole, and benzeneacetaldehyde.

On the whole, the compounds of the cocoa liquors studied represent 57.89 to 65.79% of OAV >1, and 32.61 to 42.11% represent OAV < 1. Among these compounds, fruity, floral, chocolate, buttery, and undesirable aromas are highlighted (Table 5). Several compounds account for 17 to 40% of fruit flavors with a VAO > 1, including isoamyl acetate (banana), 2-heptanol (citrus), ethyl phenylacetate, isoamyl alcohol (banana), furaneol (strawberry), pentyl 2-acetate (orange), ethyl 3-methylbutanoate, 2-nonanone, 2-heptanol, 2-heptanone (coconut), pentylacetate, and 3-methylbutyl acetate. From 14 to 25% of the floral compounds, 2-phenylethyl alcohol, linalool, 2-phenylethylacetate, acetophenone, 2-propanone, 2-phenylacetaldehyde, and β-myrcene are included. From 20 to 33% provide chocolate notes, including 2- and 3-methylbutanal, 5-ethyl-2,3-dimethylpyrazine, trimethylpyrazine, 2,3,5-trimethylpyrazine, benzaldehyde, 2-ethyl-5 and 6-methylpyrazine, 2,6-dimethylpyrazine, 2,5-dimethylpyrazine, tetramethylpyrazine, methylpyrazine, and 2-methylpropanal. From 3 to 14% provide creamy and buttery tones, such as ethylpyrazine, furfural, 3-hydroxy-2-butanone, 2,3-butanedione, 2,3-pentanedione, and 2-methylpropanoic acid. Finally, the compounds providing caramel and earthy notes in the 3% to 15% range include gamma-butyrolactone and 3-ethyl-2,5-dimethylpyrazine, respectively. Figure 4 details the set of OAV > 1 compounds described above, which is common among the studies of Streker aldehydes, linalool, isoamyl acetate, and 2-heptanol.

The compounds with an OAV < 1 include isoamyl alcohol, ethyl acetate, ethylhexanoate, limonene, acetophenone, linalool, 2,5-dimethylpyrazine, 2,6-dimethylpyrazine, 2-ethyl-5-methylpyrazine, tetramethylpyrazine, propionic acid, 2-butanone, ethanol, 2-butanol, 2-methyl-1-propanol, 2-pentylfuran, 3-hydroxy-2-butanone and ethanol, with fruity, floral, chocolate, buttery, and mainly undesirable aromas. The compounds exclusive to the study include rose oxide and β-myrcene that reveal floral notes in cocoa of the forastero and national variety, respectively.

Table 5. Odor activity values of key aroma compounds in cocoa liquor.

Variety	OAV *	Odor Contribution %								Reference
		Fruity	Floral	Chocolate	Buttery	Spices	Caramel	Undesirables	Earthy	
CCN51		31.82	13.64	22.73	13.64	-	-	18.18	-	[4]
National		4-	16.00	2-	8.00	-	-	16.00	-	
Forastero	>1	17.24	13.79	27.59	3.45	3.45	6.90	24.14	3.45	[9]
National		16.67	25.00	33.33	8.33	-	8.33	8.33	-	[12]
**		2-	2-	3-	5.00	-	-	15.00	1-	
CCN51		25.00	12.50	25.00	-	-	-	37.50	-	[4]
National		7.69	7.69	30.77	-	-	-	53.85	-	
Foastero	<1	35.29	17.65	11.76	11.76	-	5.88	17.65	-	[9]
National		25.00	12.50	37.50	-	-	12.50	12.50	-	[12]

* Odor activity values. ** African cocoa.

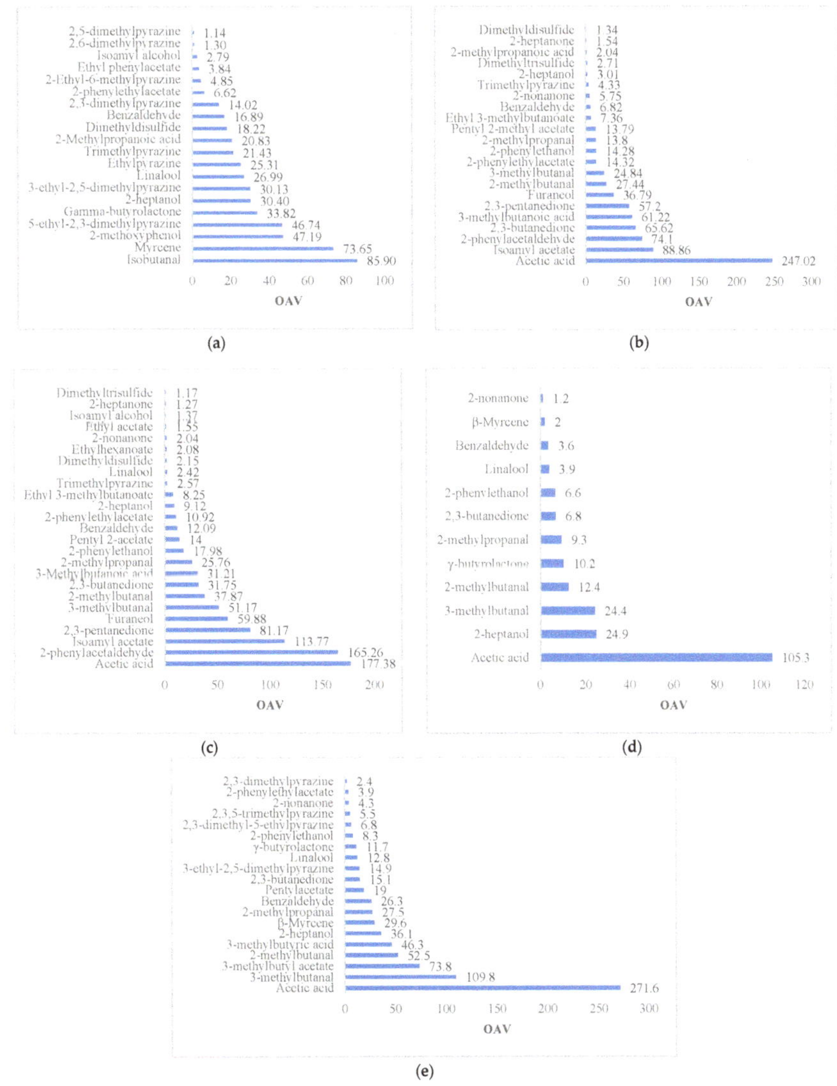

Figure 4. Studies of the volatile compounds with OAV > 1 in the liquor of the cocoa determined by (**a**) [9] in forastero cocoa beans, (**b**) [4] in CCN51 cocoa beans, (**c**) [4] in nacional cocoa beans, (**d**) [12] Ecuadorian liquor and (**e**) [12] African liquor.

Smoky, acidic, hammy, or musty aromas are the common compounds in cocoa liquors, and are present throughout the postharvest process [79]. From 8 to 24% of the total active odor aromas (OAV > 1) correspond to undesirable compounds in cocoa liquors and dried beans, such as acetic acid, 2,3-pentanedione, 2- and 3-methylbutanoic acid, and 2-methoxyphenol. In addition, 2,3-butanedione, dimethyl di- and trisulfide, methylpropanoic acid, and butanoic acid are in roasted beans and persist in cocoa liquor. During chocolate processing with these liquors, the OAV partially decreases or is undetected. For instance, in acetic acid and 2-nonanone, the OAV decreases by 5 to 14 times, and the concentration of 2,3-butanedione is undetected, so their OAV cannot be calculated [12]. The same occurs with butyric acid and 3-methylbutyric acid, which are only detected in cocoa liquors and disappear in processed chocolate. It is worth mentioning that although acids present a lower number of compounds in cocoa liquors, they have the highest VAOs, affecting the liquor's pH. Cocoa liquor with a low pH (4.75 to 5.19) is more likely to have off-flavors.

3.2.4. Chocolate

The flavor of chocolate depends on the way the series of processes described above are carried out. Conching is the last of these processes, whereby the manufacturer can obtain the flavor and aroma required for a particular product. However, this process cannot correct previous mistakes, like an unpleasant smoky or moldy flavor due to poor drying, nor can it turn an inferior cocoa taste into a perfect one [80]. Specifically, the function of conching is to evaporate volatile acids, achieve adequate viscosity, remove excess moisture, and develop a desirable color [81,82], as well as remove off-flavors and aromas while retaining desirable ones [80]. Chocolate producers often use different conching temperatures and times depending on cocoa bean varieties and the origin of the chocolate products with the desired aromatic properties [83]. In particular, levels of the most important odors decrease significantly by rising conching duration (from 6 to 10 h at 80 °C). Prolonged times reduce most pyrazines (including 2,5-dimethylpyrazine, 2-ethyl-5-methylpyrazine, and 2,3,5-trimethylpyrazine) and the levels of alcohol, acid, aldehydes, and small esters [25]. Similarly, [11] reported that prolonged conching reduces volatile acid concentration, alcohol, 3-methylbutanal, benzaldehyde, and several lesser volatile pyrazines, like trimethylpyrazine, tetramethylpyrazine, and acetylpyrrole. They also noted that this treatment increases the furfural content and does not affect the isobutanal, 2-methylbutanal, and phenylacetaldehyde levels because of the additional reaction compensation to form Strecker aldehydes during conching. Other components with significant contributions to chocolate (Figure 5) and that are highly odorous based on their low odor thresholds are 2-methylbutanal (2.2 ng/g), 3,5-diethyl-2-methylpyraniza, furaneol (27 ng/g), 2,3-diethyl-5-methylpyrazine (7.2 ng/g), ethyl 2- and 3-methylbutanoate (0.37 and 0.98 ng/g), 2-methylpropanal (3.4 ng/g), 3-isobutyl-methoxypyrazine (0.04 ng/g), 3-isopropyl-2-methoxypyrazine (0.01 ng/g), and linalool (37 ng/g) [10–12,44,84].

Other major aroma compounds remaining in chocolates include 2- and 3-methylbutanal (chocolate, malt), benzaldehyde (roasted almonds, malt), gamma-butyrolactone (sweet, caramel), 2-methylpropanal (unroasted cocoa, malt), linalool (flowery, fruity, tea-like), acetic acid (bitter, vinegar), and 2-phenylethanol (honey, rose) (Figure 5). Ref. [84] reported similar compounds, such as 3-methylbutanal, 2-methylpropanal, phenylacetaldehyde, tetramethylpyrazine, 2-acetyl-1-pyrroline, trimethylpyrazine, 3-methylbutanoic acid, acetic acid, and vanillin. The uncommon compound is vanillin, which is a highly odorous compound with OAV = 100 and was found only in chocolate with 90% cocoa (Figure 5e). In fact, the most uncommon aromatic compounds among the studies are found in this chocolate. Likewise, 2-acetyl-1-pyrroline (popcorn-like aroma) is found in roasted beans (OAV = 39) [6], cocoa liquor (OAV = 207.55) and chocolate (OAV= 396.23) [85]. Seyfried and Granvogl (2019) [10] reported OAV = 2 in 90% cocoa chocolate, and this was unidentified in 99% cocoa chocolate. This compound is highly volatile and mainly generated during roasting.

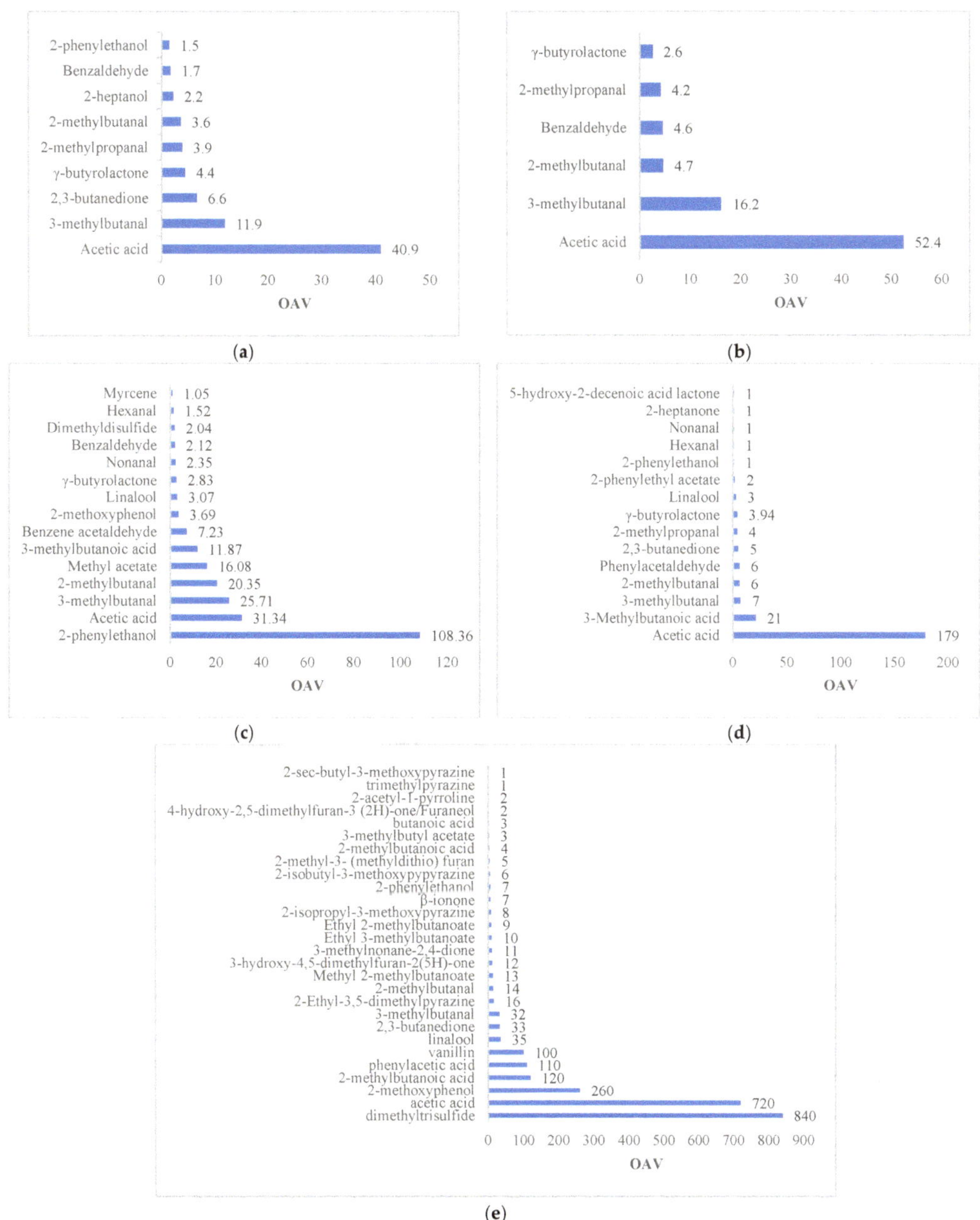

Figure 5. Studies of volatile compounds with OAV > 1 in dark chocolate determined by (**a**) [12] in chocolate from national cocoa beans, (**b**) [12] in chocolate from African cocoa beans, (**c**) [44] in chocolate from forastero cocoa beans, (**d**) [11] in Vietnamese cocoa beans and (**e**) [10] in commercial chocolate.

The greater the availability of the compounds, the more complex the chocolate aroma is due to a wide volatile matrix. The studies in this paper show that acetic acid is the most abundant compound, and the rest of the compounds represent a mixture from different families (acids, alcohols, pyrans, aldehydes, esters, furans, and pyrazines). Table 6 shows that the chocolates made with 40 and 90% cocoa have an almost complete distribution of aromas ranging from fruity to undesirable in a range of 13 to 25%.

Table 6. Odor activity values of key aroma compounds in dark chocolate.

Description	OAV	Odor Contribution %								Reference
		Fruit	Floral	Chocolate	Buttery	Spice	Caramel	Undesirable	Earthy	
Chocolate 90% cocoa		25.00	14.29	21.43	3.57	7.14	3.57	25.00	-	[10]
Chocolate 40% cocoa liquor		13.33	26.67	2.00	13.33	6.67	6.67	13.33	-	[11]
* Chocolate 51.6% cocoa		11.11	11.11	44.44	11.11	-	11.11	11.11	-	[12]
** Chocolate 51.6% cocoa	>1	-	-	66.67	-	-	16.67	16.67	-	
Chocolate 70% cocoa liquor		6.67	26.67	2.00	-	6.67	6.67	33.33	-	[44]
Chocolate 90% cocoa		23.81	4.76	9.52	23.81	9.52	14.29	14.29	-	[10]
Chocolate 40% cocoa liquor		11.11	-	66.67	11.11	-	-	11.11	-	[11]
* Chocolate 51.6%cocoa		36.36	27.27	18.18	-	-	18.18	-	-	[12]
** Chocolate 51.6% cocoa	<1	35.71	28.57	14.29	7.14	-	14.29	-	-	
Chocolate 70% cocoa liquor		40.91	13.64	18.18	-	-	4.55	18.18	4.55	[44]

* Made with Ecuadorian national cocoa liquor. ** Made from African cocoa liquor.

Many of the compounds that remain in chocolate have an increased concentration along the processing chain or remained unchanged, as has been reported; the influence of the manufacturing process is greater than that of the difference in the cocoa production area, providing a diversity of aroma profiles [86]. Particularly, criollo cocoa roasting increases acetophenone, tetramethylpyrazine, 2,3,5-trimethylpyrazine, and 2,5-dimethylpyrazine concentration, whereas 2-heptanol, phenylethyl alcohol, 2,3-butanedione, 2-phenylphenylpyrazine, 2,3-butanedione, 2,3-butanedione and 2,3-butanedione, 2,3-butanedione, 2-phenyl-2-butenal, 5-methyl-2-phenyl-2-hexanal, ethyl octanoate, ethyl phenylacetate, ethyl decanoate, and trans-linalool oxide remain stable during roasting [2]. In addition, some of the samples' volatile compounds are affected by each brand formulation, masking or enhancing a specific volatile. Ref. [26] considered three of the predominant masses of chocolates, namely, mass 33, 43, and 61, which were identified as methanol, a fragment of diverse origin, and acetic acid, respectively, and because of their high concentration, he suggested these three masses can have a huge impact on the fingerprint analysis that differentiates them by regions and brands.

4. Conclusions

Throughout the cocoa processing chain, aromatic precursor compounds give rise to characteristic aromas in each cocoa matrix. The way aroma compounds contribute to a specific matrix is estimated by the compound odor activity value. Initially, the dry fermented beans had acidic notes for which the OAV was high. The diversity of acids depends on the fermentation method used and also on bean preconditioning to reduce the fermentable sugar and, therefore, the final acidity of the bean, whereas fruity and floral aromas were characteristic of dry fermented beans. However, their concentration and abundance of compounds depends on the variety used. Compounds that exude chocolate aromas are scarce in dry fermented beans as they arise from Maillard reactions during roasting. The roasting parameters, like temperature, roasting time, and the roasting method, influence the appearance of new compounds and the preservation of those already found in the dry beans. Temperatures above 160 °C for a period of 35 min favor the appearance of pyrazines but reduce the compound concentration responsible for fruity and floral aromas, such as esters and ketones. In cocoa liquor, on average, 61.84% of the compounds represent fruity and floral aromas. The abundance of compounds in chocolate is directly related to the conching process, whereby the remaining fraction of moisture and undesirable aroma is eliminated, and the desirable aromas are concentrated.

Author Contributions: Conceptualization, O.M.Q. and D.P.H.; methodology, O.M.Q.; validation, O.M.Q., P.V.A. and A.A.B.; formal analysis, O.M.Q.; investigation, O.M.Q., D.P.H., A.A.B., P.V.A. and N.V.A.; writing—original draft preparation, D.P.H.; writing—review and editing, O.M.Q.; visualization, N.V.A.; supervision, N.V.A.; project administration, O.M.Q. All authors have read and agreed to the published version of the manuscript.

Funding: This research received no external funding.

Institutional Review Board Statement: Not applicable.

Informed Consent Statement: Not applicable.

Data Availability Statement: Not applicable.

Conflicts of Interest: The authors declare no conflict of interest.

References

1. Kongor, J.E.; Hinneh, M.; de Walle, D.V.; Afoakwa, E.O.; Boeckx, P.; Dewettinck, K. Factors influencing quality variation in cocoa (*Theobroma cacao*) bean flavour profile—A review. *Food Res. Int.* **2016**, *82*, 44–52. [CrossRef]
2. Ascrizzi, R.; Flamini, G.; Tessieri, C.; Pistelli, L. From the raw seed to chocolate: Volatile profile of Blanco de Criollo in dif-ferent phases of the processing chain. *Microchem. J.* **2017**, *133*, 474–479. [CrossRef]
3. Castro-Alayo, E.M.; Idrogo-Vásquez, G.; Siche, R.; Cardenas-Toro, F.P. Formation of aromatic compounds precursors during fermentation of Criollo and Forastero cocoa. *Heliyon* **2019**, *5*, e01157. [CrossRef]
4. Rottiers, H.; Sosa, D.A.T.; Lemarcq, V.; De Winne, A.; De Wever, J.; Everaert, H.; Jaime, J.A.B.; Dewettinck, K.; Messens, K. A multipronged flavor comparison of Ecuadorian CCN51 and Nacional cocoa cultivars. *Eur. Food Res. Technol.* **2019**, *245*, 2459–2478. [CrossRef]
5. Rottiers, H.; Sosa, D.A.T.; De Winne, A.; Ruales, J.; De Clippeleer, J.; De Leersnyder, I.; De Wever, J.; Everaert, H.; Messens, K.; Dewettinck, K. Dynamics of volatile compounds and flavor precursors during spontaneous fermentation of fine flavor Tri-nitario cocoa beans. *Eur. Food Res. Technol.* **2019**, *245*, 1917–1937. [CrossRef]
6. Frauendorfer, F.; Schieberle, P. Key aroma compounds in fermented Forastero cocoa beans and changes induced by roasting. *Eur. Food Res. Technol.* **2019**, *245*, 1907–1915. [CrossRef]
7. Qin, X.-W.; Lai, J.-X.; Tan, L.-H.; Hao, C.-Y.; Li, F.-P.; He, S.-Z.; Song, Y.-H. Characterization of volatile compounds in Criollo, Forastero, and Trinitario cocoa seeds (*Theobroma cacao* L.) in China. *Int. J. Food Prop.* **2017**, *20*, 2261–2275. [CrossRef]
8. Utrilla-Vázquez, M.; Rodríguez-Campos, J.; Avendaño-Arazate, C.H.; Gschaedler, A.; Lugo-Cervantes, E. Analysis of volatile compounds of five varieties of Maya cocoa during fermentation and drying processes by Venn diagram and PCA. *Food Res. Int.* **2020**, *129*, 108834. [CrossRef]
9. Hinneh, M.; Van de Walle, D.; Tzompa-Sosa, D.A.; De Winne, A.; Termote, S.; Messens, K.; Van Durme, J.; Afoakwa, E.O.; De Cooman, L.; Dewettinck, K. Tuning the aroma profiles of FORASTERO cocoa liquors by varying pod storage and bean roasting temperature. *Food Res. Int.* **2019**, *125*, 108550. [CrossRef]
10. Seyfried, C.; Granvogl, M. Characterization of the Key Aroma Compounds in Two Commercial Dark Chocolates with High Cocoa Contents by Means of the Sensomics Approach. *J. Agric. Food Chem.* **2019**, *67*, 5827–5837. [CrossRef] [PubMed]
11. Tran, P.D.; Van Durme, J.; Van De Walle, D.; De Winne, A.; Delbaere, C.; De Clercq, N.; Phan, T.T.Q.; Nguyen, C.P.; Dewettinck, K. Quality Attributes of Dark Chocolate Produced from Vietnamese Cocoa Liquors. *J. Food Qual.* **2016**, *39*, 311–322. [CrossRef]
12. Tuenter, E.; Delbaere, C.; De Winne, A.; Bijttebier, S.; Custers, D.; Foubert, K.; Van Durme, J.; Messens, K.; Dewettinck, K.; Pieters, L. Non-volatile and volatile composition of West African bulk and Ecuadorian fine-flavor cocoa liquor and chocolate. *Food Res. Int.* **2019**, *130*, 108943. [CrossRef]
13. Füllemann, D.; Steinhaus, M. Characterization of Odorants Causing Smoky Off-Flavors in Cocoa. *J. Agric. Food Chem.* **2020**, *68*, 10833–10841. [CrossRef]
14. Rodriguez-Campos, J.; Escalona-Buendía, H.; Orozco-Avila, I.; Lugo-Cervantes, E.; Jaramillo-Flores, M. Dynamics of volatile and non-volatile compounds in cocoa (*Theobroma cacao* L.) during fermentation and drying processes using principal com-ponents analysis. *Food Res. Int.* **2011**, *44*, 250–258. [CrossRef]
15. Koua, B.K.; Koffi, P.M.E.; Gbaha, P. Evolution of shrinkage, real density, porosity, heat and mass transfer coefficients during indirect solar drying of cocoa beans. *J. Saudi Soc. Agric. Sci.* **2019**, *18*, 72–82. [CrossRef]
16. Afoakwa, E.O.; Kongor, J.E.; Takrama, J.; Budu, A.S. Changes in nib acidification and biochemical composition during fer-mentation of pulp pre-conditioned cocoa (*Theobroma cacao*) beans. *Int. Food Res. J.* **2013**, *20*, 1843–1853.
17. Souza, C.D.S.; Block, J.M. Impact of the addition of cocoa butter equivalent on the volatile compounds profile of dark chocolate. *J. Food Sci. Technol.* **2018**, *55*, 767–775. [CrossRef] [PubMed]
18. Marty-Terrade, S.; Marangoni, A.G. Impact of Cocoa Butter Origin on Crystal Behavior. *Cocoa Butter Relat. Compd.* **2012**, 245–274. [CrossRef]
19. Afoakwa, E.O.; Paterson, A.; Fowler, M.; Ryan, A. Matrix effects on flavour volatiles release in dark chocolates varying in particle size distribution and fat content using GC–mass spectrometry and GC–olfactometry. *Food Chem.* **2009**, *113*, 208–215. [CrossRef]
20. Valle-Epquín, M.G.; Balcázar-Zumaeta, C.R.; Auquiñivín-Silva, E.A.; Fernández-Jeri, A.B.; Idrogo-Vásquez, G.; Castro-Alayo, E.M. The roasting process and place of cultivation influence the volatile fingerprint of Criollo cocoa from Amazonas, Peru. *Sci. Agropecu.* **2020**, *11*, 599–610. [CrossRef]
21. Hinneh, M.; Semanhyia, E.; Van de Walle, D.; De Winne, A.; Tzompa-Sosa, D.A.; Scalone, G.L.L.; De Meulenaer, B.; Messens, K.; Van Durme, J.; Afoakwa, E.O.; et al. Assessing the influence of pod storage on sugar and free amino acid profiles and

the implications on some Maillard reaction related flavor volatiles in Forastero cocoa beans. *Food Res. Int.* **2018**, *111*, 607–620. [CrossRef]

22. Marseglia, A.; Musci, M.; Rinaldi, M.; Palla, G.; Caligiani, A. Volatile fingerprint of unroasted and roasted cocoa beans (*Theobroma cacao* L.) from different geographical origins. *Food Res. Int.* **2020**, *132*, 109101. [CrossRef]
23. Hinneh, M.; Van de Walle, D.; Tzompa-Sosa, D.A.; Haeck, J.; Abotsi, E.E.; De Winne, A.; Messens, K.; Van Durme, J.; Afoakwa, E.O.; De Cooman, L.; et al. Comparing flavor profiles of dark chocolates refined with melanger and conched with Stephan mixer in various alternative chocolate production techniques. *Eur. Food Res. Technol.* **2019**, *245*, 837–852. [CrossRef]
24. Engeseth, N.J.; Pangan, M.F.A. Current context on chocolate flavor development—A review. *Curr. Opin. Food Sci.* **2018**, *21*, 84–91. [CrossRef]
25. Owusu, M.; Petersen, M.A.; Heimdal, H. Effect of Fermentation Method, Roasting and Conching Conditions on the Aroma Volatiles of Dark Chocolate. *J. Food Process. Preserv.* **2012**, *36*, 446–456. [CrossRef]
26. Acierno, V.; Yener, S.; Alewijn, M.; Biasioli, F.; Van Ruth, S. Factors contributing to the variation in the volatile composition of chocolate: Botanical and geographical origins of the cocoa beans, and brand-related formulation and processing. *Food Res. Int.* **2016**, *84*, 86–95. [CrossRef]
27. Caligiani, A.; Marseglia, A.; Palla, G. Cocoa: Production, Chemistry, and Use. *Encycl. Food Health* **2015**, 185–190. [CrossRef]
28. Cevallos-Cevallos, J.M.; Gysel, L.; Maridueña-Zavala, M.G.; Molina-Miranda, M.J. Time-Related Changes in Volatile Com-pounds during Fermentation of Bulk and Fine-Flavor Cocoa (*Theobroma cacao*) Beans. *J. Food Qual.* **2018**, *2018*, 1–14. [CrossRef]
29. Samaniego, I.; Espín, S.; Quiroz, J.; Ortiz, B.; Carrillo, W.; García-Viguera, C.; Mena, P. Effect of the growing area on the methylxanthines and flavan-3-ols content in cocoa beans from Ecuador. *J. Food Compos. Anal.* **2020**, *88*, 103448. [CrossRef]
30. Anecacao, 2019. Cacao Nacional | Anecacao Ecuador. Cacao Nac. Available online: http://www.anecacao.com/index.php/es/quienes-somos/cacao-nacional.html (accessed on 8 December 2020).
31. Dang, Y.K.T.; Nguyen, H.V.H. Effects of Maturity at Harvest and Fermentation Conditions on Bioactive Compounds of Cocoa Beans. *Plant Foods Hum. Nutr.* **2019**, *74*, 54–60. [CrossRef]
32. Pedan, V.; Weber, C.; Do, T.; Fischer, N.; Reich, E.; Rohn, S. HPTLC fingerprint profile analysis of cocoa proanthocyanidins depending on origin and genotype. *Food Chem.* **2018**, *267*, 277–287. [CrossRef] [PubMed]
33. Scollo, E.; Neville, D.C.; Oruna-Concha, M.J.; Trotin, M.; Cramer, R. UHPLC–MS/MS analysis of cocoa bean proteomes from four different genotypes. *Food Chem.* **2020**, *303*, 125244. [CrossRef] [PubMed]
34. Afoakwa, E.O. Cocoa Cultivation and Practices. In *Chocolate Science and Technology*; Wiley Blackwell: Singapore, 2010; pp. 16–18.
35. Rottiers, H.; Sosa, D.A.T.; Van de Vyver, L.; Hinneh, M.; Everaert, H.; De Wever, J.; Messens, K.; Dewettinck, K. Discrimination of Cocoa Liquors Based on Their Odor Fingerprint: A Fast GC Electronic Nose Suitability Study. *Food Anal. Methods* **2019**, *12*, 475–488. [CrossRef]
36. Scalone, G.L.L.; Textoris-Taube, K.; De Meulenaer, B.; De Kimpe, N.; Wöstemeyer, J.; Voigt, J. Cocoa-specific flavor components and their peptide precursors. *Food Res. Int.* **2019**, *123*, 503–515. [CrossRef]
37. Calva-Estrada, S.; Utrilla-Vázquez, M.; Vallejo-Cardona, A.; Roblero-Pérez, D.; Lugo-Cervantes, E. Thermal properties and volatile compounds profile of commercial dark-chocolates from different genotypes of cocoa beans (*Theobroma cacao* L.) from Latin America. *Food Res. Int.* **2020**, *136*, 109594. [CrossRef]
38. Hamdouche, Y.; Meile, J.C.; Lebrun, M.; Guehi, T.; Boulanger, R.; Teyssier, C.; Montet, D. Impact of turning, pod storage and fermentation time on microbial ecology and volatile composition of cocoa beans. *Food Res. Int.* **2019**, *119*, 477–491. [CrossRef]
39. Rodriguez-Campos, J.; Escalona-Buendía, H.; Contreras-Ramos, S.; Orozco-Avila, I.; Jaramillo-Flores, E.; Lugo-Cervantes, E. Effect of fermentation time and drying temperature on volatile compounds in cocoa. *Food Chem.* **2012**, *132*, 277–288. [CrossRef] [PubMed]
40. Torres-Moreno, M.; Tarrega, A.; Blanch, C. Effect of cocoa roasting time on volatile composition of dark chocolates from different origins determined by HS-SPME/GC-MS. *CyTA—J. Food* **2021**, *19*, 81–95. [CrossRef]
41. Lemarcq, V.; Van de Walle, D.; Monterde, V.; Sioriki, E.; Dewettinck, K. Assessing the flavor of cocoa liquor and chocolate through instrumental and sensory analysis: A critical review. *Crit. Rev. Food Sci. Nutr.* **2021**, *62*, 5523–5539. [CrossRef]
42. Schlüter, A.; Hühn, T.; Kneubühl, M.; Chatelain, K.; Rohn, S.; Chetschik, I. Novel Time- and Location-Independent Postharvest Treatment of Cocoa Beans: Investigations on the Aroma Formation during "Moist Incubation" of Unfermented and Dried Cocoa Nibs and Comparison to Traditional Fermentation. *J. Agric. Food Chem.* **2020**, *68*, 10336–10344. [CrossRef]
43. van Ruth, S.M. Methods for gas chromatography-olfactometry: A review. *Biomol. Eng.* **2001**, *17*, 121–128. [CrossRef] [PubMed]
44. Hinneh, M.; Abotsi, E.E.; Van de Walle, D.; Tzompa-Sosa, D.A.; De Winne, A.; Simonis, J.; Messens, K.; Van Durme, J.; Afoakwa, E.O.; De Cooman, L.; et al. Pod storage with roasting: A tool to diversifying the flavor profiles of dark chocolates produced from 'bulk' cocoa beans? (Part I: Aroma profiling of chocolates). *Food Res. Int.* **2019**, *119*, 84–98. [CrossRef]
45. Parker, J. Introduction to aroma compounds in foods. In *Flavour Development, Analysis and Perception In Food and Beverages*; Elsevier Ltd.: Amsterdam, The Netherlands, 2015. [CrossRef]
46. Servent, A.; Boulanger, R.; Davrieux, F.; Pinot, M.-N.; Tardan, E.; Forestier-Chiron, N.; Hue, C. Assessment of cocoa (*Theobroma cacao* L.) butter content and composition throughout fermentations. *Food Res. Int.* **2018**, *107*, 675–682. [CrossRef] [PubMed]
47. Afoakwa, E.O.; Quao, J.; Takrama, J.; Budu, A.S.; Saalia, F.K. Chemical composition and physical quality characteristics of Ghanaian cocoa beans as affected by pulp pre-conditioning and fermentation. *J. Food Sci. Technol.* **2013**, *50*, 1097–1105. [CrossRef]

48. Afoakwa, E.O.; Quao, J.; Budu, A.S.; Takrama, J.; Saalia, F.K. Effect of pulp preconditioning on acidification, proteolysis, sugars and free fatty acids concentration during fermentation of cocoa (*Theobroma cacao*) beans. *Int. J. Food Sci. Nutr.* **2011**, *62*, 755–764. [CrossRef] [PubMed]
49. Afoakwa, E.O.; Quao, J.; Takrama, F.S.; Budu, A.S.; Saalia, F.K. Changes in total polyphenols, o-diphenols and anthocyanin concentrations during fermentation of pulp pre-conditioned cocoa (*Theobroma cacao*) beans. *Int. Food Res. J.* **2012**, *19*, 1071–1077.
50. D'Souza, R.N.; Grimbs, A.; Grimbs, S.; Behrends, B.; Corno, M.; Ullrich, M.S.; Kuhnert, N. Degradation of cocoa proteins into oligopeptides during spontaneous fermentation of cocoa beans. *Food Res. Int.* **2018**, *109*, 506–516. [CrossRef]
51. Kumari, N.; Kofi, K.J.; Grimbs, S.; D'Souza, R.N.; Kuhnert, N.; Vrancken, G.; Ullrich, M.S. Biochemical fate of vicilin storage protein during fermentation and drying of cocoa beans. *Food Res. Int.* **2016**, *90*, 53–65. [CrossRef] [PubMed]
52. Voigt, J.; Textoris-Taube, K.; Wöstemeyer, J. pH-Dependency of the proteolytic formation of cocoa- and nutty-specific aroma precursors. *Food Chem.* **2018**, *255*, 209–215. [CrossRef] [PubMed]
53. Sunoj, S.; Igathinathane, C.; Visvanathan, R. Nondestructive determination of cocoa bean quality using FT-NIR spectroscopy. *Comput. Electron. Agric.* **2016**, *124*, 234–242. [CrossRef]
54. Afoakwa, E.; Jennifer, Q.; Agnes, S.B.; Jemmy, S.T.; Firibu, K.S. Influence of pulp-preconditioning and fermentation on fermentative quality and appearance of ghanaian cocoa (Theobroma cacao) beans. *Int. Food Res. J.* **2012**, *19*, 127–133.
55. Ilangantileke, S.G.; Wahyudi, T.; Bailon, M.G. Assessment methodology to predict quality of cocoa beans for export. *J. Food Qual.* **1991**, *14*, 481–496. [CrossRef]
56. Brunetto, M.D.R.; Gallignani, M.; Orozco, W.; Clavijo, S.; Delgado, Y.; Ayala, C.; Zambrano, A. The effect of fermentation and roasting on free amino acids profile in Criollo cocoa (*Theobroma cacao* L.) grown in Venezuela. *Braz. J. Food Technol.* **2020**, *23*, 1–12. [CrossRef]
57. Crafack, M.; Keul, H.; Eskildsen, C.E.; Petersen, M.A.; Saerens, S.; Blennow, A.; Skovmand-Larsen, M.; Swiegers, J.H.; Petersen, G.B.; Heimdal, H.; et al. Impact of starter cultures and fermentation techniques on the volatile aroma and sensory profile of chocolate. *Food Res. Int.* **2014**, *63*, 306–316. [CrossRef]
58. Reineccius, G.A.; Andersen, D.A.; Kavanagh, T.E.; Keeney, P.G. Identification and Quantification of the Free Sugars in Cocoa Beans. *J. Agric. Food Chem.* **1972**, *20*, 199–202. [CrossRef]
59. Domínguez-Pérez, L.A.; Beltrán-Barrientos, L.M.; González-Córdova, A.F.; Hernández-Mendoza, A.; Vallejo-Cordoba, B. Artisanal cocoa bean fermentation: From cocoa bean proteins to bioactive peptides with potential health benefits. *J. Funct. Foods* **2020**, *73*, 104134. [CrossRef]
60. Kumari, N.; Grimbs, A.; D'Souza, R.N.; Verma, S.K.; Corno, M.; Kuhnert, N.; Ullrich, M.S. Origin and varietal based proteomic and peptidomic fingerprinting of Theobroma cacao in non-fermented and fermented cocoa beans. *Food Res. Int.* **2018**, *111*, 137–147. [CrossRef]
61. Tchouatcheu, G.A.N.; Noah, A.M.; Lieberei, R.; Niemenak, N. Effect of cacao bean quality grade on cacao quality evaluation by cut test and correlations with free amino acids and polyphenols profiles. *J. Food Sci. Technol.* **2019**, *56*, 2621–2627. [CrossRef]
62. Bonvehí, J.S.; Coll, F.V. Factors Affecting the Formation of Alkylpyrazines during Roasting Treatment in Natural and Alkalinized Cocoa Powder. *J. Agric. Food Chem.* **2002**, *50*, 3743–3750. [CrossRef]
63. Caligiani, A.; Marseglia, A.; Prandi, B.; Palla, G.; Sforza, S. Influence of fermentation level and geographical origin on cocoa bean oligopeptide pattern. *Food Chem.* **2016**, *211*, 431–439. [CrossRef]
64. Marseglia, A.; Palla, G.; Caligiani, A. Presence and variation of γ-aminobutyric acid and other free amino acids in cocoa beans from different geographical origins. *Food Res. Int.* **2014**, *63*, 360–366. [CrossRef]
65. ICCO. (n.d.). Cultivo de cacao - Organización Internacional del Cacao. Available online: https://www.icco.org/growing-cocoa/ (accessed on 27 May 2021).
66. Niether, W.; Smit, I.; Armengot, L.; Schneider, M.; Gerold, G.; Pawelzik, E. Environmental Growing Conditions in Five Production Systems Induce Stress Response and Affect Chemical Composition of Cocoa (*Theobroma cacao* L.) Beans. *J. Agric. Food Chem.* **2017**, *65*, 10165–10173. [CrossRef] [PubMed]
67. Selamat, J.; Harun, S.M.; Ghazali, N.M. Formation of Methyl Pyrazine during Cocoa Bean Fermentation. *Pertanika* **1994**, *17*, 27–32.
68. Counet, C.; Ouwerx, C.; Rosoux, D.; Collin, S. Relationship between procyanidin and flavor contents of cocoa liquors from different origins. *J. Agric. Food Chem.* **2004**, *52*, 6243–6249. [CrossRef]
69. Rentería, O.; Pliego-Arreaga, R.; Regalado, C.; Amaro-Reyes, A.; García-Almendárez, B.E. Enhancing isoamyl acetate biosynthesis by Pichia fermentans. *Rev. Mex. Ing. Química* **2021**, *12*, 505–511. [CrossRef]
70. Schwab, W.; Davidovich-Rikanati, R.; Lewinsohn, E. Biosynthesis of plant-derived flavor compounds. *Plant J.* **2008**, *54*, 712–732. [CrossRef]
71. Ziegleder, G. Linalool contents as characteristic of some flavor grade cocoas. *Z. Lebensm. Unters. Forsch.* **1990**, *191*, 306–309. [CrossRef]
72. Maga, J.A. Pyrazine update. *Food Rev. Int.* **1992**, *8*, 479–558. [CrossRef]
73. Ziegleder, G. Composition of flavor extracts of raw and roasted cocoas. *Z. Lebensm. Unters. Forsch.* **1991**, *192*, 521–525. [CrossRef]
74. Perotti, P.; Cordero, C.; Bortolini, C.; Rubiolo, P.; Bicchi, C.; Liberto, E. Cocoa smoky off-flavor: Chemical characterization and objective evaluation for quality control. *Food Chem.* **2020**, *309*, 125561. [CrossRef] [PubMed]
75. Lemarcq, V.; Tuenter, E.; Bondarenko, A.; Van de Walle, D.; De Vuyst, L.; Pieters, L.; Sioriki, E.; Dewettinck, K. Roasting-induced changes in cocoa beans with respect to the mood pyramid. *Food Chem.* **2020**, *332*, 127467. [CrossRef]

76. Spizzirri, U.G.; Ieri, F.; Campo, M.; Paolino, D.; Restuccia, D.; Romani, A. Biogenic Amines, Phenolic, and Aroma-Related Compounds of Unroasted and Roasted Cocoa Beans with Different Origin. *Foods* **2019**, *8*, 306. [CrossRef]
77. Liu, M.; Liu, J.; He, C.; Song, H.; Liu, Y.; Zhang, Y.; Wang, Y.; Guo, J.; Yang, H.; Su, X. Characterization and comparison of key aroma-active compounds of cocoa liquors from five different areas. *Int. J. Food Prop.* **2017**, *20*, 2396–2408. [CrossRef]
78. Misnawi, J.; Ariza, B.T.S. Use of gas Chromatography-Olfactometry in combination with solid phase micro extraction for cocoa liquor aroma analysis. *Int. Food Res. J.* **2011**, *18*, 829–835.
79. Dand, R. Cocoa bean processing and the manufacture of chocolate. In *The International Cocoa Trade*, 3rd ed.; Woodhead Publishing: Sawston, UK, 2011; pp. 268–289. [CrossRef]
80. Beckett, S.T.; Paggios, K.; Roberts, I. Conching. In *Beckett's Industrial Chocolate Manufacture and Use*; John Wiley & Sons: Hoboken, NJ, USA, 2017.
81. Barišić, V.; Kopjar, M.; Jozinović, A.; Flanjak, I.; Ačkar, Đ.; Miličević, B.; Šubarić, D.; Jokić, S.; Babić, J. The Chemistry behind Chocolate Production. *Molecules* **2019**, *24*, 3163. [CrossRef] [PubMed]
82. Toker, O.S.; Palabiyik, I.; Konar, N. Chocolate quality and conching. *Trends Food Sci. Technol.* **2019**, *91*, 446–453. [CrossRef]
83. Owusu, M.; Petersen, M.A.; Heimdal, H. Relationship of sensory and instrumental aroma measurements of dark chocolate as influenced by fermentation method, roasting and conching conditions. *J. Food Sci. Technol.* **2013**, *50*, 909–917. [CrossRef]
84. Toker, O.S.; Palabiyik, I.; Pirouzian, H.R.; Aktar, T.; Konar, N. Chocolate aroma: Factors, importance and analysis. *Trends Food Sci. Technol.* **2020**, *99*, 580–592. [CrossRef]
85. Liu, J.; Liu, M.; He, C.; Song, H.; Guo, J.; Wang, Y.; Yang, H.; Su, X. A comparative study of aroma-active compounds between dark and milk chocolate: Relationship to sensory perception. *J. Sci. Food Agric.* **2015**, *95*, 1362–1372. [CrossRef]
86. Kitani, Y.; Putri, S.P.; Fukusaki, E. Investigation of the effect of processing on the component changes of single-origin chocolate during the bean-to-bar process. *J. Biosci. Bioeng.* **2022**, *134*, 138–143. [CrossRef]

 *fermentation*

Article

Pilot Scale Evaluation of Wild *Saccharomyces cerevisiae* Strains in Aglianico

Davide Gottardi [1,†], Gabriella Siesto [2,3,*,†], Antonio Bevilacqua [4], Francesca Patrignani [1], Daniela Campaniello [4], Barbara Speranza [4], Rosalba Lanciotti [1], Angela Capece [2,3] and Patrizia Romano [3,5]

[1] Department of Agricultural and Food Sciences, Campus of Food Science, Piazza Goidanich 60, 47521 Cesena, Italy

[2] School of Agricultural, Forestry, Food and Environmental Sciences, University of Basilicata, Via dell'Ateneo Lucano 10, 85100 Potenza, Italy

[3] Spin-Off StarFInn s.r.l.s., University of Basilicata, Via dell'Ateneo Lucano 10, 85100 Potenza, Italy

[4] Department of Agriculture, Food, Natural Resources and Engineering, University of Foggia, Via Napoli 25, 71122 Foggia, Italy

[5] Faculty of Economy, Universitas Mercatorum, Piazza Mattei 10, 00186 Rome, Italy

* Correspondence: gabriella.siesto@unibas.it; Tel.: +39-0971205585

† These authors contributed equally to this work.

Abstract: In winemaking, the influence of *Saccharomyces cerevisiae* strains on the aromatic components of wine is well recognized on a laboratory scale, but few studies deal with the comparison of numerous strains on a pilot scale fermentation. In this scenario, the present work aimed to validate the fermentative behavior of seven wild *S. cerevisiae* strains on pilot-scale fermentations to evaluate their impact on the aromatic profiles of the resulting wines. The strains, isolated from grapes of different Italian regional varieties, were tested in pilot-scale fermentation trials performed in the cellar in 1 hL of Aglianico grape must. Then, wines were analyzed for their microbiological cell loads, main chemical parameters of enological interest (ethanol, total sugars, fructose, glucose, total and volatile acidity, malic and lactic acids) and volatile aroma profiles by GC/MS/SPME. Seventy-six volatile compounds belonging to six different classes (esters, alcohols, terpenes, aldehydes, acids, and ketones) were identified. The seven strains showed different trends and significant differences, and for each class of compounds, high-producing and low-producing strains were found. Since the present work was performed at a pilot-scale level, mimicking as much as possible real working conditions, the results obtained can be considered as a validation of the screened *S. cerevisiae* strains and a strategy to discriminate in real closed conditions strains able to impart desired wine sensory features.

Keywords: pilot scale fermentation; *Saccharomyces cerevisiae*; wine flavor; volatile molecule fingerprinting; Aglianico

Citation: Gottardi, D.; Siesto, G.; Bevilacqua, A.; Patrignani, F.; Campaniello, D.; Speranza, B.; Lanciotti, R.; Capece, A.; Romano, P. Pilot Scale Evaluation of Wild *Saccharomyces cerevisiae* Strains in Aglianico. *Fermentation* **2023**, *9*, 245. https://doi.org/10.3390/fermentation9030245

Academic Editor: Niel Van Wyk

Received: 6 February 2023
Revised: 28 February 2023
Accepted: 1 March 2023
Published: 3 March 2023

1. Introduction

The use of indigenous strains of *Saccharomyces cerevisiae* as region-specific wine starters is increasingly attracting the interest of wine researchers and winemakers [1]. These yeasts are well adapted to micro-area conditions of a given region [2–4] and could ensure the maintenance of the typical flavor and aroma of wines obtained by grapevine cultivars of a specific geographical area [5–7]. Indeed, *S. cerevisiae* strains collected from ecologically and geographically diverse sources typically show genetic divergences associated with habitat type [8–10]. Furthermore, due to their better acclimatization to the environmental conditions of the wine-producing region and of the grape must composition, selected indigenous strains are mostly endowed with a higher level of dominance than commercial strains, exhibiting an elevated competition against yeast microbiota naturally present in the grape must [11].

It is widely recognized that *S. cerevisiae* produces different amounts of compounds affecting wine aroma profile, in the function of grape variety, fermentation conditions and must composition (availability of micronutrients, vitamins and assimilable nitrogen, etc.) [12] and that there is a high strain difference within this species due to the variation in their production of aromatic metabolites of wine [13–15]. Many studies demonstrated that indigenous *S. cerevisiae* strains with strain-specific metabolic profiles are correlated with wine chemical composition, reflecting the specificity of a *terroir* [16–18].

In general, wine aromas can be classified into varietal, fermentative, and aging aromas, but most of the compounds responsible for wine aroma are volatile molecules, and they can be classified into different chemical classes such as higher alcohols, carbonyl compounds, volatile fatty acids, esters, sulfur compounds, terpenoids and volatile phenols, that result from the yeast metabolism [19]. Overall, the aroma plays an important role in determining the organoleptic quality of the wine and, therefore, significantly affects the attractiveness of the wine [20–22].

A large amount of literature is available on the influence of the different *S. cerevisiae* strains on the aromatic components of wine, but the publications substantially describe fermentations carried out on a laboratory scale, and few studies deal with the comparison of numerous strains in pilot-scale fermentation. Indeed, laboratory-scale fermentation is an important condition to evaluate the specific metabolic traits of the different yeast strains because it allows us to compare a large number of strains under different conditions at the same time. The researchers' choice to carry out lab-scale fermentations depends mainly on the practical requirements (such as the availability of small-sized fermenters and the cost of reagents, substrates, and instruments) [23], and there are some advantages for sampling and controlling the fermentation parameters. To evaluate the fermentative behavior of selected indigenous *S. cerevisiae* starters at a laboratory scale is a useful preliminary screening, although the results often differ from those obtained from a winery level in large volumes.

Some authors compared lab- and pilot-scale fermentations showing that the small fermentations cannot mimic those on a pilot or industrial scale and that there are significant differences in the kinetics and in the production of the aromatic compounds between these scales [24].

Based on these observations, the present work focused on the study of seven different wild strains of *S. cerevisiae* in pilot-scale fermentations with the aim of evaluating their impact on the product through the aromatic profiles of the resulting wines.

2. Materials and Methods

2.1. Yeast Strains

Seven wild *Saccharomyces cerevisiae* strains from the UNIBAS Yeast Collection (UBYC), University of Basilicata (Potenza, Italy), were used to conduct the alcoholic fermentation at a pilot scale. The strains were previously isolated from grapes of different varieties, directly collected in the vineyard of different Italian regions (Table 1), and characterized for their oenological performances [25], such as high dominance and fermentative vigor, ethanol tolerance of 16–17% (v/v), medium relative nitrogen demand, low foam production, killer factor. The strains were maintained at 4 °C for short-term storage on a YPD medium (bacteriological peptone, 20 g/L; yeast extract, 10 g/L; glucose, 20 g/L; agar, 15 g/L, Oxoid, Hampshire, UK).

Table 1. Origin and grape variety of *S. cerevisiae* strains used in this study.

Strain	Origin	Grape Vine Variety
Sc7	Apulia	Bombino nero (red)
Sc22	Campania	Fiano (white)
Sc34	Basilicata	Aglianico Vulture (red)
Sc41	Sicily	Nero d'Avola (red)
Sc48	Sardinia	Vermentino (white)
Sc65	Apulia	Aglianico (red)
Sc70	Sardinia	Cannonau (red)

2.2. Pilot Scale Fermentation

Fermentation trials were carried out in a winery of the Apulia region using stainless steel 1-hL capacity vessels containing 0.9 hL of Aglianico grape must with the following characteristics: total acidity 5.37 g/L; pH 3.6; TSS 22.0; density 1.097 g/L; Yeast Assimilable Nitrogen 211.9 mg N/L. Subsequently, 50 mg/L of sulfur dioxide (Sigma, St. Louis, MO 63304, USA) was added and mixed to the grape must with the aim of inhibiting the natural microbiota possibly present in the must, in particular acetic and lactic acid bacteria. In addition, pectinolytic enzymes (1 g/hL) were added to the must in order to favor clarification and facilitate settling [26].

In the laboratory, the yeast cultures were precultured by refreshing each strain on YPD plates and incubated at 26 °C for 24 h. Then, 1 loopful of each strain was inoculated into 0.5-L flasks filled with 0.2-L of YPD broth for 24 h at 25 °C under shaking conditions (180 rpm). For the fermentation trials in the cellar, the biomass was achieved by BioFlo/CelliGen 110 bioreactor (Eppendorf, Hamburg, Germany) by inoculating the precultures of each strain in a vessel containing 5 L of YPD broth. During growth, the following parameters were controlled: a temperature of 26 °C, stirring at 200 rpm, and oxygen at 4 vvm.

After overnight growth in the bioreactor, the cells of each strain were collected by centrifugation (4700 rpm for 10 min) and cell concentration was determined by microscope counting by Thoma chamber (magnification 400×). The recovered biomass was stored for limited times at 4 °C until use. Once in the winery, each strain culture was suspended and mixed in 1 L of Aglianico grape must and left for 1 h at room temperature in order to allow the cells to adapt to the must. Then, the strain-must mixture was inoculated at a density of approximately 1×10^7 cells/mL in a 1 hL vessel containing 0.90 hL of the same Aglianico grape must. All fermentations were carried out at 25 °C in duplicate by breaking the cap twice a day by gently pressing the skins with a steel plunger. During the process, the fermentative course was monitored daily by measuring the temperature, pH and the reducing sugars by refractometric analysis (°Brix). Before the application, an aliquot of each sample was centrifuged at maximum speed for 5 min to avoid interference from the presence of yeast cells in the suspension. At the end of the fermentation (8 days), samples were collected from each wine and immediately stored at −20 °C until analysis.

Basic chemical parameters of wines, such as ethanol, total sugars, fructose, glucose, total and volatile acidity, and malic and lactic acids, were determined by Fourier transform infrared spectrometry (FTIR) using a Wine Scan analyzer (OenoFoss™, Hillerød, Denmark), calibrated according to OIV [27].

2.3. Microbiological Control

Samples for microbiological control were taken from the grape must (T0) before inoculum and SO_2 addition and at the initial (1 day, T1), tumultuous (4 days, T4) and final fermentation (8 days, T8). Samples, taken at different stages of fermentation, were used to count yeast cells and to monitor the persistence of non-*Saccharomyces* during the process.

Aliquots of 10-fold dilution of the samples were spread onto Wallerstein Laboratory Nutrient Agar medium (WL, Oxoid, Hampshire, UK) [28], and colony counting was performed after 5 days of incubation at 26 °C. The samples were also spread on Lysine agar medium (Oxoid, Hampshire, UK) in order to differentiate *Saccharomyces* from non-*Saccharomyces* yeasts, as the former cannot grow on this medium.

2.4. Analysis of Volatile Compounds

The volatile compounds of the obtained wines were analyzed using solid phase micro extraction-gas chromatography-mass spectrometry (SPME-GC/MS) using Agilent 6890 GC gas chromatograph coupled with Agilent 5973 mass spectrometry (MS) detector. DB-WAXetr column, 30 m × 0.25 mm i.d., 0.25 μm film thickness (J&W Scientific, Folsom, CA, USA), was employed and helium as the carrier gas was used with a flow rate of 1.5 mL/min. The injector temperature was 250 °C, and the oven temperature was programmed from 40 °C for 6 min to 180 °C, at 5 °C/min for 3 min, then at 7 °C/min to 240 °C for 5 min.

The fiber used for the extraction of the volatile molecules, in headspace condition, was the polydimethylsiloxane (PDMS) 100 μm. About 20 mL of each sample was placed into a 50 mL amber glass vial containing 3 g of NaCl (saturation level) and 0.5 mL of isooctane as internal standard (IS). The sample vials were equilibrated for 30 min at 40 °C in a thermostated bath. Afterward, the fiber was exposed to the headspace for 20 min, inserting the stainless-steel needle through the vial's septum and pushing the fiber into the sample headspace to collect the analytes. The fiber was then withdrawn within the needle, and the SPME device was removed from the vial and inserted into the injection port of the GC apparatus for thermal desorption. The analytes removal from the fiber was carried out in the splitless mode at 240 °C for 5 min. Detection was carried out by mass spectrometry on the total ion current obtained by electron impact at 70 eV, and the masses were scanned from m/z 29–300. The volatile compounds were identified by comparison of the mass spectra with the NIST (National Institute of Standards and Technology, Gaithersburg, MD, USA) library database and quantified in equivalent μg/L by comparison of their mass spectra and retentions time on the basis of standard compounds.

2.5. Statistic

All analyses were done in duplicate over two independent samples. Significant differences were pointed out through a 1-way Analysis of Variance (ANOVA) and LSD test as the post-hoc-test; the P-level was set to 0.05. The concentrations of the different compounds were also analyzed through Principal Component Analysis, using the method "SS/(n-1)" to compute the variance and evaluate sample distribution in the factorial space (SS is the sum of squares, and n is the number of observations).

Finally, data on sugar consumption throughout fermentation (as °Bx) were modeled through the shoulder/tail model of Geeraerd et al. [29], cast in the following form:

$$S = (S_0 - S_{res}) \exp(-k_{max}t) \frac{\exp(k_{max}SL)}{1 + [\exp(k_{max}SL)] \exp(-k_{max}t)} + S_{res}$$

where S is sugar concentration over time (°Bx), t is the time (days), S_0 and S_{res} are the initial and residual sugar concentration, respectively, k_{max} is the rate of sugar consumption (°Bx/day), and SL is the shoulder length (time before the beginning of sugar decrease, days).

Statistics were done through the software Statistica for Windows (Statsoft, Tulsa, OK, USA).

3. Results
3.1. Fermentation Kinetics and Main End-Products

The fermentation kinetics of *S. cerevisiae* strains were monitored by assessing sugar concentration to evaluate the time before the beginning of sugar consumption by yeasts (shoulder length), the rate of sugar consumption and the residual amount at the end of the experiment. Data on sugar concentrations were modeled through the shoulder/tail model by Geerared et al. [29]. The fitting lines are in Figure 1, while the kinetic parameters are reported in Table 2.

It is worth mentioning that for the data hereby reported, the tail effect represents the residual concentration at the end of the fermentation assessment in the cellar (8 days), not the final residual concentration of sugars evaluated through chemical methods and shown in Table 3.

Regardless of origin, strains experienced a shoulder length of approximately 3 days (from 2.44 to 2.78 days) without significant differences; concerning the tail, residual sugar concentration was at 2–3 °Bx, except for strain Sc34, which showed a residual sugar concentration, after 8 days of 6.13 °Bx.

The resulting wines were also characterized for some main end-products (ethanol, total and volatile acidity, malic and lactic acids, and sugar concentration at the end of fermentation; Table 3). The amount of ethanol ranged from 11.88 to 12.33% *v/v*, while the concentration of residual sugars was 1.25–1.45 g/L. Both these parameters were not affected by the kind of strains, as well as total and volatile acidity (8.90–9.65 g/L and 0.25–0.41 g/L, respectively).

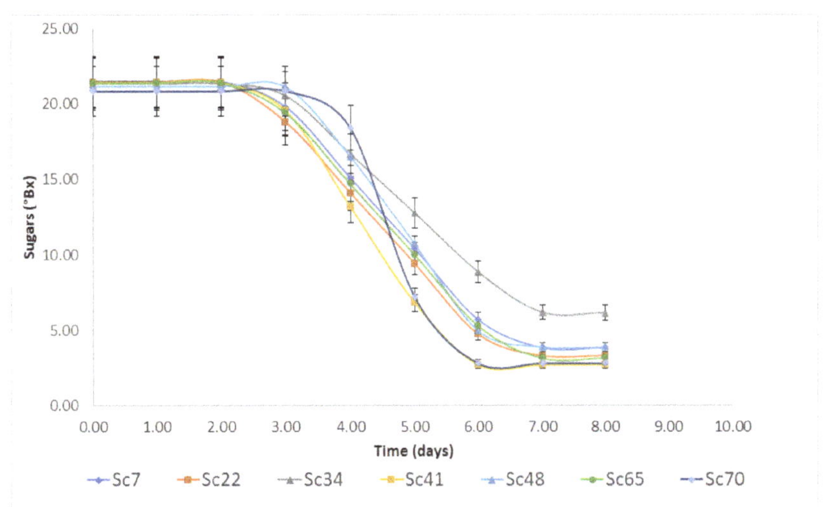

Figure 1. Kinetic of sugar consumption by the seven *S. cerevisiae* strains. Bars represent standard deviation, while lines are the best fit through the model of Geeraerd et al. [29].

Table 2. Kinetic of sugar consumptions for the seven strains of *S. cerevisiae*: fitting parameters evaluated through shoulder-tail model. SL, shoulder length (time before the beginning of sugar consumption, days). k_{max}, rate of sugar consumption (°Bx/day) S_{res}, sugar residual (°Bx). Mean values ± standard error. For each column, letters indicate significant differences (one-way ANOVA and LSD test, $p < 0.05$).

Strain	SL	k_{max}	S_{res}
Sc7	2.65 ± 0.25 A	10.82 ± 1.09 A	3.85 ± 0.74 A
Sc22	2.44 ± 0.26 A	10.81 ± 1.07 A	3.28 ± 0.72 A
Sc34	2.82 ± 0.30 A	9.03 ± 1.09 A	6.13 ± 0.71 B
Sc41	2.71 ± 0.30 A	14.65 ± 2.62 A	2.65 ± 0.89 A
Sc48	3.20 ± 0.32 A	13.34 ± 2.08 A	3.89 ± 0.87 A
Sc65	2.58 ± 0.27 A	10.85 ± 1.12 A	3.13 ± 0.77 A
Sc70	3.78 ± 0.12 A	25.79 ± 3.27 B	2.77 ± 0.58 A

Table 3. End-products and residual sugars in the wines obtained by the seven *S. cerevisiae* strains. Mean ± standard deviation. For each column, letters indicate significant differences (one-way ANOVA and LSD test, $p < 0.05$).

Strain	Ethanol (%, *v/v*)	Sugars (g/L)	Total Acidity (g/L)	Volatile Acidity (g/L)	Malic Acid (g/L)	Lactic Acid (g/L)
Sc7	11.88 ± 0.41 A	1.45 ± 0.07 A	9.65 ± 0.68 A	0.37 ± 0.04 A	3.00 ± 0.30 B	0.60 ± 0.11 C, D
Sc22	12.20 ± 0.14 A	1.25 ± 0.07 A	9.25 ± 0.59 A	0.33 ± 0.05 A	3.30 ± 0.25 B	0.60 ± 0.06 C, D
Sc34	12.04 ± 0.07 A	1.23 ± 0.11 A	9.35 ± 1.64 A	0.25 ± 0.06 A	2.90 ± 0.21 A, B	0.15 ± 0.04 A
Sc41	12.32 ± 0.10 A	1.45 ± 0.07 A	8.90 ± 0.66 A	0.41 ± 0.08 A	3.20 ± 0.21 B	0.45 ± 0.08 B, C
Sc48	11.97 ± 0.10 A	1.55 ± 0.07 A	9.55 ± 1.53 A	0.37 ± 0.06 A	2.38 ± 0.22 A	0.70 ± 0.01 D
Sc65	12.20 ± 0.11 A	1.25 ± 0.07 A	9.55 ± 0.78 A	0.35 ± 0.06 A	3.65 ± 0.35 B	0.67 ± 0.13 D
Sc70	12.33 ± 0.10 A	1.40 ± 0.14 A	9.45 ± 0.17 A	0.26 ± 0.03 A	3.00 ± 0.23 B	0.40 ± 0.03 B

The amount of malic acid was 2.38–3.65 g/L, with the lowest concentration in the wine resulting from strain Sc48. The concentration of lactic acid was 0.15–0.70 g/L, with the lowest amount in the wine from strain Sc34 and the highest for strains Sc48 and Sc65 (0.67–0.70 g/L).

The yeast cell loads were evaluated to monitor the persistence of non-*Saccharomyces* during the process and to confirm that the *S. cerevisiae* strains were dominant at the end of the fermentation. For this purpose, yeast sampling was performed at different stages of the fermentative process both on WL and Lysine agar medium. At T0 (before *S. cerevisiae* strains inoculum and the addition of SO_2), autochthonous microbiota, belonging only to non-*Saccharomyces* species, was at the level of about 3.0×10^5 UFC/mL. After the first day (T1), the colonies showing the typical *Saccharomyces* morphology were about 1.0×10^8 UFC/mL, and the number of non-*Saccharomyces* was reduced to a level of about 2.1×10^4 UFC/mL. At the middle fermentation time (T4) and at the end (T8), only *Saccharomyces* strains (3.2×10^8 UFC/mL and 1.0×10^6 UFC/mL, respectively) were found. The presence of non-*Saccharomyces* only until the second fermentation days was also confirmed by the test on lysine agar. These results underlined that the high level of inoculum (1×10^7 UFC mL) of the *S. cerevisiae* starter strains, endowed with competitive oenological characteristics, was ensured to completely overcome the presence of non-*Saccharomyces* autochthonous strains in 72–96 h, also guaranteeing the success of the guided fermentations.

3.2. Volatile Aromatic Compounds

After the assessment of strain performance under real conditions, also the VOCs (volatile organic compounds) were investigated by using a GC/MS/SPME approach, identifying 76 different compounds belonging to six different classes (esters, 42 compounds; alcohols, 16 compounds; terpenes, six compounds; aldehydes, seven compounds; acids, three compounds; ketones, three compounds).

The total amount of esters, alcohols, aldehydes and terpenes is reported in Table 4. The strains showed different trends and significant differences, and for each class of compounds, high-producing and low-producing strains were found. In the case of esters, the lowest producer was strain Sc48 (946.10 μg/L), and the highest producer was strain Sc34 (1863.55 μg/L), followed by strains Sc41, Sc65, and Sc70 (1290–1350 μg/L), while for the other classes of VOCs, strain Sc41 was always the highest producer (2374.80 μg/L of alcohols, 52.45 μg/L of terpenes and 156.44 μg/L of aldehydes). Acids and ketones were always below the detection limit or at very low concentrations (max 0.15 μg/L) (data not shown).

Table 4. Total amounts of esters, alcohols, terpenes and aldehydes produced by the seven *S. cerevisiae* strains (equivalent μg/L). Mean values ± standard deviation. For each column, letters indicate significant differences (one-way ANOVA and LSD test, $p < 0.05$).

Strain	Esters	Alcohols	Terpenes	Aldehydes
Sc7	1142.01 ± 55.87 B	1825.21 ± 81.02 D	36.61 ± 1.97 B	120.15 ± 6.59 B
Sc22	1024.21 ± 28.37 A,B	1388.67 ± 40.11 B	28.33 ± 0.68 A	85.57 ± 1.85 A
Sc34	1863.55 ± 13.33 D	1562.44 ± 49.94 B,C	45.99 ± 1.75 C	93.17 ± 11.42 A
Sc41	1345.58 ± 57.59 C	2374.80 ± 186.40 E	52.45 ± 1.99 D	156.44 ± 13.94 C
Sc48	946.10 ± 94.82 A	1413.19 ± 19.02 B,C	26.28 ± 0.38 A	83.55 ± 2.20 A
Sc65	1290.20 ± 63.24 C	1123.17 ± 15.75 A	28.69 ± 0.18 A	118.63 ± 7.43 B
Sc70	1350.42 ± 0.53 C	1592.37 ± 12.17 C	46.57 ± 2.05 C	109.20 ± 0.56 B

The amounts of esters, alcohols, aldehydes, and terpenes were used as input variables for a Principal Component Analysis to assess the global differences of the strains, as shown in Figure 2. The analysis accounted for ca. 91% of the total variability; alcohols, terpenes, and aldehydes were mainly related to component 1 (factor correlation between −0.80 to −0.94), while esters' amount was related to component 2 (factor correlation at 0.83) (Figure 2A). The strains were clustered in the factorial space in three different groups (labeled as 1, 2 and 3) (Figure 2B). Group 1 was composed of strains Sc65, Sc22, and Sc48, which were characterized by very similar trends and the low production of VOCs. On the other hand, group 3 comprised only strain Sc41, which was generally characterized by a high production of VOCs, mainly alcohols, aldehydes, and terpenes.

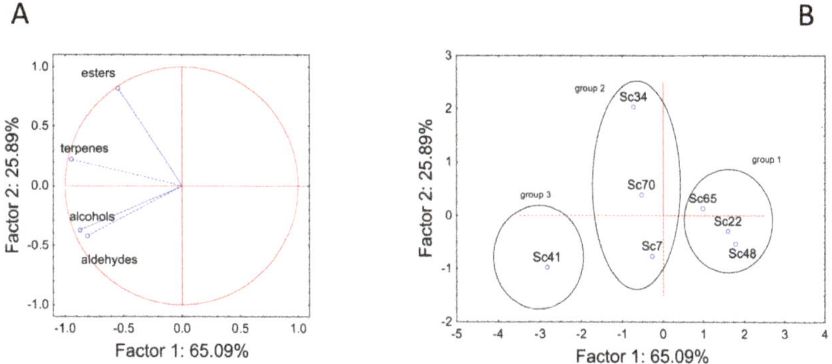

Figure 2. Principal Component Analysis run on the total amounts of esters, terpenes, alcohols, and aldehydes. (**A**) variables' projection; (**B**) cases' projection.

Finally, an intermediate group (medium producers) was found (strains Sc7, Sc70, and Sc34). The strains belonging to this group showed similar trends for alcohols, aldehydes, and terpenes but not for esters, as strain Sc34 was an ester/high-producer.

Then, a second series of PCA was run using the amounts of the different esters (Figure 3) and alcohols (Figure 4). The compounds with very low concentrations or below the detection limit were excluded (for example, in the case of esters, methyl butyrate, isobutyl hexanoate, methyl octanoate, ethyl dodecanoate, ethyl dihydrocinnamate, ethyl cinnamate, ethyl 2-hydroxyisovalerate, ethyl 3-hydroxybutyrate and others). In addition, the analysis was not done for aldehydes and terpenes due to the low number of compounds. The actual amount of the different compounds, the odor and threshold values are summarized in Supplementary Table S1.

PCA run on esters is reported in Figure 3. The analysis accounted for 63% of the total variability, and the input variables were either related to component 1 or component 2. The compounds **1** (ethyl propanoate, threshold level (TL) 2.1 mg/L), **2** (ethyl isobutyrate, TL 0.6 mg/L), **4** (propyl acetate, TL 0.2 mg/L), **14** (ethyl hexanoate, TL 0.05 mg/L), **17** (ethyl heptanoate, TL 0.6 mg/L), **21** (ethyl octanoate, TL 0.02 mg/L), and **29** (ethyl phenylacetate, TL 0.25 mg/L) were related to the component 2, while the others to the component 1 (Figure 3A).

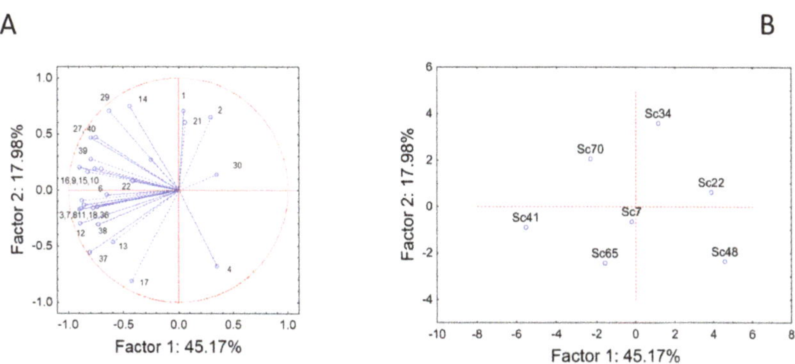

Figure 3. Principal Component Analysis run on esters. (**A**) variables' projection; (**B**) cases' projection.

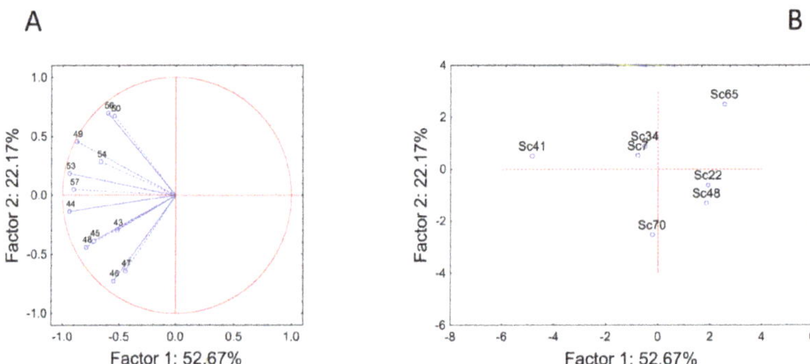

Figure 4. Principal Component Analysis run on alcohols. (**A**) variables' projection; (**B**) cases' projection.

Strain distribution in the factorial space is complex, and it was not possible to point out groups with homogeneous trends (Figure 3B). Strain Sc41 was characterized by the high amount of many esters, like isoamyl acetate (compound **11**), ethyl hexanoate (compound **14**), and ethyl butanoate (compound **3**). Strain Sc34 (the highest ester producer, at least in terms of the total amount) was characterized by the highest production of ethyl propanoate (compound **1**), ethyl isobutyrate (compound **2**), and ethyl octanoate (compound **21**).

The other strains showed the highest production for some specific compounds, but their concentration was generally lower than the molecules reported for strains Sc41 and Sc34.

PCA run on alcohols accounted for 75% of the total variability. The input variables generally were related to component 1, except for the compounds **46** (1-pentanol, TL 80 mg/L), **47** (4-methyl-1-pentanol, TL 50 mg/L), 50 (cis-3-hexen-1-ol, TL 0.4 mg/L) and **56** (benzyl alcohol), which showed a correlation with the factor 2 (factor correlations between 0.64 and 0.73) (Figure 4A). The factorial distribution of the strains reproduced the distribution in three classes, as reported for the PCA with the total amounts of the different classes. In fact, strain Sc41 showed a completely different trend from all other strains, as it generally produced a high amount of all alcohols. In the other groups, strains Sc34 and Sc7 showed a similar trend, as well as strains Sc22 and Sc48 (Figure 4B).

4. Discussion

The first aim of the present research was to select *S. cerevisiae* strains able to work at an up-scale level and to impart to the Aglianico wine-specific volatile molecule fingerprinting in relation to the strain used. In the present experimental work, the seven tested strains were evaluated at pilot-scale mimicking conditions closed as much as possible to the real conditions, and, according to our knowledge, this is one of the few pieces of experimental evidence performed at the pilot scale also assessing the final wine aroma profile. Although lab-scale fermentations offer numerous options for sampling and control of fermentation conditions, they are unable to mimic pilot- or industrial-scale fermenters, especially in relation to secondary metabolism, such as aroma production by yeasts [24]. Instead, pilot-scale fermentations in 100 L tanks are described as well-adapted for mimicking industrial fermentations [30].

The seven wild strains were chosen according to parameters that make them highly competitive due to a combination of properties [31], such as fast growth, efficient glucose consumption, good ability to produce ethanol, high tolerance to ethanol and medium relative demand of nitrogen, have exhibited fermentative profiles comparable to the commercial strain generally employed for Aglianico wine production such as *S. cerevisiae* FE (Fermol Elegance) [32]. In fact, the tested *S. cerevisiae* strains were able to efficiently convert the grape sugars to alcohol, fitting the primary selection criteria of yeast strain selection [5]. In particular, strains Sc41 and Sc70 showed the highest rate of sugar consumption, even if the residual sugars were at an acceptable level for all the tested strains. Also, the volatile acidity detected for all the final wines responded to the required selection criteria according to strain ability for volatile acidity, which needs to be lower than 0.45 g/L. As regards the microbial cell loads detected on WL and lysine medium, the results confirmed the dominance of *S. cerevisiae* strains on non-*Saccharomyces* ones during the evolution of the guided fermentations, confirming the data present in the literature [11,33]. Furthermore, the high inoculum level of the strains (1.0×10^7 UFC/mL) ensured the predominance over non-*Saccharomyces* yeasts, considering that their development is strongly inhibited by the initial concentration of the starter culture and by its growth rate during vinification.

Until a few decades ago, wine yeasts were selected basically on their ability to quickly transform grape sugars into ethanol, their resistance to sulfur dioxide and their low acetic acid production. Actually, their role has been significantly enlarged by the advent of modern oenological microbiology, and thus, their selection has involved the addition of criteria that take into account the improvement of wine quality in terms of color, aroma, structure, technological, and also healthy properties [34,35]. The importance of these criteria depends on the type and style of wine to be made as well as the requirements of the winery. In particular, the selection of yeast strains according to the generated volatile aroma profiles has been largely considered by several authors since the yeast strain's ability to impart good sensory features is one of the most important criteria. In fact, although the complexity of wine aroma can also vary depending on many variables such as the type of grape variety, *terroir*, microbial starter, fermentation process, aging, and bottling, the selection of new wild strains within the species *S. cerevisiae*, able to satisfy conventional selection criteria and impart new aroma volatile profile could be a useful tool to enlarge the gamma of sensory properties of wine products [32,36]. In fact, consumers, who are increasingly more demanding, are looking for distinctive characteristics in wines, and this encourages the producers and researchers to develop biotechnological strategies (including new strain selection) to improve the aromatic complexity of wines. The challenge is actually to find new wild strains able to work on winery conditions. In fact, the literature shows a huge amount of research works based on lab trials, but few are focused on up-scale conditions [37].

The volatile wine profiles of the present work were obtained by GC/MS/SPME, a technique widely used for this type of sample due to the high volatility of esters and alcohols, particularly. Moreover, the volatile fingerprinting obtained throughout the headspace analysis technique results in profiles closed to the perception of the panelist [35]. The results showed that the tested *S. cerevisiae* strains influenced and drove wine volatile molecule profiles in a strain-dependent way, imparting specific features to the obtained wines. In particular, in our experimental conditions, the inoculation with *S. cerevisiae* Sc41 resulted in wine characterized by large amounts of alcohols, terpenes and aldehydes able to impart specific flavor notes to the final wine.

Differently, strain Sc34 was the highest producer of esters in terms of amount, followed by strain Sc41. Since different authors have found great variability in wine volatile molecule profile in relation to the *S. cerevisiae* strain tested [38], emphasizing the potential role of this parameter as a trait for starter culture selection also to enlarge the potential of diversification of the wine sector, the selection of the best-performing yeasts could be done according to the traits to impart to the final product. In general, esters are formed by yeasts during the alcoholic fermentation, and they are responsible for the fruity odor, while terpenic and nor-isoprenoid compounds are the most important constituent of the varietal aroma of grapes and confer a flowery odor to the wine [39].

From a technological point of view, it could be interesting to find yeast strains able to increase linalool, a-terpineol, and citronellol, able to impart citrus and peach flavor notes, throughout their B-glucosidase activity, a trait not very common among *S. cerevisiae* strains [40]. Particularly, volatile esters constitute one of the most important classes of aroma compounds and are largely responsible for the fruity aromas associated with wine and other fermented beverages [19,41,42]. Their formation differs widely among yeast strains, and other external factors such as fermentation temperature, nutrient availability, pH, unsaturated fatty acid/sterol levels, and oxygen levels could contribute, all playing an important part in determining the end levels of esters in a wine.

Our research suggested that, in real conditions, *S. cerevisiae* Sc41 was able to produce high amounts of many esters, like isoamyl acetate (TL 0.03 mg/L), ethyl hexanoate (TL 0.05 mg/L), ethyl butanoate (TL 0.6 mg/L), which are largely described to contribute to fruity aromas (pineapple, apricot). On the other hand, strain Sc34 (the highest ester producer, at least for the total amount) was characterized by the highest production of ethyl propanoate (TL 2.1 mg/L), ethyl isobutyrate (TL 0.6 mg/L) and ethyl octanoate (TL 0.02 mg/L).

Regarding alcohols, which play a fundamental role since they usually have a strong, pungent smell, particularly when exceeding 400 mg/L, all the samples contained 2-phenyl-ethanol responsible for honey, spice, rose, and lilac nuances even if below the concentration threshold (0.75 mg/L). Strain Sc41 produced the highest amount of 2-phenyl-ethanol and benzyl alcohol recognized for their rose and fruity note, respectively.

Sc41 produced also the highest amount of hexanol (TL 4 mg/L) able to impart grass note.

Nevertheless, the ratio among the volatile molecules plays an important role in defining the final wine flavor and taste and not only the individual compounds. Also, the interaction with non-volatile compounds needs to be taken into consideration.

According to the final features intended to impart to the product in terms of alcohols or esters, a proper selection can be defined. In fact, although all the strains showed the potential to ferment Aglianico must, different profiles for volatile compounds were identified, and this trait is fundamental for strain selection.

5. Conclusions

The selection of starter cultures is a complex process, and many researchers stop their experiments after a preliminary laboratory validation, while in this article, a fermentation with 100 L (with independent batches for each strain) was carried out. Although lab-scale fermentations offer numerous options for sampling and control of fermentation conditions, they are unable to mimic pilot- or industrial-scale fermenters, especially in relation to secondary metabolism, such as aroma production by yeasts. At the same time, pilot-scale fermentations in 100 L tanks are described as suitable for mimicking industrial fermentations, producing even reproducible results.

In conclusion, since the present work was performed at a pilot-scale level, mimicking as much as possible real working conditions, the results obtained can be considered as a validation of the screened *S. cerevisiae* strains and a strategy to discriminate a strain, among others, to impart desired wine sensory "features," also providing to the experts of the sector exploitable information for operating in cellars.

Supplementary Materials: The following supporting information can be downloaded at: https://www.mdpi.com/article/10.3390/fermentation9030245/s1, Table S1: Amounts of esters, alcohols, terpenes (equivalent µg/L) and their odor and threshold values. The numbers in the column "code" are those used as a symbol of the molecules in the PCA (Figures 3 and 4). In a row, values with different letters are significantly different (one-way ANOVA and LSD test, $p < 0.05$).

Author Contributions: Conceptualization, P.R., F.P. and A.B.; methodology, P.R., F.P. and A.B.; software, D.G. and G.S.; validation, F.P., A.B. and P.R.; formal analysis, all; investigation, all; resources, P.R., F.P. and A.C.; data curation, D.G. and G.S.; writing—original draft preparation, P.R., A.B. and F.P.; writing—review and editing, all; visualization, all; supervision, P.R., A.B., F.P. and G.S.; project administration, A.B., F.P. and P.R. All authors have read and agreed to the published version of the manuscript.

Funding: DC received a grant from the Apulia Region (funder) for her researcher position through the call REFIN (funding number: UNIFG283).

Institutional Review Board Statement: Not applicable.

Informed Consent Statement: Not applicable.

Data Availability Statement: The data presented in this study are available in the article.

Acknowledgments: A.C. was supported by the project PSR Regione Basilicata 2014–2020, sottomisura 16.2 IN.VINI.VE.RI.TA.S (Innovare la viti-VINIcoltura lucana: VErso la RIgenerazione varieTAle, la Selezione di vitigni locali e proprietà antiossidanti dei vini), N. 976. JRU MIRRI-IT is greatly acknowledged for scientific support.

Conflicts of Interest: The authors declare no conflict of interest or any role of the funding sponsors in the choice of the research project; design of the study; in the collection, analyses or interpretation of data; in the writing of the manuscript; or in the decision to publish the results.

References

1. Pretorius, I.S. Tasting the terroir of wine yeast innovation. *FEMS Yeast Res.* **2020**, *20*, foz084. [CrossRef] [PubMed]
2. Van der Westhuizen, T.J.; Augustyn, O.P.H.; Pretorius, I.S. Geographical distribution of indigenous *Saccharomyces cerevisiae* strains isolated from vineyards in the coastal regions of the Western Cape in South Africa. *S. Afr. J. Enol. Vitic.* **2000**, *21*, 3–10. [CrossRef]
3. Torija, M.J.; Rozès, N.; Poblet, M.; Guillamón, J.M.; Mas, A. Yeast population dynamics in spontaneous fermentations: Comparison between two different wineproducing areas over a period of three years. *Antonie Van Leeuwenhoek* **2001**, *79*, 345–352. [CrossRef] [PubMed]
4. Lopes, C.A.; van Broock, M.; Querol, A.; Caballero, A.C. *Saccharomyces cerevisiae* wine yeast populations in a cold region in Argentinean Patagonia. A study at different fermentation scales. *J. Appl. Microbiol.* **2002**, *93*, 608–615. [CrossRef] [PubMed]
5. Pretorius, I.S. Tailoring wine yeast for the new millennium: Novel approaches to the ancient art of winemaking. *Yeast* **2000**, *16*, 675–729. [CrossRef]
6. Nikolaou, E.; Soufleros, E.H.; Bouloumpasi, E.; Tzanetakis, N. Selection of indigenous *Saccharomyces cerevisiae* strains according to their oenological characteristics and vinification results. *Food Microbiol.* **2006**, *23*, 205–211. [CrossRef]
7. Capece, A.; Romaniello, R.; Pietrafesa, R.; Romano, P. Indigenous *Saccharomyces cerevisiae* yeasts as a source of biodiversity for the selection of starters for specific fermentations. *BIO Web Conf.* **2014**, *3*, 02003. [CrossRef]
8. Legras, J.L.; Merdinoglu, D.; Cornuet, J.M.; Karst, F. Bread, beer and wine: *Saccharomyces cerevisiae* diversity reflects human history. *Mol. Ecol.* **2007**, *16*, 2091–2102. [CrossRef]
9. Borneman, A.R.; Forgan, A.H.; Pretorius, I.S.; Chambers, P.J. Comparative genome analysis of *Saccharomyces cerevisiae* wine strain. *FEMS Yeast Res.* **2008**, *8*, 1185–1195. [CrossRef]
10. Liti, G.; Carter, D.M.; Moses, A.M.; Warringer, J.; Parts, L.; James, S.A.; Davey, R.P.; Roberts, I.N.; Burt, A.; Koufopanou, V.; et al. Population genomics of domestic and wild yeasts. *Nature* **2009**, *458*, 337–341. [CrossRef]
11. Capece, A.; Pietrafesa, R.; Siesto, G.; Romaniello, R.; Condelli, N.; Romano, P. Selected Indigenous *Saccharomyces cerevisiae* Strains as Profitable Strategy to Preserve Typical Traits of Primitivo Wine. *Fermentation* **2019**, *5*, 87. [CrossRef]
12. Carrau, F.M.; Medina, K.; Farina, L.; Boido, E.; Henschke, P.A.; Dellacassa, E. Production of fermentation aroma compounds by *Saccharomyces cerevisiae* wine yeasts: Effects of yeast assimilable nitrogen on two model strains. *FEMS Yeast Res.* **2008**, *8*, 1196–1207. [CrossRef] [PubMed]
13. Romano, P.; Caruso, M.; Capece, A.; Lipani, G.; Paraggio, M.; Fiore, C. Metabolic diversity of *Saccharomyces cerevisiae* strains from spontaneously fermented grape musts. *World J. Microbiol. Biotechnol.* **2003**, *19*, 311–315. [CrossRef]
14. Callejon, R.; Clavijo, A.; Ortigueira, P.; Troncoso, A.M.; Paneque, P.; Morales, M.L. Volatile and sensory profile of organic red wines produced by different selected autochthonous and commercial *Saccharomyces cerevisiae* strains. *Anal. Chim. Acta* **2010**, *660*, 68–75. [CrossRef]
15. Orlić, S.; Vojvoda, T.; Babić, K.H.; Arroyo-López, F.N.; Jeromel, A.; Kozina, B.; Iacumin, L.; Comi, G. Diversity and oenological characterization of indigenous *Saccharomyces cerevisiae* associated with Žilavka grapes. *World J. Microbiol. Biotechnol.* **2010**, *26*, 1483–1489. [CrossRef]
16. Knight, S.; Klaere, S.; Fedrizzi, B.; Goddard, M.R. Regional microbial signatures positively correlate with differential wine phenotypes: Evidence for a microbial aspect to terroir. *Sci. Rep.* **2015**, *5*, 14233. [CrossRef]
17. Bokulich, N.A.; Collins, T.S.; Masarweh, C.; Allen, G.; Heymann, H.; Ebeler, S.E.; Mills, D.A. Associations among wine grape microbiome, metabolome, and fermentation behaviour suggest contribution to regional wine characteristics. *mBio* **2016**, *7*, 1–12. [CrossRef]
18. Anagnostopoulos, D.A.; Kamilari, E.; Tsaltas, D. Contribution of the Microbiome as a Tool for Estimating Wine's Fermentation Output and Authentication. In *Advances in Grape and Wine Biotechnology*; Morata, A., Loira, I., Eds.; IntechOpen: London, UK, 2019; pp. 1–21.
19. Romano, P.; Braschi, G.; Siesto, G.; Patrignani, F.; Lanciotti, R. Role of Yeasts on the Sensory Component of Wines. *Foods* **2022**, *11*, 1921. [CrossRef]
20. Bruwer, J.; Alant, K.; Li, C.; Bastian, S. Wine Consumers and Makers: Are They Speaking the Same Language? *Aust. N. Z. Grapegrow. Winemak.* **2005**, *496*, 80–84.
21. Jackson, R.S. *Wine Science. Principles and Applications*, 3rd ed.; Academic Press: San Diego, CA, USA, 2008.

22. Petropulos, V.I.; Bogeva, E.; Stafilov, T.; Stefova, M.; Siegmund, B.; Pabi, N.; Lankmayr , E. Study of the Influence of Maceration Time and Oenological Practices on the Aroma Profile of Vranec Wines. *Food Chem.* **2014**, *165*, 506–514. [CrossRef]
23. Cadiere, A.; Aguera, E.; Caille, S.; Ortiz-Julien, A.; Dequin, S. Pilot-scale evaluation the enological traits of a novel, aromatic wine yeast strain obtained by adaptive evolution. *Food Microbiol.* **2012**, *32*, 332–337. [CrossRef] [PubMed]
24. Casalta, E.; Aguera, E.; Picou, C.; Rodriguez Bencomo, J.J.; Salmon, J.-M.; Sablayrolles, J.-M. A comparison of laboratory and pilot-scale fermentations in winemaking conditions. *Appl. Microbiol. Biotechnol.* **2010**, *7*, 1665–1673. [CrossRef] [PubMed]
25. Romano, P.; Fiore, C. Influenza del ceppo di lievito autoctono sulle qualità organolettiche del vino. In Proceedings of the Workshop POM A01project on Raccolta Meccanica Delle Uve da Vino in Ambienti Meridionali ed Insulari Italiani, Senorbi, Italy, 8 May 2001.
26. Kashyap, D.R.; Vohra, P.K.; Chopra, S.; Tewari, R. Applications of pectinases in the commercial sector: A review. *Bioresour. Technol.* **2001**, *77*, 215–227. [CrossRef] [PubMed]
27. Anonymous. *Guidelines on Infrared Analysers in Oenology. OIV/OENO Resolution 390/2010*; International Organization of Vine and Wine General Assembly (OIV): Tbilisi, Georgia, 2010.
28. Pallmann, C.L.; Brown, J.A.; Olineka, T.L.; Cocolin, L.; Mills, D.A.; Bisson, L.F. Use of WL Medium to Profile Native Flora Fermentations. *Am. J. Enol. Vitic.* **2001**, *52*, 3. [CrossRef]
29. Geeraerd, A.H.; Herremans, C.H.; Van Impe, J.F. Structural model requirements to describe microbial inactivation during mild heat treatment. *Int. Food Microbiol.* **2000**, *56*, 185–209. [CrossRef]
30. Aguera, E.; Sablayrolles, J.M. Vinification à l'échelle pilote (100 L). II Caractérisation—Intérêt. *Wine Internet Techn. J.* **2005**, *7*, 1–9.
31. Piskur, J.; Rozpedowska, E.; Polakova, S. How did *Saccharomyces* evolve to become a good brewer? *Trends Genet.* **2006**, *22*, 183–186. [CrossRef]
32. Testa, B.; Coppola, F.; Lombardi, S.; Iorizzo, M.; Letizia, F.; Di Renzo, M.; Succi, M.; Tremonte, P. Influence of *Hanseniaspora uvarum* AS27 on Chemical and Sensorial Characteristics of Aglianico Wine. *Processes* **2021**, *9*, 326. [CrossRef]
33. Salvadò, Z.; Arroyo-Lòpez, F.N.; Barrio, E.; Querol, A.; Guillamòn, J.M. Quantifying the individual effect of ethanol and temperature on the fitness advantage of *Saccharomyces cerevisiae*. *Food Microbiol.* **2011**, *28*, 1155–1161. [CrossRef]
34. Fleet, G.H. Wine yeasts for the future. *FEMS Yeast Res.* **2008**, *8*, 979–995. [CrossRef]
35. Patrignani, F.; Chinnici, F.; Serrazanetti, D.I.; Vernocchi, P.; Ndagijimana, M.; Riponi, C.; Lanciotti, R. Production of Volatile and Sulfur Compounds by 10 *Saccharomyces cerevisiae* Strains Inoculated in Trebbiano Must. *Front. Microbiol.* **2016**, *7*, 243. [CrossRef] [PubMed]
36. Belda, I.; Ruiz, J.; Esteban-Fernández, A.; Navascués, E.; Marquina, D.; Santos, A.; Moreno-Arribas, M.V. Microbial Contribution to Wine Aroma and Its Intended Use for Wine Quality Improvement. *Molecules* **2017**, *22*, 189. [CrossRef] [PubMed]
37. Rossouw, D.; Jacobson, D.; Bauer, F.F. Transcriptional regulation and the diversification of metabolism in wine yeast strains. *Genetics* **2012**, *190*, 251–261. [CrossRef] [PubMed]
38. Capece, A.; Romano, P. Yeasts and their metabolic impact on wine flavour. In *Yeasts in the Production of Wine*; Romano, P., Ciani, M., Fleet, G.H., Eds.; Springer: Berlin, Germany, 2019; pp. 43–80.
39. Vararu, F.; Moreno-García, J.; Zamfir, C.I.; Cotea, V.V.; Moreno, J. Selection of aroma compounds for the differentiation of wines obtained by fermenting musts with starter cultures of commercial yeast strains. *Food Chem.* **2016**, *197*, 373–381. [CrossRef]
40. Vernocchi, P.; Patrignani, F.; Ndagijimana, M.; Lopez, C.C.; Suzzi, G.; Gardini, F.; Lanciotti, R. Trebbiano wine produced by using *Saccharomyces cerevisiae* strains endowed with β-glucosidase activity. *Ann. Microbiol.* **2015**, *65*, 1565–1571. [CrossRef]
41. Sumby, K.M.; Grbin, P.R.; Jiranek, V. Microbial modulation of aromatic esters in wine: Current knowledge and future prospects. *Food Chem.* **2010**, *121*, 1–16. [CrossRef]
42. Parpinello, G.P.; Ricci, A.; Folegatti, B.; Patrignani, F.; Lanciotti, R.; Versari, A. Unraveling the potential of cryotolerant *Saccharomyces eubayanus* in Chardonnay white wine production. *LWT* **2020**, *134*, 110183. [CrossRef]

fermentation

Article

Investigation of Geraniol Biotransformation by Commercial *Saccharomyces* Yeast Strains by Two Headspace Techniques: Solid-Phase Microextraction Gas Chromatography/Mass Spectrometry (SPME-GC/MS) and Proton Transfer Reaction-Time of Flight-Mass Spectrometry (PTR-ToF-MS)

Rebecca Roberts [1,2], Iuliia Khomenko [2], Graham T. Eyres [1], Phil Bremer [1], Patrick Silcock [1], Emanuela Betta [2] and Franco Biasioli [2,*]

[1] Department of Food Science, University of Otago, P.O. Box 56, Dunedin 9054, New Zealand
[2] Sensory Quality Unit, Research and Innovation Centre, Fondazione Edmund Mach, 38098 Trento, Italy
*   Correspondence: franco.biasioli@fmach.it

Citation: Roberts, R.; Khomenko, I.; Eyres, G.T.; Bremer, P.; Silcock, P.; Betta, E.; Biasioli, F. Investigation of Geraniol Biotransformation by Commercial *Saccharomyces* Yeast Strains by Two Headspace Techniques: Solid-Phase Microextraction Gas Chromatography/Mass Spectrometry (SPME-GC/MS) and Proton Transfer Reaction-Time of Flight-Mass Spectrometry (PTR-ToF-MS). *Fermentation* 2023, 9, 294. https://doi.org/10.3390/fermentation9030294

Academic Editor: Niel Van Wyk

Received: 28 February 2023
Revised: 8 March 2023
Accepted: 9 March 2023
Published: 17 March 2023

Abstract: Hop-derived volatile organic compounds (VOCs) and their transformation products significantly impact beer flavour and aroma. Geraniol, a key monoterpene alcohol in hops, has been reported to undergo yeast-modulated biotransformation into various terpenoids during fermentation, which impacts the citrus and floral aromas of the finished beer. This study monitored the evolution of geraniol and its transformation products throughout fermentation to provide insight into differences as a function of yeast species and strain. The headspace concentration of VOCs produced during fermentation in model wort was measured using Solid-Phase Microextraction Gas Chromatography/Mass Spectrometry (SPME-GC/MS) and Proton Transfer Reaction-Time of Flight-Mass Spectrometry (PTR-ToF-MS). In the absence of yeast, only geraniol was detected, and no terpenoid compounds were detected in geraniol-free ferments. During fermentation, the depletion of geraniol was closely followed by the detection of citronellol, citronellyl acetate and geranyl acetate. The concentration of the products and formation behaviour was yeast strain dependent. SPME-GC/MS provided confidence in compound identification. PTR-ToF-MS allowed online monitoring of these transformation products, showing when formation differed between *Saccharomyces cerevisiae* and *Saccharomyces pastorianus* yeasts. A better understanding of the ability of different yeast to biotransform hop terpenes will help brewers predict, control, and optimize the aroma of the finished beer.

Keywords: beer; fermentation; geraniol; biotransformation; SPME-GC/MS; PTR-ToF-MS

1. Introduction

Monoterpenoids are volatile organic compounds (VOCs) which can strongly impact food flavour. They are found in the essential oils of various plants, including hops [1,2]. Monoterpene alcohols, such as geraniol, linalool, nerol and α-terpineol, provide a "floral" or "citrus" aroma to beer. Interestingly monoterpenes such as citronellol and geranyl acetate have been detected in beer but not in hops [3]. While the generation of such compounds is not fully understood due to the complex chemical, physical and biochemical changes that occur throughout brewing and fermentation, *Saccharomyces* yeast have been reported to biotransform aroma compounds [4–7]. Specifically, geraniol has been reported to be a precursor for many of the monoterpenoids (via biotransformation) present in wine [8–10] or beer [4–6,11,12].

The complexity of hop essential oils and the various transformation reactions during fermentation make it challenging to determine the origin of VOCs produced in beer [3]. This challenge is particularly true for terpenoid compounds, which are responsible for much of the flavour and aroma in beer. Terpenoids can be present in the form of glycosides.

The cleavage of glycosides by yeast enzymes during fermentation can lead to the release of free terpenoids in beer, further contributing to the challenge of identifying the origin of aroma in beer [13]. Investigating the biotransformation of individual compounds in a model system could provide a better understanding of potential reaction pathways and the impact of different yeast strains on terpene production during fermentation. To date, research on biotransformation has mostly relied on techniques such as Solid Phase Micro Extraction (SPME) coupled with Gas Chromatography/Mass Spectrometry (GC/MS) to determine the VOC composition of beer. A limitation to this approach is that the analysis is typically performed only at the end of the fermentation, which does not provide all the information required to understand the dynamics of the biochemical reactions. Steyer et al. 2013, were among the first to evaluate the transformations of terpenes over time using Stir Bar Sorptive Extraction-Liquid Desorption (SBSE-LD) and GC/MS. The findings of their study indicated that geraniol underwent a transformation during fermentation by *S. cerevisiae*, resulting in the production of citronellol, linalool, nerol, citronellyl acetate, and geranyl acetate [14]. Alternative high throughput techniques, such as Proton Transfer Reaction-Time of Flight-Mass Spectrometry (PTR-ToF-MS) have been used to follow VOC development during fermentation [15]. PTR-ToF-MS provides dynamic measurements, increasing the understanding of volatile compounds' generation dynamics or reaction pathways, which can be useful for identifying the impact of different brewing conditions on the flavour and aroma [16].

Due to its high concentration in fresh hops, geraniol was an appropriate choice as the initial compound to investigate using model ferments. Previous studies have proposed several pathways for the biotransformation from geraniol. Still, limitations of these studies are the use of a complex starting material (whole hop cones) as well as the use of different microorganisms that are not commonly used in beer fermentation: *Cyanobacterium* [17], *Aspergillus niger* [18], *Castellaniella defragrans* and *Pseudomonas aeruginoa* [19]. An overview of the current literature related to beer and wine on the biotransformation of geraniol by *S. cerevisiae* is displayed in Figure 1 [5,6,20–22].

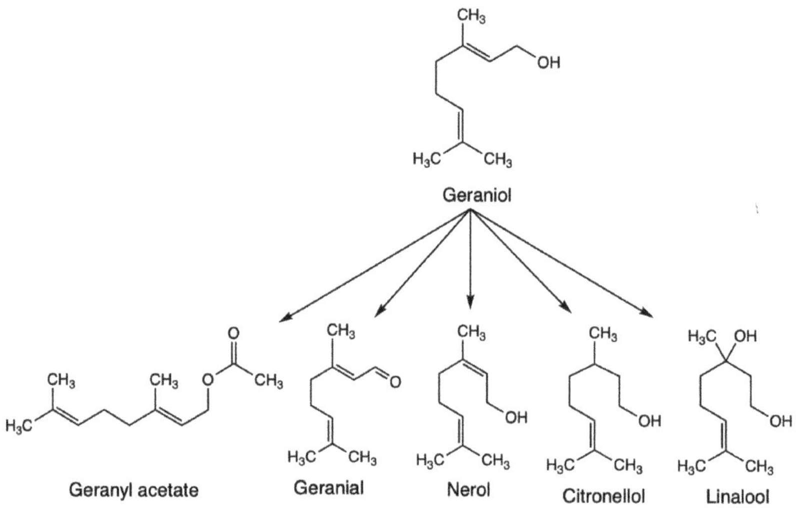

Figure 1. Previously identified biotransformation reactions of geraniol [5,6,20–22].

The current "gold standard" for the identification and off-line monitoring of VOCs is Gas Chromatography/Mass Spectrometry (GC/MS) which is a widely used analytical technique due to its high sensitivity, selectivity, wide linear range, versatility, and precision. GC/MS can detect VOCs at very low concentrations, even in the presence of other compounds. It can separate and identify individual VOCs in a sample, and it can measure a wide range of VOC concentrations, from low parts-per-billion (ppb) to high parts-per-million (ppm) levels [23]. However, drawbacks of GC/MS include the long time required for a single analysis due to its labour-intensive sample preparation and quantitative analysis requiring reference standards [24]. In contrast, PTR-ToF-MS suits rapid quantitative analysis of VOCs in complex mixtures. It is an ultra-high sensitive technique that allows for the online analysis of VOCs based on their mass-to-charge ratio and can detect trace gases at ultra-low concentration levels (low ppt). A limitation of PTR-ToF-MS is that isomers are not distinguishable. Therefore, identification should be complemented by another analytical technique, such as GC/MS [25–27] or by implementing additional tools, such as fast-GC [28]. These two techniques have been previously used to measure VOCs in cheese, potatoes, infant formula, blueberries, milk, olive oil and truffles [26,29–31] and can support the identification of spectrometric peaks used for rapid monitoring over time [26,32]. The rapid PTR-MS-based methods also allow for the measurement of a larger number of replicates, making results statistically more robust. Therefore, two separate analyses were carried out at the same time in this study: one using SPME-GC/MS of a few select time points to identify the VOCs present, and the other using PTR-ToF-MS at more time points to monitor the generation of the VOC overtime more accurately.

The overall aim of the current study was to monitor the dynamic changes of geraniol during beer fermentation to understand and quantify in real time the point at which differences between *S. cerevisiae* and *S. pastorianus* yeast occurred and to provide new information on the biotransformation of geraniol during beer production.

2. Materials and Methods

2.1. Yeast Hydration and Model Wort Preparation

Commercially available yeast strains supplied by Fermentis (Lilles, France) were *Saccharomyces cerevisiae* strains SafAle US-05 and SafAle WB-06 and *Saccharomyces pastorianus* strains SafLager W-34/70 and SafLager S-23. Table 1 provides an overview of the samples measured with SPME-GC/MS and PTR-ToF-MS. The SPME-GC/MS samples were measured once every 24 h over a 5-day period, while the PTR-ToF-MS samples were measured once every 6 h over the same 5-day period. Each dried yeast strain was rehydrated separately in model wort. The model wort was prepared by dissolving 260 g of spray-dried malt extract (Briess Golden light) into 2 L deionized water (18 MΩ cm). For pH correction, 166 mg of calcium chloride ($CaCl_2$) was added [33]. In place of bittering hops, 76.7 mg of Iso-α-acids (ICS—I4 Iso Standard; American Society of Brewing Chemists, St. Paul, MN, USA) was added to provide an international bitterness unit (IBU) of 20. The model wort was heated to 90 °C using a water bath and held for 10 min, then decreased to 20 °C using an ice bath. The main analytical characteristic of the model wort was pH 5.2, with a specific gravity of 12 °P. Each yeast strain starting density was 10×10^7 cells per mL, which was in line with manufacturing recommendations and best brewing practices.

Table 1. An overview of the samples measured with SPME-GC/MS and PTR-ToF-MS.

Yeast Species	Yeast Strain	SPME-GC/MS	Measurement Frequency (h)	PTR-ToF-MS	Measurement Frequency (h)
S. cerevisiae	SafAle US-05	✓	24	✓	6
S. cerevisiae var. Diastaticus	SafAleWB-06	✓	24	✓	6
S. pastorianus	SafLager W-34/70	-	-	✓	6
S. pastorianus	SafLager S-23	-	-	✓	6

2.2. Micro-Fermentations

Each 3 mL micro-fermentation consisted of model wort, yeast and 5 ppm of geraniol. In addition, samples without geraniol and samples without yeast served as blank controls. The samples were added into 20 mL glass head-space vials, sealed then placed into a thermostatic autosampler tray (set to 20 °C) in a randomized order (CTC CombiPAL, CTC Analytics, Zwingen, Switzerland).

2.3. HS- SPME-GC/MS Analytical Conditions

VOCs were extracted using Head Space Solid Phase Microextraction on (HS-SPME-GC/MS) with 2-cm fibre coated with 50/30-µm divinyl benzene/carboxen/polydimethylsiloxane (DVB/CAR/PDMS, Supelco, Bellefonte, PA, USA). The fibre was exposed to the headspace for 40 min. The compounds absorbed on the SPME fibre were desorbed at 250 °C in the GC/MS injection port. The mass detector operated in electron ionization mode (EI, internal ionization source; 70 eV) with a scan range from m/z 33 to 350. Analysis was carried out using Perkin Elmer Clarus 500 GC/MS equipped with an HP-INNOWax fused silica capillary column (30 m, 0.32-mm ID, 0.5-µm film thickness; Agilent Technologies, Palo Alto, CA, USA). The oven temperature was initially set at 40 °C for 1 min, then increased to 220 °C at 4 °C/min, increased to 250 °C at 15 °C/min and maintained for 2 min. Helium was used as carrier gas with a flow rate of 1.5 mL/min. Compound identification was based on mass spectra matching with NIST14/Wiley98 libraries. Linear retention indices were calculated under the same chromatographic conditions after the injection of a C7–C30 n-alkane series (Supelco).

2.4. PTR-ToF-MS Measurement

Headspace measurements were performed with a commercial PTR-ToF-MS 8000 apparatus from Ionicon Analytik GmbH (Innsbruck, Austria) in a standard configuration (V mode). The ionization conditions were as follows: 500 V drift voltage, 110 °C drift temperature, and 2.80 mbar drift pressure resulting in an E/N ratio of 130 Townsend (1 Td = 10^{-17} cm^2 V^{-1} s^{-1}). Sample handling, headspace flushing and sampling were carried out using an autosampler (MPS MultiPurpose Sampler, Gerstel, Germany) specially adapted for PTR-ToF-MS [34]. The autosampler moved the sample from the incubation tray to the temperature-controlled purging site, connected to the PTR-ToF-MS inlet. Dynamic headspace analysis took place for 60 s with an acquisition rate of one mass spectrum per second between m/z 15 and 349. Due to the high ethanol concentration, argon was added to the inlet system at a flow rate of 120 sccm, with the total flow rate of the system at 160 sccm. This prevented primary ion depletion and the formation of ethanol clusters that might affect the final quantification of volatiles [35]. The argon flow rate was controlled by a multi-gas controller (MKS Instruments, Inc., Andover, MA, USA). After measurement, the vial was moved back to the same position as the incubation tray, and the cycle was repeated on the following sample. During fermentation, the measurement was repeated every 6 h to monitor the fermentation process.

Deadtime correction, internal calibration of mass spectral data, and peak extraction were performed according to previously described procedures [36,37]. The peak intensity in ppb/v (parts per billion per volume) was estimated using the formula described in the literature [38]. The formula uses a constant value for the reaction rate coefficient (k = 2.10^{-9} cm^3 s^{-1}).

Systematic errors can arise due to various factors, such as the use of a constant reaction coefficient, humidity, and fragmentation. However, in most cases, the error associated with measuring the absolute concentration of each compound is less than 30% and can be corrected post-analysis [36]. Certain mass peaks, such as those associated with isotopologues of ^{13}C, ^{18}O, and ^{27}S, as well as water and ethanol clusters, were excluded from the dataset to minimise errors. Tentative compound identification was conducted by comparing the measured mass to the theoretical mass in the literature (Table 2). SPME-GC/MS was employed in conjunction with PTR-TOF-MS to confirm the tentative identification of compounds through the comparison of their mass spectra and chromatographic retention times.

Table 2. List of the peaks identified with PTR-ToF-MS. The measured mass, the identified mass, the sum formula and a tentative identification are given.

Theoretical *m/z*	Measured *m/z*	Sum Formula	Chemical Class	Tentative Identification
28.0062	28.006	$C_2H_5^+$	Alcohols	Ethanol Fragment
33.0339	33.034	CH_4OH^+	Alcohols	Methanol
48.0529	48.053	$C_2H_5OH^+$	Alcohols	Ethanol (isotopologue)
59.0491	59.049	$C_3H_6OH^+$	Aldehydes/ketones	Propanol/acetone
62.0317	62.031	$C_2H_4O_2H^+$	Esters and acids	Acetic acid
64.0292	64.029	$C_2H_6SH^+$	Sulphur compounds	Dimethylsulfide
69.0697	69.069	$C_5H_8H^+$	Terpene	Terpene fragment
76.047	75.043	$C_3H_6O_2H^+$	Esters and acids	Propionic acid
81.0699	81.07	$C_6H_8H^+$	Terpene	Terpene fragment
83.0783	83.084	$C_6H_{10}H^+$	Terpene	Terpene fragment
85.0654	85.064	$C_5H_8OH^+$	Aldehydes/Ketones	Pentanal/pentenone
87.0439	87.043	$C_4H_6O_2H^+$	Ketones	Butanedione
87.0803	87.08	$C_5H_{10}OH^+$	Alcohols	Pentanol
94.0952	93.068	$C_7H_7^+$	Terpene	Terpene fragment
95.0492	95.046	$C_6H_6OH^+$	Phenols	Phenol
95.096	95.09	$C_7H_{10}H^+$	Terpenes	Terpene fragment
97.0284	97.027	$C_5H_4O_2H^+$	Aldehydes	Furfural
97.0642	97.057	$C_6H_8OH^+$	Aldehydes/Furans	Hexadienal/ethylfuran
99.0802	99.079	$C_6H_{10}OH^+$	Aldehydes	Hexenal/methylpentenone
101.0951	101.091	$C_6H_{12}OH^+$	Alcohols	Hexanol
103.0749	103.074	$C_5H_{10}O_2H^+$	Esters and acids	Methylbutanoic acid
107.0705	107.07	$C_7H_6OH^+$	Aldehydes	Benzaldehyde
107.1071	107.102	$C_8H_{10}H^+$	Aromatic hydrocarbons	Xylene/ethylbenzene
109.0712	109.059	$C_7H_8OH^+$	Phenols	Benzyl alcohol (cresol)
111.0463	111.042	$C_6H_6O_2H^+$	Furans	Acetyl furan
111.0804	111.076	$C_7H_{10}OH^+$	Aldehydes	Heptadienal
113.0965	113.096	$C_7H_{12}OH^+$	Aldehydes	Heptanal
115.1109	115.111	$C_7H_{14}OH^+$	Ketones	Heptanone
121.0691	121.067	$C_8H_8OH^+$	Aldehydes	Methylbenzaldehyde-coumaran
127.1117	127.112	$C_8H_{14}OH^+$	Ketones	Octenone/methylheptenone
129.0911	129.091	$C_7H_{12}O_2H^+$	Esters and acids	Hexenyl formate
129.1272	129.125	$C_8H_{16}OH^+$	Ketones	Octanone/Dimethylcyclohexanol
131.1062	131.107	$C_7H_{14}O_2H^+$	Esters and acids	Heptanoic acid/hexyl formate
135.1032	135.109	$C_{10}H_{14}H^+$	Aromatic hydrocarbons	Methylpropylbenzene
136.1073	136.112	$C_9H_{13}NH^+$	Heterocyclic compounds	Butyl-pyridine/ethyl-propylpyridine
137.132	137.133	$C_{10}H_{16}H^+$	Terpenes	Various monoterpenes
141.1357	141.127	$C_9H_{16}OH^+$	Aldehydes	Nonanal
143.1443	143.148	$C_9H_{18}OH^+$	Ketones/Aldehydes	Nonanone/nonanal
151.1108	151.112	$C_{10}H_{14}OH^+$	Terpenes	Carvacrol/safranal
153.0615	153.063	$C_8H_8O_3H^+$	Aldehydes	Vanillin, methyl salicylate
153.1234	153.126	$C_{10}H_{16}OH^+$	Aldehydes	Citral
155.1424	155.143	$C_{10}H_{18}OH^+$	Alcohols	Linalool/geraniol/a-terpineol/nerol
157.1576	157.158	$C_{10}H_{20}OH^+$	Alcohols	Citronellol/dihydrolinalool
171.1373	171.137	$C_{10}H_{18}O_2H^+$	Terpenes	Linalool oxide/Citronellic acid
199.1677	199.169	$C_{12}H_{23}O_2H^+$	Terpenes	Citronellyl acetate
201.1819	201.184	$C_{12}H_{24}O_2H^+$	Terpenes	Dihydrocitronellyl acetate
205.1878	205.200	$C_{12}H_{23}O_2H^+$	Terpenes	Humulene

Fragmentation Pattern Measurement

To improve the confidence in m/z used to monitor terpenes, the fragmentation patterns of pure standards were also measured (Table 3). Terpenoid standards; linalool, geraniol, α-terpineol, citral, citronellal, citronellal acetate, limonene, β-pinene, nerol, geranyl acetate, dihydrocitronellyl acetate, dihydrolinalool, myrcene, and caryophyllene, were diluted to a final concentration of 5 ppm through serial dilutions. These diluted standards were then analyzed with PTR-ToF-MS to obtain their fragmentation patterns. Preliminary experiments determined that the headspace concentrations were suitable and not below the detection limit. Compound identification was then carried out by comparing spectral data with fragmentation data. However, it is important to mention the identification from PTR-ToF-MS remained tentative as isomeric product ions (both molecular ions and fragments) of different compounds can overlap at a given m/z.

Table 3. Terpenes, their molecular weight (MW), and formular and relative abundances of their different fragments determined by PTR-ToF-MS.

	MW (g/mol)	m/z	81	83	93	95	135	137	139	155	157	199
Geraniol	154.25	$C_{10}H_{18}O$	82.03		100	51.25		46.06		0.27		
Citronellol	156.27	$C_{10}H_{20}O$	78.52		100	64.75		26.60			20.32	
Geranyl acetate	196.29	$C_{12}H_{22}O_2$	100			10.45	1.6	42.53				
Citronellyl acetate	198.30	$C_{12}H_{22}O_2$	40.47	100		11.10			41.32			25.51

2.5. Data Analysis

Table 1 provides an overview of the samples measured with SPME-GC/MS and PTR-ToF-MS. Multivariate statistical analysis was carried out using R 3.2.0 (R Foundation for Statistical Computing, Vienna, Austria) internal statistical functions and external packages, specifically: ggplot2 and ANOVA. Principal component analysis (PCA) was carried out using the R package "mixomics" [39] on the log-transformed and mean-centred data. A two-way ANOVA (yeast strain and time, $p < 0.001$) was used to determine the mass peaks with significant differences between yeast strains. When a monoisotopic mass peak was saturated, its isotopologue was considered a substitute ion.

3. Results and Discussion

3.1. SPME-GC/MS Results

The VOCs detected during the fermentation by the different yeasts with SPME-GC/MS are presented in Table 4. *S. cerevisiae* strains SafAle US-05 and SafAle WB-06 were selected to be measured using PTR-ToF-MS because previous experiments (data not shown) indicated that they were the most different in terms of their VOC composition. A preliminary data exploration was made using a principal component analysis (PCA), each point representing a distinct measurement (Figure 2). The first two principal components accounted for 88.44% of the total variability. Time-dependent evolutions are observed with different colours, representing the time points from day 1 to day 5. The first time point for all samples was similar, irrespective of yeast strain or compound, and clustered together close to the top left quadrant. After day 1, different evolutions were evident when comparing the samples with or without geraniol and separating by yeast strain. The loadings plot (Figure 2) shows the contribution of each VOC to the principal components at different time points and illustrates the differences in VOC evolution between yeast strains. On the other hand, the correlation circle plot (Figure 3) identifies which VOCs are most strongly associated with each principal component. In this study, terpenoid compounds such as geraniol (18), geranyl acetate (15), citronellol (16), and citronellyl acetate (13) played a crucial role in differentiating samples with added geraniol from the samples without added geraniol.

The time evolution of the detected terpenes (as the area under the curve) is displayed in Figure 4. In the samples with geraniol spiked, regardless of the yeast strain, the peak area of geraniol decreased over the first two days of fermentation and remained constant. This initial loss could result from several factors, including the removal of the compound from the solution due to CO_2 production by the yeast (stripping) during fermentation, as well as loss during sample measurement (purging). The ability of CO_2 to "blow off" the linalool during fermentation was investigated by Ferreira et al. (1996), who observed a reduction of 7.5% after 24 h [40].

When geraniol was not added, no terpenoids were detected. The terpenoids, geraniol, geranyl acetate, citronellol and citronellyl acetate were only detected in samples containing yeast to which geraniol had been added, with the concentration varying greatly depending on the yeast strain. The peak area of geranyl acetate on the second day of fermentation was more than three times higher with SafAle WB-06 than in the samples fermented with SafAle

US-05. Interestingly, geranyl acetate concentration in SafAle WB-06 decreased dramatically over time but remained relatively constant for SafAle US-05.

For the first two days of fermentation, citronellyl acetate was not detected in samples fermented with SafAle US-05. It then increased and plateaued until the final measurement on day 5. The formation of citronellyl acetate with SafAle WB-06 was comparable to the pattern for geranyl acetate; a dramatic increase followed by a decline. At the end of fermentation, the mean peak area of citronellyl acetate was similar for both yeast strains (SafAle US-05: $4.49 \times 10^6 \pm 5.53 \times 10^5$ and SafAle WB-06: $4.39 \times 10^6 \pm 9.80 \times 10^5$). The production of citronellol by SafAle US-05 and SafAle WB-06 followed a similar pattern, with similar abundance at each time point. Studies using GC/MS generally only provided a snapshot of the volatile profile of beer at a single time point, often at the end of fermentation. As a result, this approach has previously missed important dynamic changes in volatile production that occurred over the course of fermentation. Taking dynamic measurements throughout the fermentation provides a more comprehensive understanding of yeast metabolism and strain-dependent differences.

Table 4. Compounds detected at the end of fermentation with SPME-GC/MS.

Number	Compound	Formula	CAS
1	Ethyl Acetate	$C_4H_8O_2$	141-78-6
2	Ethanol	C_2H_6O	200-578-6
3	Ethyl propanoate	$C_5H_{10}O_2$	105-37-3
4	Ethyl butanoate	$C_6H_{12}O_2$	105-54-4
5	Isobutyl alcohol	$C_4H_{10}O$	78-83-1
6	Isoamyl acetate	$C_7H_{14}O_2$	123-92-2
7	Isoamyl alcohol	$C_5H_{12}O$	123-51-3
8	Ethyl hexanoate	$C_8H_{16}O_2$	123-66-0
9	Ethyl octanoate	$C_{10}H_{20}O_2$	106-32-1
10	Acetic acid	CH_3COOH	64-19-7
11	Ethyl decanoate	$C_{12}H_{24}O_2$	110-38-3
12	Isoamyl octanoate	$C_{13}H_{26}O_2$	2035-99-6
13	Citronellyl acetate	$C_{12}H_{22}O_2$	150-84-5
14	Ethyl 9-decenoate	$C_{12}H_{22}O_2$	67233-91-4
15	Geranyl acetate	$C_{12}H_{20}O_2$	105-87-3
16	Citronellol	$C_{10}H_{20}O$	106-22-9
17	Ethyl dodecanoate	$C_{14}H_{28}O_2$	106-33-2
18	Geraniol	$C_{10}H18O$	106-24-1
19	Phenylethyl alcohol	$C_8H_{10}O$	60-12-8
20	Octanoic acid	$C_8H_{16}O_2$	124-07-2

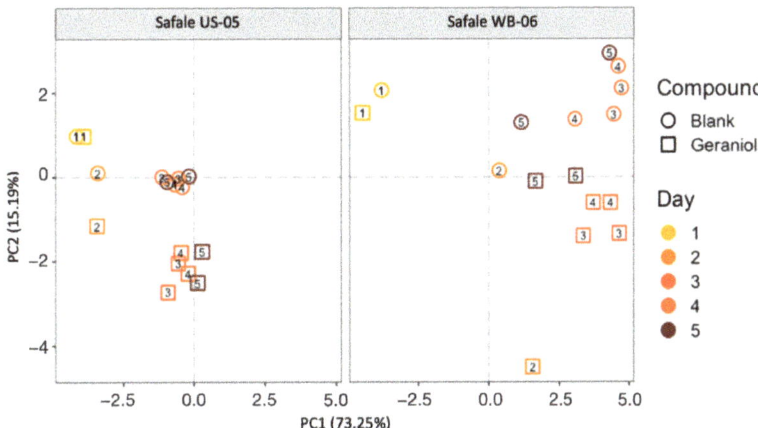

Figure 2. Score plot of the first two principal component analyses (PCA) of VOC produced during fermentation (5 days) by commercially available yeast: *S. cerevisiae* strain SafAle US-05 and *S. cerevisiae var Diastaticus* strain SafAle WB-06 (n = 2). Different colours represent the different days of fermentation.

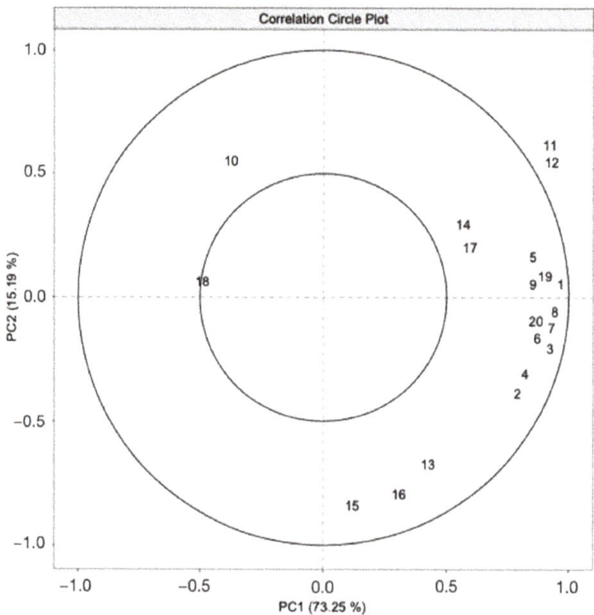

Figure 3. Correlation circle plots from the PCA applied to GC/MS data. Correlation circle plots display the correlation structure between compounds produced throughout fermentation in the space spanned by PC1 and PC2. The numbers represent the name of the compound displayed in Table 4.

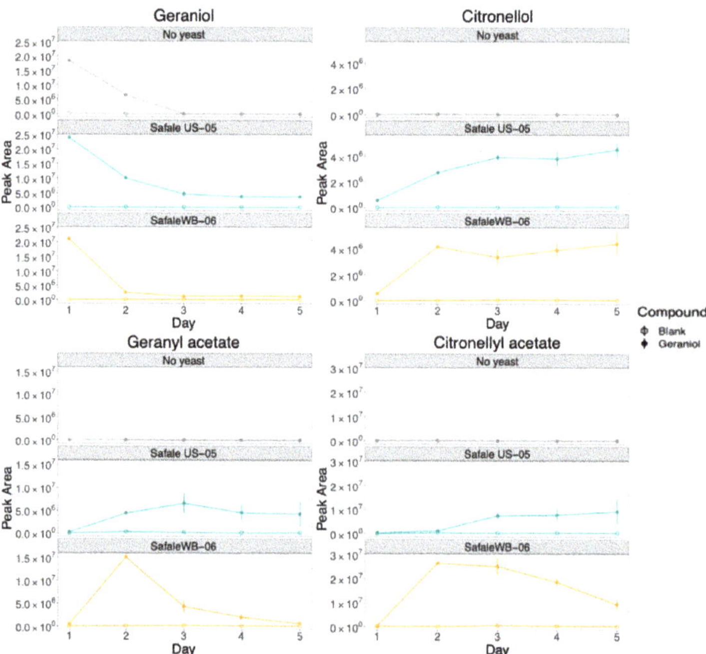

Figure 4. Time evolution of VOCs determined with SPME-GC/MS GC–MS. Quantification is given in the peak area detected under the curve, which can be quantitatively compared between the different measurements. Data presented as the mean peak area ± standard deviation of two replicates.

Steyer et al. (2013) [14] investigated terpene production between two yeast strains, *S. cerevisiae* strain S288c and a haploid strain 59a derived from a wine strain EC1118. Each yeast was added to a synthetic must medium (MS300) with geraniol (1 mg/L). Stir bar sorptive extraction (SBSE) and GC–MS analysis showed a rapid disappearance of geraniol from the medium, followed by the appearance of citronellol, nerol and linalool, and finally, geranyl- and citronellyl acetate were synthesized by both yeast stains. Neither nerol nor linalool was observed in the current study. Commercial yeast strains have been developed to meet the specific requirements of brewers with regard to stress resistance, brewing performance, enzyme release and the profile of aromantic compounds produced. Comparing the results from the current study highlights the impact of terpenoid addition and yeast on the compounds produced. Even yeast strains classified as the same species often show a high level of genetic divergence [41]. This genetic variability affects metabolism, and this information is not generally provided to brewers even though it could have a large impact on the VOC profile and flavour of the beer.

3.2. PTR-ToF-MS Results

S. cerevisiae strains SafAle US-05 and SafAle WB-06 and *S. pastorianus* strains SafLager W-34/70 and SafLager S-23 were selected for measurement with PTR-ToF-MS. Of the measured mass range between 15–245 m/z, 345 peaks were observed. After peak extraction and filtering, data calibration and filtration (eliminating isotopologues, water and ethanol clusters), 47 peaks were assigned to a sum formula and tentatively assigned to one or more compounds based on GC/MS identification and literature (Table 4). These tentatively identified compounds belonged to various chemical classes, many derived from yeast metabolism. PTR-ToF-MS measurement of four individual terpenoids produced fragment ions of masses 81 and 95 as non-isotopic ions (only ^{12}C and ^{1}H, not ^{13}C or ^{2}H) (Table 3). Mass 81 and 95 have previously been reported as terpene fragments [42–44]. Holzinger et al. (2000) proposed calculating the total monoterpene concentration from the signal of masses 67, 81, 95, 137 and 156 [42].

Development of Volatiles during Fermentation

The emission of ethanol and CO_2 during fermentation is directly associated with yeast activity as carbohydrates are converted into CO_2, ethanol, and hundreds of other secondary metabolites. Monitoring the evolution of ethanol (m/z 47.049) and carbon dioxide (m/z 44.999) was easily achieved as their protonated molecular ions are the predominant peaks [45]. No significant difference ($p < 0.001$) in the concentration of CO_2 between yeast strains was observed during fermentation (Figure 5). In contrast, the ethanol concentration in the samples produced by yeast strain WB-06 was significantly higher in the second, third and fourth measurement. There was no significant difference in the concentration of ethanol between yeast strains for the remainder of fermentation.

Secondary metabolites generated by yeast at the same time as CO_2 and ethanol are formed, which can influence the aroma and taste of beer. Variation in the metabolites across different yeast strains is what allows yeast to impart characteristic flavours to beer [46]. The selected 11 peaks were the protonated molecular ions of each terpenoid and their fragments identified from the SPME-GC/MS data: geraniol (m/z 155.143, 137.132, 95.089, 93.952, 81.073), geraniol acetate (m/z 135.109), citronellol (m/z 157.158), citronellyl acetate (m/z 199.169, 139.141, 83.084) and a fragment which is used to demonstrate total terpene concentration (m/z 67.056) as mentioned by Holzinger et al. (2000). The identification of the compound by m/z was done using standards, literature (if available) and by comparison to SPME-GC/MS data. Figure 6 displays the concentration (ppbV) of geranyl acetate (m/z 135.109), geraniol (m/z 155.143), citronellol (m/z 157.158) and citronellyl acetate (m/z 199.169) and measured by PTR-ToF-MS for four commercial yeast strains. The remaining 7 peaks (m/z 139.141, 137.132, 95.089, 93.952, 83.084 and 81.073) are displayed in the supplementary material.

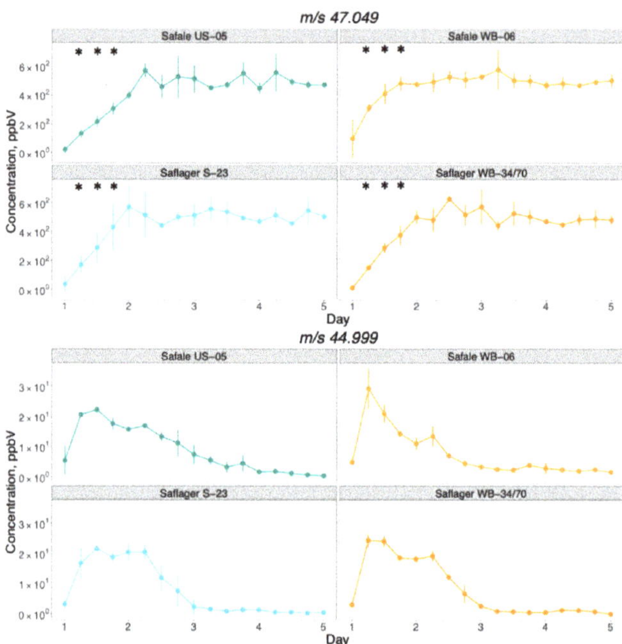

Figure 5. Mean concentration (ppbV) of ethanol (*m/z 47.049*) and carbon dioxide (*m/z 44.999*) during fermentation by commercially available yeast: *Saccharomyces cerevisiae* strain SafAle US-05, *Saccharomyces cerevisiae var Diastaticus* strain SafAle WB-06 and *Saccharomyces pastorianus* strains SafLager S-23 and SafLager W-34/70. Data presented as mean ± standard deviation of seven independent measurements. Asterisk (*) reflects statistically significant differences between strains with a *p*-value < 0.001, one-way ANOVA.

The signal evolution pattern for *m/z 157.158* is associated with citronellol as confirmed with pure standards and literature [47]. The gradual increase in citronellol over time was comparable to the results of SPME-GC/MS (Figure 4). The rapid analysis of samples with PTR-ToF-MS enabled the inclusion of additional yeast strains, such as *S. pastorianus* strains SafLager S-23 and SafLager WB-34/70, in the analysis. This increased sample throughput enabled the identification of species-specific differences that had not previously been observed. A significant increase in citronellol around the last two days of fermentation was identified, with the final concentration being highest in samples produced by *S. pastorianus* yeast. Species-dependent differences in the final concentration of citronellol from geraniol have been previously reported by Haslbeck et al. (2018). Unhopped wort with 70 µg/L of geraniol produced between 0.7–0.9 µg/L and 0.4 ug/L of citronellol by *S. cerevisiae* and *S. pastorianus*, respectively [48]. The old yellow enzyme (OYE) has been postulated as the enzyme responsible for this reduction of geraniol into citronellol [14]. The authors demonstrated this by fermenting using strains with the OYE2 gene either overexpressed or deleted. Deletion of the gene resulted in considerably less citronellol, and overexpression of the gene resulted in considerably more citronellol. The reduction of geraniol to citronellol may change the floral character of the beer to a more citrus-like character [4].

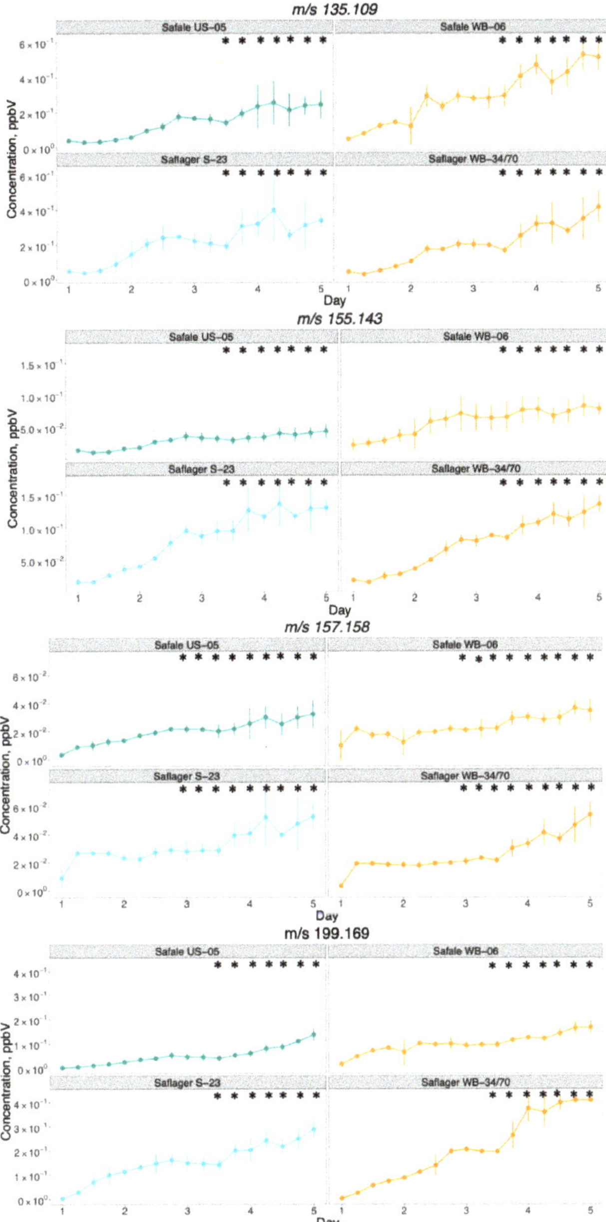

Figure 6. Mean concentration (ppbV) of geranyl acetate (*m/z 135.109*), geraniol (*m/z 155.143*), citronellol (*m/z 157.158*) and citronellyl acetate (*m/z 199.169*) during fermentation by commercially available yeast: *Saccharomyces cerevisiae* strain SafAle US-05, *Saccharomyces cerevisiae var Diastaticus* strain SafAle WB-06 and *Saccharomyces pastorianus* strains SafLager S-23 and SafLager W-34/70. Data presented as mean ± standard deviation of seven independent measurements. Asterisk (*) reflects statistically significant differences between strains with a *p*-value < 0.001, one-way ANOVA.

The unique signals *of m/z 83.084, m/z 139.141* and *m/z 199.169* were associated with citronellyl acetate. Monitoring the protonated molecule (*m/z 199.169*) showed increased concentration throughout fermentation by all yeast strains. Consistent with the SPMS GC/MS results, an initial lag in the concentration of citronellyl acetate by SafAle US-05 was observed. All four yeast strains produced similar concentrations throughout fermentation until the end of day four, where a significantly higher concentration was produced by both *S. cerevisiae* strains. The unique signal of *m/z 135.109* was associated with geranyl acetate, which was determined using a pure reference standard. A significant difference in the concentration began halfway through day 3 and continued until the end of fermentation, with the highest concentration from samples produced by SafAle WB-06. The lowest concentration was from samples produced by SafAle US-05. SafLager S-23 and W 34/70 had a similar formation pattern throughout fermentation. Alcohol acetyltransferase (ATF) is the enzyme involved in the esterification of geraniol to citronellyl- and geranyl acetate [49,50]. Steyer et al. (2013) used a BY4741 strain with ATF1, or ATF2 genes deleted derived from *S. cerevisiae* S288C to display the involvement of ATF1 and ATF2 in the formation of terpenyl acetates through overexpression and deletion. When ATF1 and ATF2 were deleted, there was a drastic reduction in the formation of geranyl- and citronellyl acetate from geraniol. When the same gene was overexpressed, an increase in formation was observed. In the current study, the level of OYE and ATF expression in the different strains is likely to explain the differences observed. The loss of geraniol by biotransformation is mainly due to its reduction to citronellol catalyzed by OYE and acetylation to citronellyl acetate and geranyl acetate catalyzed by ATF. Brewers could utilize this knowledge to choose yeast strains that express these genes at desired levels, thereby achieving the desired concentrations of citronellol, citronellyl acetate and geranyl acetate in their beer.

3.3. Comparison between PTR-ToF-MS and GC/MS to Monitor the Formation of Compounds throughout Beer Fermentation

PTR-ToF-MS can measure more time points compared to GC/MS (four times a day vs. once a day for each micro-fermentation in our case). This is important, especially at the beginning of the fermentation, when many changes in VOCs are occurring, as evident in the initial 24 h for *m/z 157.158* (citronellol), *m/z 199.169* (citronellyl acetate) and *m/z 135.109* (geranyl acetate). An initial rapid increase in the concentration of citronellol was observed after 6 h. The concentration of citronellyl acetate started to increase, and finally, after 18 h, the concentration of geranyl acetate increased. A summary of this formation is shown in Figure 7. The delay of acetylation is likely due to the repression of ATF gene expression when oxygen (dissolved in the wort) is present [51]. The concentration of oxygen in wort gradually decreases during the first hours of fermentation. By 210 min, complete oxygen depletion is typically observed [52].

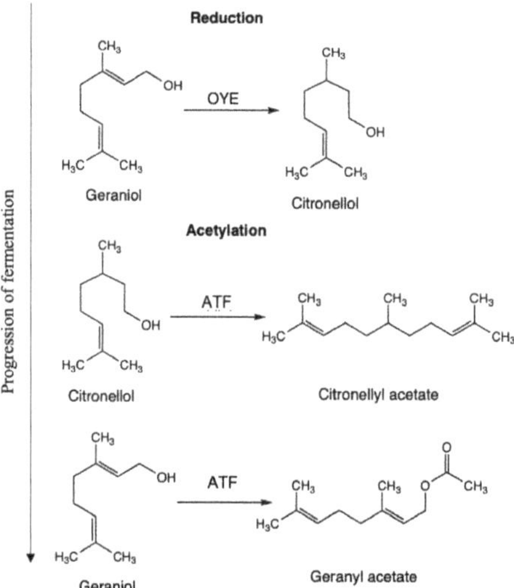

Figure 7. Proposed formation of citronellol, citronellyl acetate and geranyl acetate during fermentation. Enzymes (OYE and ATF) are based on literature [14,50,51].

4. Conclusions

The fate of geraniol during beer fermentation by *S. cerevisiae* or *S. pastorianus* was investigated as an example of biotransformation of hop flavour compounds occurring during beer fermentation. The reduction in the concentration of geraniol was closely followed by the detection of citronellol, citronellyl acetate and geranyl acetate. The ability of yeast to transform geraniol into these compounds has previously been identified. However, this is the first study to monitor the changes in real-time by direct injection mass spectrometry. The use of PTR-ToF-MS enabled differences between yeast strains to be identified during fermentation with high temporal resolution. The implementation of two separate techniques allowed for the identification of compounds (GC/MS), online monitoring and quantitative determination (PTR-MS). The results show that the concentrations of terpenoids detected throughout fermentation were impacted by the yeast species and strain. In general, a higher concentration of the detected terpenoids was produced by *S. pastorianus*. An example of strain-dependent differences was shown on the initial day of fermentation, where the increase in the concentration of citronellyl acetate (m/z *199.169*) was slower in *S. cerevisiae* SafAle US-05 and WB-06 when compared to *S. pastorianus* SafLager S-23 and W-34/74. Biotransformation terpenoids increased in concentration at a similar rate for *S. pastorianus* strains, whereas *S. cerevisiae* strains SafAle- US-05 and WB-06 differed greatly. The difference between yeasts may be due to the diverse level of OYE and ATF expression, impacting the concentration of citronellol, citronellyl acetate and geranyl acetate, respectively. The concentration of some of the VOCs detected in the micro-fermentations (3 mL) may have different magnitudes when compared to industrial-sized fermentation. Still, the yeast differences and proposed pathways are expected to be comparable. Using this developed method to investigate terpenoids important for beer aroma is essential and might be used to investigate the effect of biological and technological parameters. To expand the current understanding of aroma generation during fermentation, analyzing more terpenoids with a similar experimental design will provide more valuable information on yeast production and transformation reactions. There is also a need to analyze different yeast strains, which would give the brewers additional information to manage and change the aroma of the beer to meet consumers' preferences.

Supplementary Materials: The following supporting information can be downloaded at: https://www.mdpi.com/article/10.3390/fermentation9030294/s1, Figure S1: Mean concentration (ppbV) of m/z *139.141*, m/z *137.132*, m/z *95.089*, m/z *93.952*, m/z *83.084* and m/z *81.073* during fermentation by commercially available yeast: *Saccharomyces cerevisiae* strain SafAle US-05, *Saccharomyces cerevisiae var Diastaticus* strain SafAle WB-06 and *Saccharomyces pastorianus* strains SafLager S-23 and SafLager W-34/70. Data presented as mean ± standard deviation of seven independent measurements.

Author Contributions: Conceptualization G.T.E., P.S., P.B. and F.B.; Methodology, R.R., G.T.E., P.S., P.B., I.K. and F.B.; Investigation, R.R.; Formal analysis, R.R., I.K. and E.B.; Data Curation, R.R., I.K. and E.B.; Writing—original draft, R.R.; Writing—review and editing, F.B., G.T.E., P.S., P.B. and I.K.; Supervision, G.T.E., P.S., P.B. and F.B. All authors have read and agreed to the published version of the manuscript.

Funding: This research was funded by Provincia Autonoma di Trento (ADP FEM 2022).

Institutional Review Board Statement: Not applicable.

Informed Consent Statement: Not applicable.

Data Availability Statement: The data presented in this study are available on request from the corresponding author.

Acknowledgments: Rebecca Roberts is grateful to receive funding from the University of Otago (Doctorial Scholarship) and Fondazione Edmund Mach.

Conflicts of Interest: The authors declare no conflict of interest.

References

1. Tisserand, R.; Young, R. (Eds.) 2-Essential oil composition. In *Essential Oil Safety*, 2nd ed.; Churchill Livingstone: St. Louis, MO, USA, 2014; pp. 5–22.
2. Buttery, R.; Ling, L. The chemical composition of the volatile oil of hops. *Brew. Dig.* **1966**, *41*, 71–77.
3. Eyres, G.; Dufour, J.-P. 22-Hop Essential Oil: Analysis, Chemical Composition and Odor Characteristics. In *Beer in Health and Disease Prevention*; Preedy, V.R., Ed.; Academic Press: San Diego, CA, USA, 2009; pp. 239–254.
4. Takoi, K.; Itoga, Y.; Koie, K.; Kosugi, T.; Shimase, M.; Katayama, Y.; Nakayama, Y.; Watari, J. The Contribution of Geraniol Metabolism to the Citrus Flavour of Beer: Synergy of Geraniol and β-Citronellol Under Coexistence with Excess Linalool. *J. Inst. Brew.* **2010**, *116*, 251–260. [CrossRef]
5. King, A.; Dickinson, R. Biotransformation of monoterpene alcohols by Saccharomyces cerevisiae, Torulaspora delbrueckii and Kluyveromyces lactis. *Yeast* **2000**, *16*, 499–506. [CrossRef]
6. King, A.; Dickinson, R. Biotransformation of hop aroma terpenoids by ale and lager yeasts. *FEMS Yeast Res.* **2003**, *3*, 53–62. [CrossRef]
7. Richter, T.M.; Silcock, P.; Algarra, A.; Eyres, G.T.; Capozzi, V.; Bremer, P.J.; Biasioli, F. Evaluation of PTR-ToF-MS as a tool to track the behavior of hop-derived compounds during the fermentation of beer. *Food Res. Int.* **2018**, *111*, 582–589. [CrossRef]
8. Carrau, F.M.; Medina, K.; Boido, E.; Farina, L.; Gaggero, C.; Dellacassa, E.; Versini, G.; Henschke, P.A. De novo synthesis of monoterpenes by Saccharomyces cerevisiae wine yeasts. *FEMS Microbiol. Lett.* **2005**, *243*, 107–115. [CrossRef]
9. Gamero, A.; Manzanares, P.; Querol, A.; Belloch, C. Monoterpene alcohols release and bioconversion by Saccharomyces species and hybrids. *Int. J. Food Microbiol.* **2011**, *145*, 92–97. [CrossRef]
10. Ugliano, M.; Genovese, A.; Moio, L. Hydrolysis of wine aroma precursors during malolactic fermentation with four commercial starter cultures of Oenococcus oeni. *J. Agric. Food Chem.* **2003**, *51*, 5073–5078. [CrossRef]
11. Takoi, K.; Itoga, Y.; Takayanagi, J.; Kosugi, T.; Shioi, T.; Nakamura, T.; Watari, J. Screening of Geraniol-Rich Flavor Hop and Interesting Behavior of β-Citronellol during Fermentation under Various Hop-Addition Timings. *J. Am. Soc. Brew. Chem.* **2014**, *72*, 22–29. [CrossRef]
12. Sharp, D.; Qian, Y.; Shellhammer, G.; Shellhammer, T.H. Contributions of Select Hopping Regimes to the Terpenoid Content and Hop Aroma Profile of Ale and Lager Beers. *J. Am. Soc. Brew. Chem.* **2017**, *75*, 93–100. [CrossRef]
13. Sharp, D.C.; Vollmer, D.M.; Qian, Y.; Shellhammer, T.H. Examination of Glycoside Hydrolysis Methods for the Determination of Terpenyl Glycoside Contents of Different Hop Cultivars. *J. Am. Soc. Brew. Chem.* **2017**, *75*, 101–108. [CrossRef]
14. Steyer, D.; Erny, C.; Claudel, P.; Riveill, G.; Karst, F.; Legras, J.-L. Genetic analysis of geraniol metabolism during fermentation. *Food Microbiol.* **2013**, *33*, 228–234. [CrossRef] [PubMed]
15. Richter, T.; Algarra Alarcon, A.; Silcock, P.; Eyres, G.; Bremer, P.; Capozzi, V.; Biasioli, F. Tracking of Hop-Derived Compounds in Beer During Fermentation with PTR-TOF-MS. In Proceedings of the 15th Weurman Flavour Research Symposium, Graz, Austria, 18–22 September 2017; Technische Universität: Graz, Austria, 2017; pp. 171–174.
16. Berbegal, C.; Khomenko, I.; Russo, P.; Spano, G.; Fragasso, M.; Biasioli, F.; Capozzi, V. PTR-ToF-MS for the Online Monitoring of Alcoholic Fermentation in Wine: Assessment of VOCs Variability Associated with Different Combinations of Saccharomyces/Non-Saccharomyces as a Case-Study. *Fermentation* **2020**, *6*, 55. [CrossRef]
17. Noma, Y.; Asakawa, Y. Biotransformation of monoterpenoids by microorganisms, insects, and mammals. In *Handbook of Essential Oils*; CRC Press: Boca Raton, FL, USA, 2020; pp. 613–767.
18. Demyttenaere, J.C.R.; del Carmen Herrera, M.; De Kimpe, N. Biotransformation of geraniol, nerol and citral by sporulated surface cultures of Aspergillus niger and *Penicillium* sp. *Phytochemistry* **2000**, *55*, 363–373. [CrossRef] [PubMed]
19. Holt, S.; Miks, M.H.; de Carvalho, B.T.; Foulquié-Moreno, M.R.; Thevelein, J.M. The molecular biology of fruity and floral aromas in beer and other alcoholic beverages. *FEMS Microbiol. Rev.* **2018**, *43*, 193–222. [CrossRef] [PubMed]
20. Sharp, D.C.; Steensels, J.; Shellhammer, T.H. The effect of hopping regime, cultivar and β-glucosidase activity on monoterpene alcohol concentrations in wort and beer. *J. Inst. Brew.* **2017**, *123*, 185–191. [CrossRef]
21. Sharp, D.C. Factors that Influence the Aroma and Monoterpene Alcohol Profile of Hopped Beer. Ph.D. Thesis, Oregon State University, Corvallis, OR, USA, 2016.
22. Slaghenaufi, D.; Indorato, C.; Troiano, E.; Luzzini, G.; Felis, G.E.; Ugliano, M. Fate of Grape-Derived Terpenoids in Model Systems Containing Active Yeast Cells. *J. Agric. Food Chem.* **2020**, *68*, 13294–13301. [CrossRef]
23. da Silva, G.C.; da Silva, A.A.S.; da Silva, L.S.N.; Godoy, R.L.d.O.; Nogueira, L.C.; Quitério, S.L.; Raices, R.S.L. Method development by GC–ECD and HS-SPME–GC–MS for beer volatile analysis. *Food Chem.* **2015**, *167*, 71–77. [CrossRef]
24. Jeleń, H.H.; Majcher, M.; Dziadas, M. Microextraction techniques in the analysis of food flavor compounds: A review. *Anal. Chim. Acta* **2012**, *738*, 13–26. [CrossRef]
25. Cappellin, L.; Karl, T.; Probst, M.; Ismailova, O.; Winkler, P.M.; Soukoulis, C.; Aprea, E.; Märk, T.D.; Gasperi, F.; Biasioli, F. On quantitative determination of volatile organic compound concentrations using proton transfer reaction time-of-flight mass spectrometry. *Environ. Sci. Technol.* **2012**, *46*, 2283–2290. [CrossRef]
26. Farneti, B.; Khomenko, I.; Grisenti, M.; Ajelli, M.; Betta, E.; Algarra, A.A.; Cappellin, L.; Aprea, E.; Gasperi, F.; Biasioli, F. Exploring blueberry aroma complexity by chromatographic and direct-injection spectrometric techniques. *Front. Plant Sci.* **2017**, *8*, 617. [CrossRef] [PubMed]

27. Majchrzak, T.; Wojnowski, W.; Lubinska-Szczygeł, M.; Różańska, A.; Namieśnik, J.; Dymerski, T. PTR-MS and GC/MS as complementary techniques for analysis of volatiles: A tutorial review. *Anal. Chim. Acta* **2018**, *1035*, 1–13. [CrossRef] [PubMed]
28. Romano, A.; Fischer, L.; Herbig, J.; Campbell-Sills, H.; Coulon, J.; Lucas, P.; Cappellin, L.; Biasioli, F. Wine analysis by FastGC proton-transfer reaction-time-of-flight-mass spectrometry. *Int. J. Mass Spectrom.* **2014**, *369*, 81–86. [CrossRef]
29. Boscaini, E.; Van Ruth, S.; Biasioli, F.; Gasperi, F.; Märk, T.D. Gas chromatography-olfactometry (GC-O) and proton transfer reaction-mass spectrometry (PTR-MS) analysis of the flavor profile of grana padano, parmigiano reggiano, and grana trentino cheeses. *J. Agric. Food Chem.* **2003**, *51*, 1782–1790. [CrossRef] [PubMed]
30. Blasioli, S.; Biondi, E.; Samudrala, D.; Spinelli, F.; Cellini, A.; Bertaccini, A.; Cristescu, S.M.; Braschi, I. Identification of volatile markers in potato brown rot and ring rot by combined GC/MS and PTR-MS techniques: Study on in vitro and in vivo samples. *J. Agric. Food Chem.* **2014**, *62*, 337–347. [CrossRef]
31. van Ruth, S.M.; Floris, V.; Fayoux, S. Characterisation of the volatile profiles of infant formulas by proton transfer reaction-mass spectrometry and gas chromatography–mass spectrometry. *Food Chem.* **2006**, *98*, 343–350. [CrossRef]
32. Pollien, P.; Lindinger, C.; Yeretzian, C.; Blank, I. Proton transfer reaction mass spectrometry, a tool for on-line monitoring of acrylamide formation in the headspace of Maillard reaction systems and processed food. *Anal. Chem.* **2003**, *75*, 5488–5494. [CrossRef]
33. Marconi, O.; Rossi, S.; Galgano, F.; Sileoni, V.; Perretti, G. Influence of yeast strain, priming solution and temperature on beer bottle conditioning. *J. Sci. Food Agric.* **2016**, *96*, 4106–4115. [CrossRef]
34. Capozzi, V.; Yener, S.; Khomenko, I.; Farneti, B.; Cappellin, L.; Gasperi, F.; Scampicchio, M.; Biasioli, F. PTR-ToF-MS coupled with an automated sampling system and tailored data analysis for food studies: Bioprocess monitoring, screening and nose-space analysis. *JoVE* **2017**, *123*, e54075.
35. Khomenko, I.; Stefanini, I.; Cappellin, L.; Cappelletti, V.; Franceschi, P.; Cavalieri, D.; Märk, T.D.; Biasioli, F. Non-invasive real time monitoring of yeast volatilome by PTR-ToF-MS. *Metabolomics* **2017**, *13*, 118. [CrossRef]
36. Cappellin, L.; Biasioli, F.; Granitto, P.M.; Schuhfried, E.; Soukoulis, C.; Costa, F.; Märk, T.D.; Gasperi, F. On data analysis in PTR-TOF-MS: From raw spectra to data mining. *Sens. Actuators B Chem.* **2011**, *155*, 183–190. [CrossRef]
37. Cappellin, L.; Biasioli, F.; Fabris, A.; Schuhfried, E.; Soukoulis, C.; Märk, T.D.; Gasperi, F. Improved mass accuracy in PTR-TOF-MS: Another step towards better compound identification in PTR-MS. *Int. J. Mass Spectrom.* **2010**, *290*, 60–63. [CrossRef]
38. Lindinger, W.; Hansel, A.; Jordan, A. On-line monitoring of volatile organic compounds at pptv levels by means of proton-transfer-reaction mass spectrometry (PTR-MS) medical applications, food control and environmental research. *Int. J. Mass Spectrom. Ion Process.* **1998**, *173*, 191–241. [CrossRef]
39. Rohart, F.; Gautier, B.; Singh, A.; Lê Cao, K.-A. mixOmics: An R package for 'omics feature selection and multiple data integration. *PLoS Comput. Biol.* **2017**, *13*, e1005752. [CrossRef]
40. Ferreira, I.M.P.L.V.O.; Jorge, K.; Nogueira, L.C.; Silva, F.; Trugo, L.C. Effects of the Combination of Hydrophobic Polypeptides, Iso-α Acids, and Malto-oligosaccharides on Beer Foam Stability. *J. Agric. Food Chem.* **2005**, *53*, 4976–4981. [CrossRef] [PubMed]
41. Borneman, A.R.; Desany, B.A.; Riches, D.; Affourtit, J.P.; Forgan, A.H.; Pretorius, I.S.; Egholm, M.; Chambers, P.J. Whole-genome comparison reveals novel genetic elements that characterize the genome of industrial strains of Saccharomyces cerevisiae. *PLoS Genet.* **2011**, *7*, e1001287. [CrossRef]
42. Holzinger, R.; Sandoval-Soto, L.; Rottenberger, S.; Crutzen, P.; Kesselmeier, J. Emissions of volatile organic compounds from Quercus ilex L. measured by Proton Transfer Reaction Mass Spectrometry under different environmental conditions. *J. Geophys. Res. Atmos.* **2000**, *105*, 20573–20579. [CrossRef]
43. Warneke, C.; Karl, T.; Judmaier, H.; Hansel, A.; Jordan, A.; Lindinger, W.; Crutzen, P.J. Acetone, methanol, and other partially oxidized volatile organic emissions from dead plant matter by abiological processes: Significance for atmospheric HOx chemistry. *Glob. Biogeochem. Cycles* **1999**, *13*, 9–17. [CrossRef]
44. Tani, A.; Hayward, S.; Hewitt, C. Measurement of monoterpenes and related compounds by proton transfer reaction-mass spectrometry (PTR-MS). *Int. J. Mass Spectrom.* **2003**, *223*, 561–578. [CrossRef]
45. Soukoulis, C.; Cappellin, L.; Aprea, E.; Costa, F.; Viola, R.; Märk, T.D.; Gasperi, F.; Biasioli, F. PTR-ToF-MS, a novel, rapid, high sensitivity and non-invasive tool to monitor volatile compound release during fruit post-harvest storage: The case study of apple ripening. *Food Bioprocess Technol.* **2013**, *6*, 2831–2843. [CrossRef]
46. Dzialo, M.C.; Park, R.; Steensels, J.; Lievens, B.; Verstrepen, K.J. Physiology, ecology and industrial applications of aroma formation in yeast. *FEMS Microbiol. Rev.* **2017**, *41*, S95–S128. [CrossRef] [PubMed]
47. Amelynck, C.; Mees, B.; Schoon, N.; Bultinck, P. FA-SIFT study of the reactions of $H_3O+\cdot(H_2O)n$ (n = 0, 1, 2), NO+ and O_2+ with the terpenoid aldehydes citral, citronellal and myrtenal and their alcohol analogues. *Int. J. Mass Spectrom.* **2015**, *379*, 52–59. [CrossRef]
48. Haslbeck, K.; Bub, S.; von Kamp, K.; Michel, M.; Zarnkow, M.; Hutzler, M.; Coelhan, M. The influence of brewing yeast strains on monoterpene alcohols and esters contributing to the citrus flavour of beer. *J. Inst. Brew.* **2018**, *124*, 403–415. [CrossRef]
49. Kimura, M.; Ito, M. Bioconversion of essential oil components of Perilla frutescens by Saccharomyces cerevisiae. *J. Nat. Med.* **2020**, *74*, 189–199. [CrossRef] [PubMed]
50. Verstrepen, K.J.; Van Laere, S.D.M.; Vanderhaegen, B.M.P.; Derdelinckx, G.; Dufour, J.-P.; Pretorius, I.S.; Winderickx, J.; Thevelein, J.M.; Delvaux, F.R. Expression Levels of the Yeast Alcohol Acetyltransferase Genes ATF1, Lg-ATF1, and ATF2 Control the Formation of a Broad Range of Volatile Esters. *Appl. Environ. Microbiol.* **2003**, *69*, 5228–5237. [CrossRef]

51. Mason, A.B.; Dufour, J.P. Alcohol acetyltransferases and the significance of ester synthesis in yeast. *Yeast* **2000**, *16*, 1287–1298. [CrossRef]
52. Kucharczyk, K.; Tuszyński, T. The effect of wort aeration on fermentation, maturation and volatile components of beer produced on an industrial scale. *J. Inst. Brew.* **2017**, *123*, 31–38. [CrossRef]

 *fermentation*

Article

New Online Monitoring Approaches to Describe and Understand the Kinetics of Acetaldehyde Concentration during Wine Alcoholic Fermentation: Access to Production Balances

Charlie Guittin [1,2,*], Faïza Maçna [1], Christian Picou [1], Marc Perez [1], Adeline Barreau [2], Xavier Poitou [2], Jean-Marie Sablayrolles [1], Jean-Roch Mouret [1] and Vincent Farines [1]

[1] SPO, Université de Montpellier, INRAE, Institut Agro, 34060 Montpellier, France
[2] R&D Department, Jas Hennessy & Co., 16100 Cognac, France
*   Correspondence: charlie.guittin@inrae.fr; Tel.: +33-4-9961-2274

Abstract: The compound acetaldehyde has complex synthesis kinetics since it accumulates during the growth phase and is consumed by yeast during the stationary phase, as well as evaporating (low boiling point) throughout the process. One recurrent question about this molecule is: can temperature both increase and decrease the consumption of the molecule by yeast or does it only promote its evaporation? Therefore, the main objective of this study was to describe and analyze the evolution of acetaldehyde and shed light on the effect of temperature, the main parameter that impacts fermentation kinetics and the dynamics of acetaldehyde synthesis. Thanks to new online monitoring approaches, anisothermal temperature management and associated mathematical methods, complete acetaldehyde production balances during fermentation made it possible to dissociate biological consumption from physical evaporation. From a biological point of view, the high fermentation temperatures led to important production of acetaldehyde at the end of the growth phase but also allowed better consumption of the molecule by yeast. Physical evaporation was more important at high temperatures, reinforcing the final decrease in acetaldehyde concentration. Thanks to the use of production balances, it was possible to determine that the decrease in acetaldehyde concentration during the stationary phase was mainly due to yeast consumption, which was explained by the metabolic links found between acetaldehyde and markers of metabolism, such as organic acids.

Keywords: acetaldehyde; online monitoring; alcoholic fermentation; consumption; production balance

Citation: Guittin, C.; Maçna, F.; Picou, C.; Perez, M.; Barreau, A.; Poitou, X.; Sablayrolles, J.-M.; Mouret, J.-R.; Farines, V. New Online Monitoring Approaches to Describe and Understand the Kinetics of Acetaldehyde Concentration during Wine Alcoholic Fermentation: Access to Production Balances. *Fermentation* **2023**, *9*, 299. https://doi.org/10.3390/fermentation9030299

Academic Editor: Niel Van Wyk

Received: 3 March 2023
Revised: 16 March 2023
Accepted: 16 March 2023
Published: 18 March 2023

1. Introduction

Alcoholic fermentation is the central stage of the wine-making process using the yeast *Saccharomyces cerevisiae*. The main reaction in this biological process is the bioconversion of sugars into ethanol and carbon dioxide, as well as many other compounds responsible for the organoleptic profile of the wine. Acetaldehyde, also referred to as ethanal or acetic aldehyde, is a powerful aromatic compound that can be found in many food matrices: apple juice, spirits, beer, cider, wine, cheese, yogurt and butter [1]. In wine, free acetaldehyde can form more or less stable combinations with other molecules to produce combined or bound acetaldehyde; the sum of the free and combined acetaldehyde corresponds to the total acetaldehyde. Typically, in the presence of sulfite, the combination rate is 50–60% at the end of fermentation [2]. The concentration of total acetaldehyde in wines generally varies between 10 and 200 mg/L, with a sensory perception threshold of around 100 to 125 mg/L for free acetaldehyde. At low concentrations, acetaldehyde has a pleasant and fruity aroma, while at high levels, its odor becomes irritating and pungent, which depreciates the organoleptic qualities of wines for consumers. These high concentrations of acetaldehyde give undesired organoleptic properties to wines, such as green apple, freshly cut grass and nutty aromas, which, however, are sometimes sought after; in particular, in wines such as "vins jaunes". The acetaldehyde in wines has microbiological and/or chemical origins. Indeed, during alcoholic fermentation under reducing conditions, acetaldehyde is the most important carbonyl compound produced by yeasts after ethanol. Acetaldehyde accumulation differs according to yeast species and strain, ranging from 0.5 for the least productive ones to more than 700 mg/L for the most productive ones [1]. Moreover, under oxidative conditions, acetaldehyde can also be produced after oenological fermentation through ethanol chemical oxidation when the wine is exposed to air. The presence of the carbonyl group makes acetaldehyde very reactive and sensitive to oxidation phenomena. The main physical properties of acetaldehyde are: a boiling point of 20.1 °C [3], 120 kPa vapor pressure, and a Henry's law constant in water at 298.15 K of about 15 $mol \cdot kg^{-1} \cdot bar^{-1}$ [4]. This molecule is also very polar, with water solubility of 2.568×10^5 mg/L at 298.15 K and log Kow = −0.34.

Acetaldehyde is a key compound of yeast metabolism and an intermediate of glycolysis produced during alcoholic fermentation. Its metabolic pathway thus begins with glycolysis, which generates pyruvate as the final product. Pyruvate can then be converted to acetaldehyde and carbon dioxide by pyruvate decarboxylase (PDC); then, acetaldehyde can be reduced to ethanol through the action of alcohol dehydrogenase (ADH). These steps of alcoholic fermentation take place in the cell cytoplasm. Acetaldehyde has a central role in yeast metabolism as the precursor of many molecules, such as ethanol, acetate, acetoin, α-acetohydroxybutyrate and α-acetolactate. Acetaldehyde is also used to generate the cytoplasmic acetyl-CoA required for lipid biosynthesis. The reduction of acetaldehyde to ethanol is highly dependent on ADH activity in close connection with the cofactor NADH. Compared to its high production, acetaldehyde accumulation in the fermentation medium is negligible, as it is directly transformed into ethanol (the main flux of alcoholic fermentation). However, acetaldehyde accumulation could be one way to orient the metabolism; in particular, by inducing the synthesis of certain by-products (organic acids) while maintaining the redox balance [5]. The dynamics of acetaldehyde production during alcoholic fermentation can be divided into three phases. Early formation of this compound is observed during the lag phase at the beginning of fermentation before any detectable yeast growth [6]. Acetaldehyde accumulation continues throughout the growth phase. Finally, its concentration decreases during the stationary phase until the end of alcoholic fermentation [2,7,8]. Comparison of acetaldehyde kinetics and sugar consumption kinetics shows a possible relationship between the moment when the maximum acetaldehyde concentration is reached and the divergence of glucose and fructose degradation rates [8,9]. Thus, sugar concentration influences acetaldehyde kinetics [10]. These kinetics are also highly dependent on the redox balance of the cell. At the beginning of fermentation, glycerol production allows NADH/NAD+ recycling. This production phase ensures the maintenance of the cell redox balance. Then, in the stationary phase, glycerol mainly plays a role related to coping with osmotic stress, so its synthesis is continuous [11]. Afterwards, as soon as acetaldehyde is present in a sufficient quantity, it is able to play its role of reducer [12]. Thus, through its catabolism, acetaldehyde serves as a terminal electron acceptor in yeast redox balance and has the ability to create energy through glycolysis [1]. Although sugar is the main substrate for acetaldehyde formation, the metabolism of amino acids, such as alanine, also contributes to the formation of this compound [13]. In this case, acetaldehyde is produced through the Ehrlich pathway: alanine is catabolized to 2-oxopropanoate, which can then be decarboxylated to acetaldehyde [14].

Online monitoring of molecules, which involves dozens or hundreds of measurements during fermentation, brings new information of direct interest for the study of metabolism and for a better understanding of the kinetics of the production of compounds [15]. With manual sampling, the dynamics of acetaldehyde synthesis, including the accumulation peak at the end of the growth phase and its synchronization with the fermentation kinetics, would have been difficult results to obtain due to the compiling of precision errors. Moreover, employing this online monitoring tool, a recent study evaluated the physical properties of acetaldehyde during alcoholic fermentation by characterizing the partition coefficient of this molecule, thus making it possible to determine production balances by estimating the share of acetaldehyde losses attributable to evaporation [7]. However, to date, acetaldehyde accumulation, including its metabolic and physicochemical aspects, remains poorly described and demonstrated. Using an online monitoring tool for the production of this volatile molecule during oenological fermentation, the atypical synthesis kinetics of acetaldehyde in the presence of SO_2 have recently been investigated [2]. However, much work remains to be undertaken, particularly on the impact of fermentation parameters on the synthesis and re-consumption of this compound by yeast, since the hypotheses in the literature diverge, notably with respect to the fermentation temperature. Indeed, although otherwise contradictory, studies agree in considering the fermentation temperature a key point in acetaldehyde production [8,16–20]. For Amerine and Ough (1964), the fermentation temperature has little effect on the final acetaldehyde content, while other authors correlate the increase in acetaldehyde content with increasing fermentation temperature [18].

To obtain a general understanding of acetaldehyde dynamics, this work employed a combination of original approaches, including (a) the use of an online acetaldehyde monitoring system in the gas phase during alcoholic fermentation; (b) the implementation of complete production balances, including for physicochemical and metabolic aspects; and, finally, (c) the application of different isothermal and anisothermal temperature profiles in order to understand the impact of this parameter on the synthesis of the molecule using three *Saccharomyces cerevisiae* yeast strains in a natural must.

2. Materials and Methods

2.1. Schematic Overview of the Experimental Program

The experimental process in this study relied on different techniques: (a) the online monitoring of fermentation kinetics; (b) the control of the anisothermal temperature during fermentation; (c) the online monitoring of volatile compounds, such as acetaldehyde; and (d) as a result of this monitoring, the implementation of a complete production balance using mathematical models allowing the dissociation of the biological effects (consumption by yeast) and physical effects (evaporation) on the decrease in acetaldehyde during the stationary phase (Figure 1).

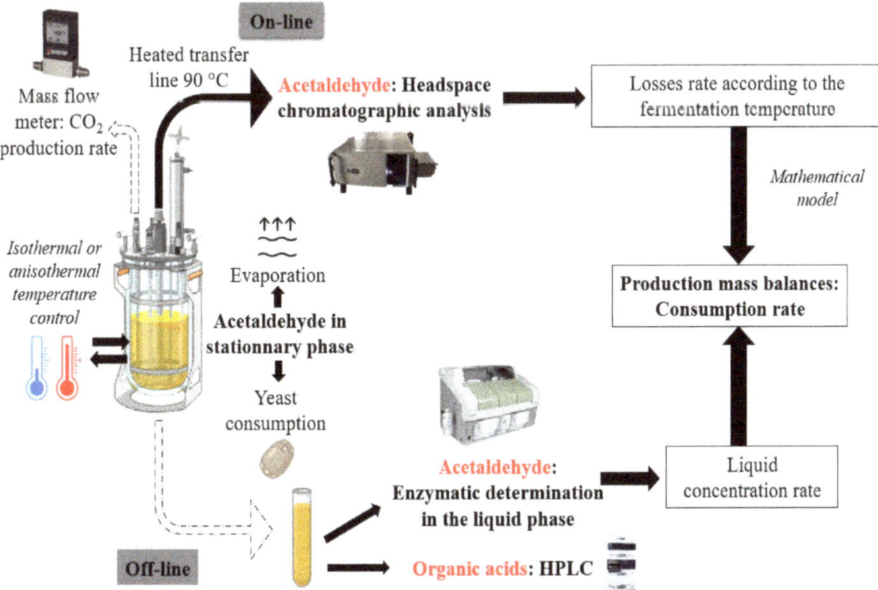

Figure 1. Schematic overview of the experimental design: acetaldehyde production balance during alcoholic fermentation.

2.2. Yeast Strains

Fermentations were carried out with the commercial *Saccharomyces cerevisae* strains Lalvin FC9® (Lallemand SA, Montreal, QC, Canada), Fermivin 7013® and Fermivin SM102® (Erbslöh S.A.S., Servian, France). Fermenters were inoculated with 20 g/hL active dry yeast previously rehydrated for 30 min at 37 °C in a 50 g/L glucose solution (1 g of dry yeast diluted in 10 mL of this solution).

2.3. Natural Must

The must was prepared from the Ugni blanc grape variety (2020) harvested in the Cognac region, France. Grapes were pressed and the must was settled for 24 h at 4 °C in the presence of 3 mL/hL enzymes (MYZYM SPIRIT, Institut Oenologique de Champagne, Epernay, France). A final turbidity of 95 NTU (measured with a 2100 N turbidimeter (Hach®, Lognes, France)) was achieved for the must and the sludge was collected separately. A correlation between sludge concentration and turbidity was previously calculated (R^2 = 0.9924) by adding different amounts of solid particles to final volumes (100 mL) of must:

$$\text{Turbidity} = 37.05 \times [\text{solids particles}] + 47.32 \tag{1}$$

where the concentration of added solid particles ([solid particles]) is in percentage points (v/v) and the turbidity in NTU.

The must characteristics were as follows: 185 g/L total sugar, 111 mg/L assimilable nitrogen (35 mg/L and 76 mg/L mineral and organic nitrogen, respectively) and pH 3.02.

Based on the relationship determined between turbidity and the percentage of solid particles (Equation (1)), 216 mL of sludge was added to a final volume of 1.8 L must to obtain a turbidity of 500 NTU. The corresponding volume of must was previously removed from the fermenter to be substituted with the sludge.

Assimilable nitrogen was adjusted to 200 mg Nass/L (=assimilable nitrogen/L) with a solution of amino acids and NH_4Cl, respecting the proportions of 30% mineral nitrogen (NH_4^+) and 70% organic nitrogen (a mix of amino acids) found in the initial Ugni blanc must. The free amino acid content of the initial Ugni blanc must was determined using cation-exchange chromatography (see Section 2.5). The composition of the amino acid stock solution added was as follows (in g/L): tyrosine, 0.29; tryptophan, 0.22; isoleucine, 0.33; aspartate, 4.33; glutamate, 6.02; arginine, 9.46; leucine, 0.49; threonine, 0.95; glycine, 0.06; glutamine, 3.89; alanine, 2.19; valine, 1.16; methionine, 0.09; phenylalanine, 1.02; serine, 1.66; histidine, 0.38; lysine, 0.15; asparagine, 0.38; and proline, 9.38.

A solution of NH_4Cl (21.45 g/L) was used as an ammonium source. To obtain 200 mg Nass/L in must, 18 mL of amino acid stock solution and 9 mL of NH_4Cl solution were added to 1.8 L of must, respectively.

2.4. Fermentation Conditions

The fermentations were performed in four 2 L jacketed glass autoclavable bioreactors (Applikon®, Delft, the Netherlands). Bioreactors were equipped with direct drive stirring systems (150 rpm) coupled with Rushton impellers (diameter 45 mm). Temperature regulation in each bioreactor was ensured with a Huber cryostat (Offenburg, Deutschland) with coolant liquid circulation in the double jacket. Each cryostat was coupled with a Pt 100 sensor to ensure temperature control. Online measurement of the rate of CO_2 production (dCO_2/dt) was undertaken automatically with a gas thermal mass flow controller (Brooks® Instrument, Hatfield, PA, USA). Different temperature profiles were applied. For the three strains, three isothermal temperature profiles were selected: 18, 24 and 30 °C.

For strains FC9® and SM102®, anisothermal temperature profiles were additionally obtained with a temperature control slope of 0.2 °C/gCO$_2$ released and a trigger level for the temperature change applied just after 20 g of CO_2 was released. This slope corresponds to the average cooling or heating slope observed in wineries' industrial tanks [21]. The first anisothermal profile was an ascending one from 18 °C to 30 °C. The second profile decreased from 30 °C to 18 °C. The third profile was 24–18 °C and, finally, the fourth was 24–30 °C. Each isothermal and anisothermal fermentation operation was performed in duplicate.

2.5. Quantification of Sterols and Fatty Acids in Grape Solids

2.5.1. Dry Matter

A total of 200 mL of must was centrifuged for 10 min at 10,000 rpm to concentrate grape solids. Supernatant was removed and grape solids were washed three times consecutively with a NaCl solution (10 mM) to remove sugars. The final pellet was freeze-dried overnight to recover dry matter (DM).

2.5.2. Lipid Composition

Total lipids from lyophilized grape solids (aliquot of 1 g) were extracted overnight with methanol/chloroform (2:1, v/v), and the solid residue was then extracted over 2 h with methanol/chloroform/water (2:1:0.8, v/v/v). The organic extracts were dried over Na_2SO_4 and concentrated to dryness using a rotatory evaporator. Phytosterols (campesterol, stigmasterol and β-sistosterol) and main fatty acids were determined in the remaining solids with the method described by Grison et al. (2015), as adapted by [22]. Total fatty acid concentration was 28.06 mg/g DM. C18 unsaturated acids represented approximately 57% of the total fatty acid content, and the most abundant saturated fatty acid was palmitic acid (23%). The concentration of phytosterols was 6.93 mg/g DM; i.e., within the limits of the concentrations described for other grape varieties [22]. The main phytosterol was β-sitosterol (84%), while campesterol and stigmasterol accounted for approximately 10.5% of total phytosterols.

Using the lipid composition of the solid particles' dry matter and Equation (1) (information on the amount of solid particles as a function of turbidity), it was calculated that the 500 NTU turbidity corresponded to 26.3 mg/L of sterols and 106.43 mg/L of fatty acids.

2.6. Quantification of Assimilable Nitrogen

Ammonium concentration was determined enzymatically (R-Biopharm AG™, Darmstadt, Germany). The free amino acid content of the must was measured using cation-exchange chromatography with post-column ninhydrin derivatization (Biochrom 30, Biochrom™, Cambridge, UK), as described by [23].

2.7. Determination of Concentration of Metabolites of Central Carbon Metabolism

With every 10 g/L of CO_2 released, samples were analyzed to determine the concentrations of ethanol, glycerol, succinate, α-cetoglutarate and acetate using HPLC (HPLC 1260 Infinity, Agilent™ Technologies, Santa Clara, CA, USA) on a Phenomenex Rezex ROA column (Phenomenex™, Le Pecq, France) at 60 °C. The column was eluted with 0.005 N H_2SO_4 at a flow rate of 0.6 mL/min. Organic acids were analyzed with a UV detector (Agilent™ Technologies, Santa Clara, CA, USA) at 210 nm; the concentrations of the other compounds were quantified with a refractive index detector (Agilent™ Technologies, Santa Clara, CA, USA). Analyses were carried out with the Agilent™ OpenLab CDS 2.x software package (Santa Clara, CA, USA).

2.8. Online Acetaldehyde Monitoring

Our online measurement device consisted of a GC Compact 4.0 (Interscience, Breda, The Netherlands) gas phase chromatography system for carbon and sulfur volatile compounds analysis.

For volatile compounds (acetaldehyde) analysis, headspace gas was sequentially pumped from each 2 L bioreactor for 1 min at a flow rate of 15 mL/min through a dedicated heated transfer line at 90 °C. The gas phase passed through a sampling loop of 50 μL to feed the analysis channel of the sulfur compounds, while carbon compounds were concentrated in a cold trap (Tenax™) at 5 °C.

After this concentration step, carbon compounds were desorbed from Tenax™ at 280 °C for 1 min. The injector temperature was 250 °C with a split flow at 2 mL/min. The GC Compact 4.0 was equipped with a programmable oven containing a MXT-Wax column (30 m × 0.25 mm ID, 0.5 μm film thickness) from Restek™ (Lisses, France). Helium at a constant pressure of 80 kPa was used as the carrier gas. The oven temperature program was: 40 °C for 2 min, increase to 160 °C at a rate of 8 °C/min, hold at 160 °C for 10 s, increase to 220 °C at a rate of 15 °C/min, and hold at 220 °C for 3 min. The Flame Ionization Detector (Thermo Fisher Scientific™, Toulouse, France) was set at 150 °C.

The areas of the acetaldehyde peaks were acquired with Chromeleon™ Chromatography Data System (CDS) software (Thermo Fisher Scientific™, Toulouse, France). The analysis frequency was once every 2 h for each bioreactor.

2.9. Determination of Acetaldehyde

2.9.1. Calibration in the Gas Phase

The GC Compact 4.0 (Interscience, Breda, the Netherlands) gas phase chromatography system was calibrated with an ATIS Adsorbent Tube Injector System 230V (Supelco®, Sigma-Aldrich, Saint-Louis, MO, USA) supplied with CO_2. The liquid calibration standard (1 μL) was injected with a microsyringe through a replaceable septum in the center of the injection glassware, which was heated to 100 °C. Each flash-vaporized sample was pumped in a CO_2 atmosphere into the gas phase chromatography system for analysis. A stock solution of acetaldehyde (CAS no. 75-07-0) (10.03 g/L) was prepared from pure acetaldehyde (Acetaldehyde, PESTANAL®, analytical standard, Sigma, St Quantin Fallavier, France). The different calibration points were determined for this stock solution with the following liquid concentrations: 1.00, 2.00, 4.01, 6.01, 8.02 and 10.03 g/L.

2.9.2. Losses during Fermentation

During fermentation, the losses in the gas $L_{(t)}$ in mg per liter of must were calculated according to the following equation:

$$L_{(t)} = \int_0^t C_{g(t)} \times Q_t \times dt \qquad (2)$$

where $C_{g(t)}$ is the concentration of acetaldehyde in milligrams per liter of CO_2 measured online in the gas phase and Q_t is the CO_2 flow rate at time t expressed in liters of CO_2 per liter of must and per hour.

The rate of losses in the gas phase $R_{L_{(t)}}$ in mg per liter of must and per hour was calculated as follows:

$$R_{L_{(t)}} = \frac{L_{(t)} - L_{(t-i)}}{t - (t-i)} \qquad (3)$$

where t and $(t-i)$ correspond to two successive sampling points in an hour.

2.9.3. Production Mass Balances

Offline liquid sampling was performed with every 10 g/L of CO_2 released until the end of fermentation and total acetaldehyde concentrations were precisely measured using a commercial enzymatic test kit (Ref 984347, ThermoFischer scientific™).

The kinetics of total acetaldehyde in the liquid phase were thus smoothed with a polynomial function of degree two for the accumulation phase in the medium (growth phase) and with logarithmic smoothing for the decrease phase (stationary phase) to obtain the concentration of acetaldehyde in the liquid $C_{l_{(t)}}$ in milligrams per liter of must.

The rate of accumulation in the liquid phase $R_{C_{l(t)}}$ in mg per liter of must and per hour was calculated as follows:

$$R_{C_{l(t)}} = \frac{C_{l(t)} - C_{l(t-i)}}{t - (t-i)} \qquad (4)$$

where t and $(t-i)$ correspond to two successive sampling points in an hour.

The rate of acetaldehyde consumption by yeasts, expressed as $R_{Co(t)}$ in milligrams per liter of must per hour, was calculated as the difference between the absolute value of the acetaldehyde accumulation rate in the liquid, expressed as $Abs\left(R_{C_{l(t)}}\right)$ in milligrams per liter of must per hour, and the absolute value of the rate of losses in the gas phase, expressed as $Abs\left(R_{L_{(t)}}\right)$ in milligrams per liter of must per hour (Equation (5)):

$$R_{Co(t)} = Abs\left(R_{C_{l(t)}}\right) - Abs\left(R_{L_{(t)}}\right) \tag{5}$$

2.10. Statistical Analysis

All the experiments were carried out in biological duplicates. Statistical analyses were performed with R version 3.6.2 (R Development Core Team 2016). The ANOVA was realized using the aov function from the R package agricolae (v1.3.5) and Tukey's test was used for the separation of means.

3. Results and Discussion

3.1. Fermentation Kinetics

In all the fermentation experiments, sugars were almost exhausted; i.e., the residual sugar content was lower than or equal to 2 g/L. As the limiting nutrient, assimilable nitrogen was completely consumed at the end of the growth phase in all experiments. Anisothermal profiles applied during the stationary phase changed the overall shape of the fermentation kinetics and the fermentation parameters, such as fermentation duration. When the temperature was increased during fermentation (from 18 to 30 °C) (Figure 2A), the fermentation rate was maintained at a high value, which resulted in faster fermentation of about 100 h compared to the initial isothermal fermentation performed at 18 °C. The temperature increase of 0.2 °C/g of CO_2 released allowed yeasts to maintain their fermentative activity [24–26]. On the other hand, when the temperature was decreased (30 °C to 18 °C) during the stationary phase, it was observed that yeast cellular activity was slowed down, which resulted in an additional 100 h of fermentation compared to the isothermal control fermentation at 30 °C (Figure 2B). Similar trends were observed as previously with the two other anisothermal temperature profiles, even though fermentation only started at 24 °C (Figure 2C). However, when the profile was raised from 24 to 30 °C, the fermentation time was 40 h shorter instead of the 100 h reduction when fermentation was started at 18 °C. Thus, isothermal and anisothermal profiles highlighted the importance of the effects of temperature on yeast metabolic activity, which resulted in major modifications in fermentation kinetics.

3.2. Synthesis of Primary Carbon Metabolites

3.2.1. Glycerol Production

Glycerol is quantitatively the most important by-product, and its production during fermentation with *Saccharomyces cerevisiae* is associated with ethanol and carbon dioxide production. In wines, levels between 1 and 15 g/L are frequently encountered; higher levels are thought to contribute to a wine's smoothness and viscosity [27] but do not contribute directly to its aroma due to glycerol's non-volatile nature. The two most important functions of glycerol synthesis in yeast are related to redox balancing and the hyperosmotic stress response. Osmotic stress is one of the most common types of stress imposed on yeasts, making it necessary for the cell to survive under these conditions. The role of NADH-consuming glycerol formation is to maintain the cytosolic redox balance, especially under anaerobic conditions, compensating for cellular reactions that produce NADH [28]. Figure 3 shows the kinetics of glycerol production as a function of CO_2 throughout the alcoholic fermentation. Glycerol synthesis was very high during the growth phase and until the middle of the stationary phase. With 50 g/L of CO_2 released and more, a decrease in the slope for glycerol synthesis was observed, indicating a tendency towards a slowing down of production. This result was observed for the three fermentation temperatures: 18, 24 and 30 °C. The application of non-isothermal temperatures did not induce any changes in the synthesis of this compound.

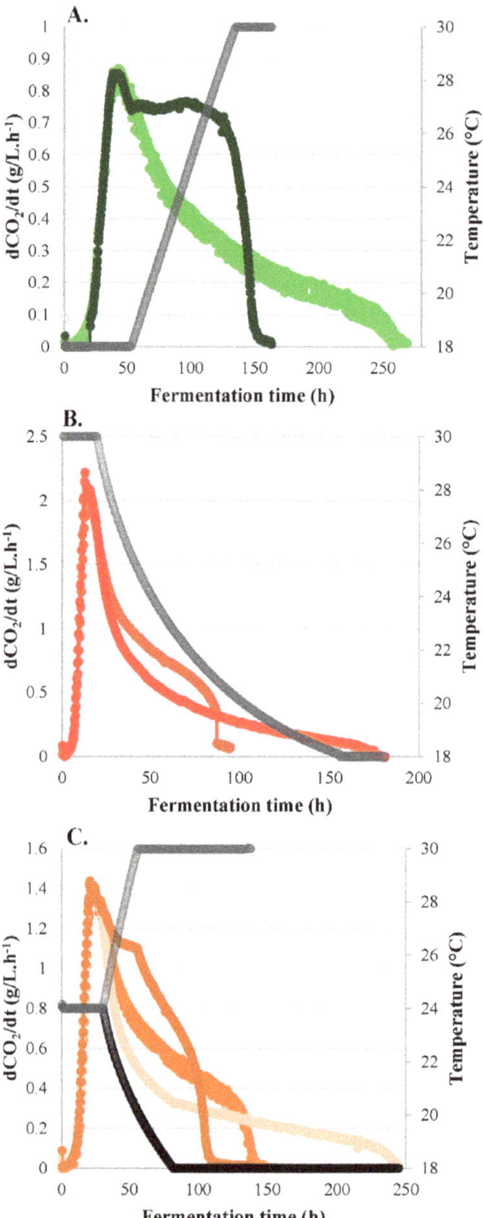

Figure 2. Fermentation kinetics analyzed at (**A**) 18 °C (light green) and 18–30 °C (dark green), (**B**) 30 °C (dark red) and 30–18 °C (light red) and (**C**) 24 °C (orange), 24–18 °C (light orange) and 24–30 °C (dark orange) with anisothermal temperature profiles (gray and black).

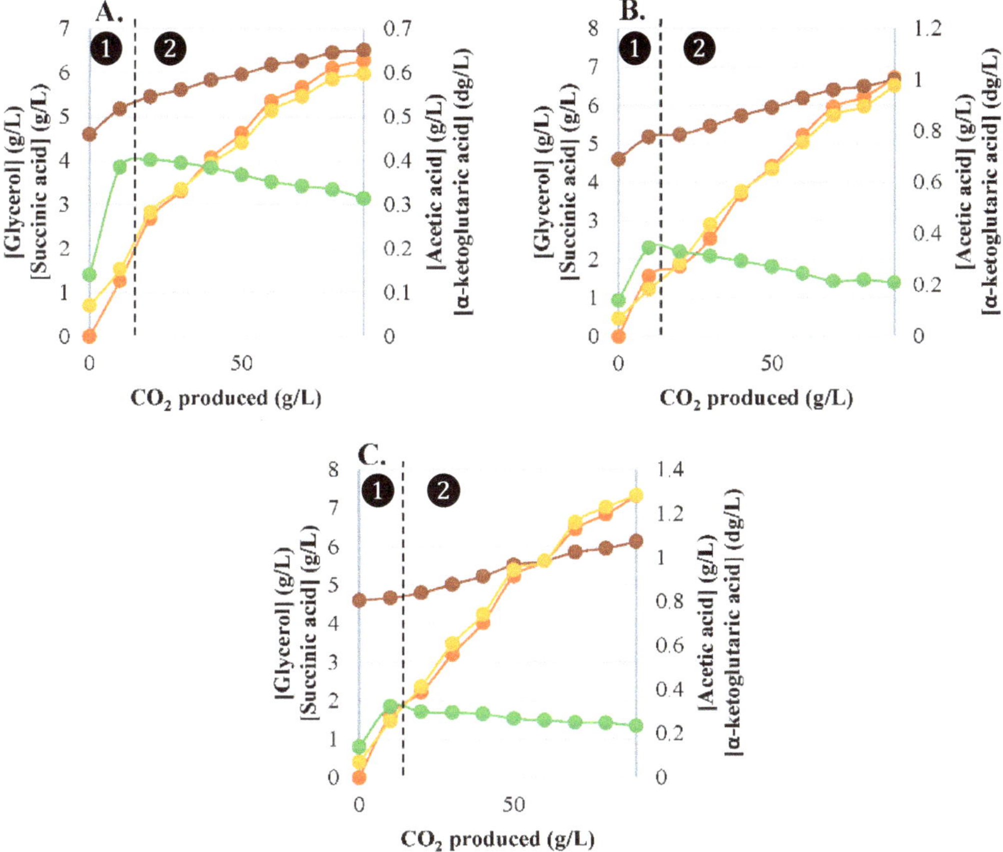

Figure 3. Production kinetics for organic acids at 18 °C (**A**), 24 °C (**B**) and 30 °C (**C**): glycerol (orange), succinate (brown), α-ketoglutarate (yellow) and acetic acid (green). ❶ growth phase and ❷ stationary phase.

Therefore, glycerol maintained the redox balance throughout the fermentation and, especially, up to the middle of the stationary phase (i.e., 60 g of CO_2 released), where a change in the slope was observed. During this phase, glycerol acted purely as a safety valve for the elimination of excess reducing power and not as a protective agent against increases in the osmotic pressure of the medium [29]. After that, one hypothesis could be that acetaldehyde served as an electron acceptor for the novel oxidation of NADH to NAD^+. This hypothesis is detailed in the following sections. Temperature favored the production of final amounts of glycerol at concentrations of 6.3 g/L at 18 °C, 6.8 g/L at 24 °C and 7.2 g/L at 30 °C (ANOVA *p*-value < 0.001), as observed in other studies [30–33]. This temperature-related overproduction could be explained by higher glycerol-3-phosphate deshydrogenase (GPDH) activity at approximately 25 °C [11].

3.2.2. α-Ketoglutaric Acid Production

α-Ketoglutarate (also known as αKG, 2-oxoglutarate and 2-oxoglutaric acid) is a key tricarboxylic acid (TCA) cycle intermediate and an important intermediate in many catabolic and anabolic processes. Under fermentative conditions with Saccharomyces cerevisiae, TCA does not function in a cyclic form, as under aerobic conditions [34], but with an oxidative and a reductive branch with succinate as the final compound [35]. Camarasa et al. (2003) showed that, under oenological conditions, about two thirds of the succinate were synthesized via the reducing branch but that the oxidative pathway was also active during the growth phase and allowed the synthesis of α-ketoglutarate, a key component of nitrogen metabolism (redirected to amino acid synthesis).

It was found in this study that α-ketoglutarate was produced throughout the alcoholic fermentation and in a very similar way as glycerol (Figure 3) for all three isothermal temperatures (no impact on the synthesis at anisothermal temperatures). When glycerol synthesis was plotted against α-ketoglutarate during alcoholic fermentation, very high correlations were obtained at 18 °C (r^2 = 0.9980), 24 °C (r^2 = 0.9916) and 30 °C (r^2 = 0.9961). These correlations clearly showed a metabolic link between the syntheses of these two compounds. Indeed, the α-ketoglutarate synthesis pathway through the TCA cycle was responsible for the reduction of NAD^+ to NADH via isocitrate conversion to α-ketoglutarate in contrast to the glycerol synthesis pathway that oxidized excess NADH to NAD^+.

Regarding the effect of temperature on α-ketoglutarate synthesis, it was observed that higher temperatures induced higher concentrations of this compound. Indeed, final α-ketoglutarate concentrations of 0.59, 0.97 and 1.28 g/L were found for 18, 24 and 30 °C, respectively (ANOVA p-value < 0.001).

Two hypotheses explaining this result could be proposed. The first hypothesis is that, although cell activity was higher at high temperature [24,32,36–38], the higher flux of α-ketoglutarate was not used by the glutamate pathway to form amino acids [39,40]. Indeed, in the present work employing a high lipid concentration (linked to a high turbidity of 500 NTU), the flux within the acetic acid synthesis pathway (precursor of lipid synthesis) was low [31,32]; therefore, so was NADPH availability. The low availability of this co-factor thus resulted in very limited α-ketoglutarate conversion to glutamate [40], leading to α-ketoglutarate accumulation in the extra-cellular medium. The second hypothesis is related to the fact that α-ketoglutarate synthesis was highly correlated with that of glycerol. When the production of the latter was enhanced, the cell needed to reduce high quantities of NAD^+ (generated by the glycerol-3-phosphate pathway) to maintain its intracellular redox balance. Therefore, the reductive pathway of the TCA cycle was activated, resulting in increased accumulation of α-ketoglutarate.

3.2.3. Succinic Acid Production

Succinic acid is a by-product in the alcoholic fermentation of yeasts, with amounts ranging from a few mg up to 2 g/L in all products of fermentation [41]. As the main acid produced by yeast, it significantly influences organoleptic balance by providing acidity and a desirable salty–bitter taste. Succinate synthesis started at the beginning of fermentation [42] and continued during the growth and the stationary phases regardless of the temperature applied (Figure 3).

Succinic acid can be formed from pyruvate or aspartate via the reductive branch of the TCA cycle or from pyruvate via the oxidative branch. In the reducing pathway of the TCA cycle, the precursor of succinate is L-malate. In the present work, it is important to note that, depending on the state of advancement of the fermentation, the correlation behavior of these two compounds diverged. Indeed, during the growth phase and for up to 50% of the fermentation process, a negative correlation existed at the three temperatures (18 °C, r^2 = 0.9913; 24 °C, r^2 = 0.9759; and 30 °C, r^2 = 0.9747): malate (present in the must at 4 g/L) was consumed to synthesize succinate. On average, at 18 and 24 °C, the decrease of 0.3 g/L of malate induced the production of 1.3 g/L for succinate, while at 30 °C, there was a decrease of 0.25 g/L for malate with an increase of 0.9 g/L of succinic acid. During the second part of the fermentation—i.e., during the stationary phase—re-accumulation of malate in the fermentation medium was associated with succinate production, with positive correlations at 18 °C (r^2 = 0.9243), 24 °C (r^2 = 0.8886) and 30 °C (r^2 = 0.9470). Indeed, during this phase, succinate was produced by the TCA reductive pathway but the malate concentration of natural musts could also have stimulated succinate synthesis in the first phase [43].

Finally, establishing correlations between succinate and other intermediates of the TCA cycle makes it possible to better understand the production kinetics of this acid. At first, citrate synthesis (precursor of succinate through the oxidative pathway) was only observed during the growth phase (positive correlation with succinate) and up to the beginning of the stationary phase; i.e., for up to 40% of the fermentation process for 18 °C ($r^2 = 0.8792$), 24 °C ($r^2 = 0.9796$) and 30 °C ($r^2 = 0.9864$). In contrast, α-ketoglutarate, which is also a precursor of succinate, was always correlated with succinate throughout the fermentation for all three temperatures ($r^2 = 0.9848$, 0.9823 and 0.9739 for 18, 24 and 30 °C, respectively). This can be explained by the fact that succinic acid can also be produced from glutamate through glutamate oxidation to α-ketoglutarate followed by the conversion of α-ketoglutarate to succinic acid via this step of the TCA cycle (a reverse pathway from glutamate). All these results confirmed that the TCA reductive branch (reductive pathway) is the main pathway operative during anaerobic fermentation [29,43].

3.2.4. Acetic Acid Production

In *Saccharomyces cerevisiae*, acetate is produced as an intermediate of the pyruvate dehydrogenase (PDH) bypass, which converts pyruvate to acetyl-CoA in a series of reactions catalyzed by pyruvate decarboxylase (PDC), acetaldehyde dehydrogenase (ACDH) and acetyl-CoA synthetase. This pathway is the sole source of cytosolic acetyl-CoA, which is required for anabolic processes, such as lipid biosynthesis [44]. The reaction catalyzed by ACDH also generates reducing equivalents, which are required for a variety of synthetic pathways and redox reactions, in the form of NADPH [45].

The monitoring of acetic acid synthesis showed that acetic acid was essentially produced during the growth phase and until the maximum CO_2 production rate was reached. Afterwards, stabilization and a slight decrease in acetate concentration in the medium were observed during the stationary phase. In the present work, the decrease was even more significant at high temperatures, with final concentrations of 0.32 g/L of acetic acid at 18 °C compared to 0.21 and 0.23 g/L at 24 and 30 °C, respectively. The volatility of the molecule could explain these results, but the data in the literature remain contradictory to date [46,47].

Acetaldehyde, a precursor of acetate through the activity of aldehyde dehydrogenase, has similar kinetics to acetate, showing accumulation during the growth phase until a maximum concentration is reached, as observed by other authors [8,18]. This was also illustrated by the positive correlation between acetaldehyde and acetate synthesis observed at 18 °C ($r^2 = 0.8230$), 24 °C ($r^2 = 0.8803$) and 30 °C ($r^2 = 0.9253$).

3.3. Acetaldehyde and Metabolism

The kinetics of acetaldehyde depend strongly on the redox balance of the cell. At the beginning of the winemaking process, glycerol synthesis allows the recycling of NADH/NAD$^+$ and ensures the maintenance of the redox balance of the cell system. Then, in the stationary phase, the slower but continuous synthesis of glycerol is generally used to fight against osmotic stress, as acetaldehyde is present in sufficient quantities to play its reducing role with an important flux [12]. A previous study stated that acetaldehyde is unable to serve as an electron acceptor for cytosolic NADH and that the accumulated NADH is instead oxidized by the reduction of dihydroxyacetone phosphate to glycerol-3-phosphate [48]. However, addition of acetaldehyde at the beginning of fermentation has been shown to reduce the lag phase of the yeast. Therefore, the acetaldehyde molecule seems to act as an activator of NADH-consuming metabolic pathways and, in particular, of the lower part of the glycolysis pathway [6].

These results were corroborated by the strong correlation between glycerol and acetaldehyde synthesis observed in our work. Figure 4A,D,G demonstrate the metabolic link between the syntheses of these two metabolic markers involved in redox balance maintenance. Additionally, an inverse correlation was noted between the final concentrations of acetaldehyde and glycerol. Thus, during its catabolism, acetaldehyde serves as a terminal electron acceptor for the redox balance of yeast and its creation of energy through glycolysis [1], while glycerol is always produced so that it can also participate in the maintenance of the redox balance, on the one hand, and for resistance against osmotic stress, on the other hand [29] Yeast consumption of extracellular acetaldehyde may hypothetically be connected with hexose transporter activity. Indeed, Saccharomyces cerevisiae harbors several hexose transporters that transport glucose and fructose through facilitated diffusion. It has been shown that, under enological conditions, the activity of the low-affinity hexose transport system starts to decrease when the fermenting cells are in the early stationary phase, limiting sugar transport [49,50]. This decreased activity reduces the flux in the earlier part of the glycolysis pathway, resulting in a lower intracellular concentration for acetaldehyde, therefore enabling the consumption of extracellular acetaldehyde to regenerate NADH (Figure 5A).

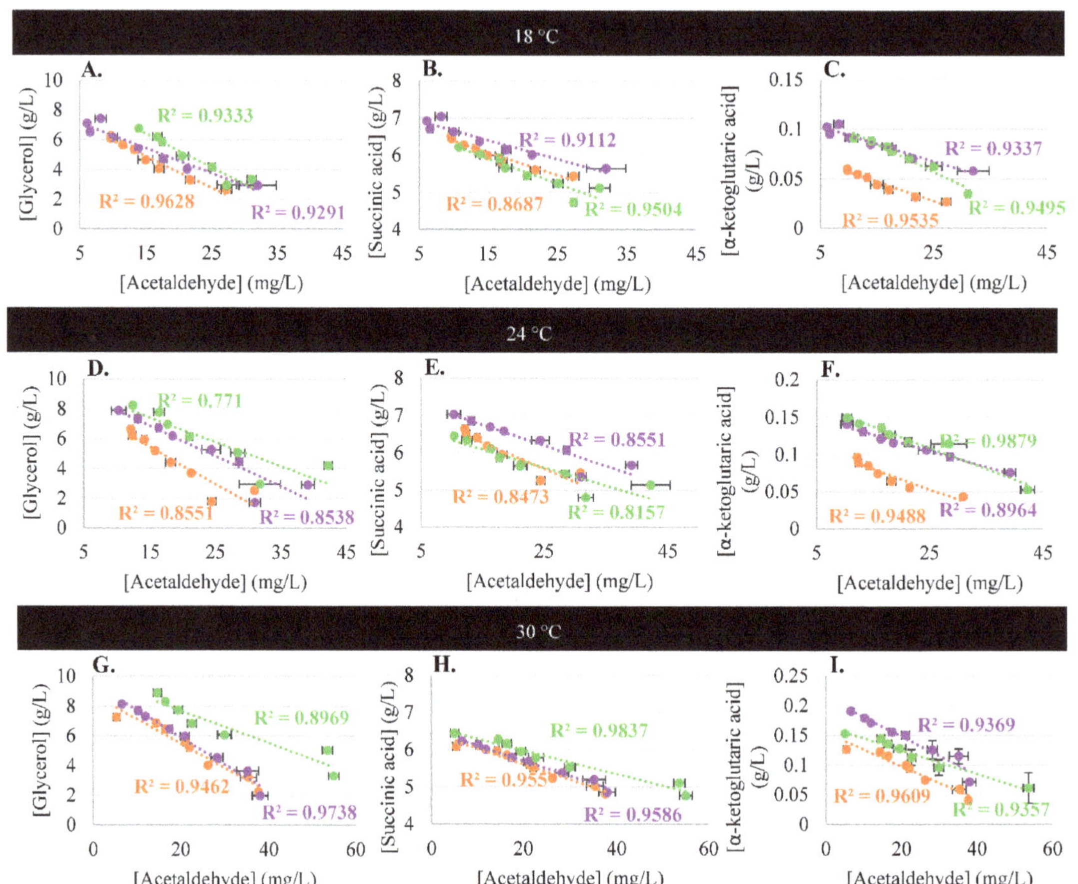

Figure 4. Correlation between the production of glycerol (**A,D,G**) and the organic acids succinate (**B,E,H**) and α-ketoglutarate (**C,F,I**), with acetaldehyde synthesis for the three isothermal temperatures (18, 24 and 30 °C) for strains FC9® (orange), SM102® (purple) and 7013® (green).

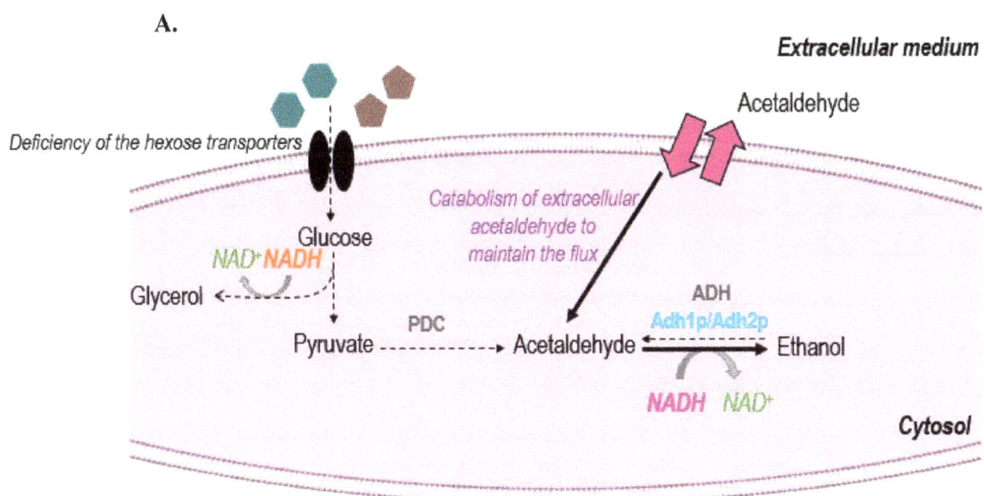

Figure 5. Hypotheses regarding the biological phenomenon of acetaldehyde reconsumption (**A**) through the decreased activity of hexose transporters and (**B**) through the principle of the acetaldehyde–ethanol Adh3p (Ac-Et) shuttle and the importance of the amount of NADH for yeast metabolism.

An alternative hypothesis regarding acetaldehyde consumption by yeast can be put forward in this study. Acetaldehyde synthesis is strongly correlated with the synthesis of TCAs, such as succinate and α-ketoglutarate, as markers of metabolism involved in the reduction of NAD+ to mitochondrial NADH (Figure 4B,C,E,F,H,I). Such a correlation between acetaldehyde consumption and organic acid synthesis by yeast during the stationary phase has already been described [51] This metabolic link can be verified by considering the acetaldehyde–ethanol shuttle that allows the exchange of mitochondrial NADH with cytosolic NADH [52]. Coupling of the mitochondrial alcohol dehydrogenases Adh3p and cytosolic Adh2p can be ensured by the diffusion of acetaldehyde and ethanol between the two compartments. However, this shuttle is not functional in oenological fermentations: if Adh3p is indeed active during fermentation, Adh2p is inactive because of repression by glucose [53]. The action of the Adh3p enzyme with the entry of acetaldehyde into mitochondria and the exit of ethanol from this compartment thus distorts the cytosolic $NADH/NAD^+$ balance: the entry of one mole of acetaldehyde into mitochondria generates a surplus of cytosolic NADH since this molecule cannot be used to form ethanol. In order to equilibrate the cytosolic $NADH/NAD^+$ balance, the yeast cell again consumes the acetaldehyde previously excreted in the medium to compensate for this redox imbalance (Figure 5B). A recent study also underlined the importance of Adh3p in the oxidization of mitochondrial NADH to NAD^+ and the central role of the major TCA-contributing metabolites (malate, α-ketoglutarate, succinate and citrate), showing a strong correlation between organic acid production and acetaldehyde consumption [5,54]. The production of these acids helps maintain the mitochondrial redox balance by reducing NAD^+ (produced by Adh3p) to NADH. The evolution of accumulated and consumed acetaldehyde over time indicates that it correlates directly with markers of interest for redox balance, thus allowing its maintenance thanks to its role as electron acceptor.

3.4. Impact of Temperature on Acetaldehyde

Temperature can impact both (a) acetaldehyde synthesis and consumption by yeast and (b) evaporation of this volatile molecule. However, to date, the relative contributions of these two phenomena to the evolution kinetics of acetaldehyde concentration in the liquid phase remain unknown.

First, high temperatures resulted in higher accumulation of acetaldehyde in the medium (Table 1) when the maximum population was reached, as previously demonstrated by [8]. For strain FC9[®], maximum acetaldehyde concentrations varied around 27.3 mg/L at 18 °C versus 35.8 mg/L at 30 °C (p-value < 0.001). With strain SM102[®], the impact of temperature was limited, and the strain appeared to produce a similar amount of acetaldehyde (around 40 mg/L, p-value < 0.05) regardless of the temperature chosen. Finally, for strain 7013[®], a major impact from temperature was observed, with a variation of 31.0 mg/L at 18 °C versus 53.5 mg/L at 30 °C (p-value < 0.001). While results concerning the levels of acetaldehyde accumulated in the medium at the end of the growth phase remain contradictory, authors agree in considering the fermentation temperature a key point in the production of acetaldehyde [8,16–20] and that high temperatures seem to promote high concentrations of acetaldehyde at the end of the growth phase [17], in addition to higher cellular activity [36]. The effect of the strain on acetaldehyde production was also highly significant (Table 1) (p-value < 0.001), with strain 7013[®] being the strain that resulted in the highest accumulation of acetaldehyde. Indeed, acetaldehyde production is a "strain-dependent" trait, which complicates comparisons with other studies [10].

A strain effect was also observed for residual acetaldehyde (Figure 6), but this time, strain 7013[®] achieved the lowest final concentration. For the other two strains, final acetaldehyde concentrations were similar with isothermal temperature profiles, while strain FC9[®] led to the lowest concentrations with anisothermal profiles. The impact of temperature on residual acetaldehyde was different from that on accumulated acetaldehyde; indeed, high temperatures induced a decrease in final acetaldehyde content. At 30 °C, the concentrations for the three strains were around 5 mg/L in the wines compared to around 9 mg/L (Tukey test < 0.001) at 18 °C. Furthermore, descending anisothermal temperature profiles (30–18 °C and 24–18 °C) tended to resulted in lower residual acetaldehyde concentrations than the ascending profiles (18–30 °C and 24–30 °C). Thus, an inverse correlation was observed for all temperatures between accumulated and residual acetaldehyde concentrations.

Table 1. Concentration of acetaldehyde accumulated in the extracellular medium at the end of the growth phase. ANOVA statistical analysis: effect of temperature and strain.

		Acetaldehyde Concentration (mg/L)		
		FC9®	SM102®	7013®
Isothermal temperature profiles	18 °C	27.3 ± 0.7	32.0 ± 0.1	31.0 ± 0.7
	24 °C	30.9 ± 0.2	39.1 ± 0.5	37.2 ± 7.7
	30 °C	35.8 ± 0.8	40.0 ± 1.4	53.5 ± 1.2
Anisothermal temperature profiles	18–30 °C	28.2 ± 0.5	37.9 ± 1.0	
	30–18 °C	36.7 ± 0.6	40.3 ± 0.3	
	24–18 °C	29.1 ± 2.6	39.5 ± 2.1	
	24–30 °C	30.1 ± 0.5	39.4 ± 1.0	
Temperature effect ANOVA (*p*-value)		***	*	***
Strain effect ANOVA (*p*-value)			***	

(***): ANCOVA $p < 0.001$, (*): ANCOVA $p < 0.05$.

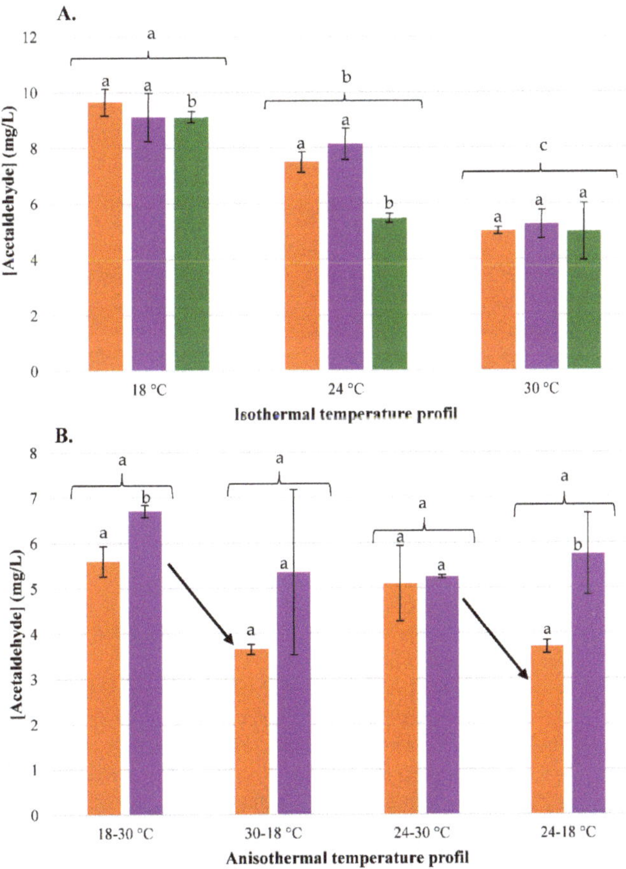

Figure 6. Residual acetaldehyde concentration at the end of alcoholic fermentation for isothermal (**A**) and anisothermal (**B**) temperature profiles with the three yeast strains: FC9® (orange), SM102® (purple) and 7013® (green). The letters (a, b, c) indicate homogeneous groups for acetaldehyde contration at the 95% confidence level, as tested by Tukey's statistical test.

To understand the impact of temperature on acetaldehyde synthesis, various hypotheses can be proposed: (a) From a biological point of view, yeast metabolic activity is lower at low temperatures and can be correlated with lower accumulation of acetaldehyde in the medium. Alternatively, poor cellular activity may be associated with poor consumption of acetaldehyde by yeast, leading to high residual concentrations [8]. In contrast, it can also result in a longer time for yeasts to consume acetaldehyde (low fermentation rate), which is associated with low residual concentrations [8–10,19]. (b) Another situation is when fermentation takes place at high temperatures: metabolic activity is higher with high acetaldehyde accumulation in the medium but there is better consumption by yeast and, thus, a lower residual concentration. Too short a fermentation duration can also reduce the time available for yeast consumption of acetaldehyde [10], thus inducing high final concentrations. (c) From a physical perspective, acetaldehyde evaporation at low temperatures will be lower and acetaldehyde will be present in higher concentrations at the end of fermentation. The reverse reasoning can be applied for high temperature. A question of interest thus emerges: what is the major phenomenon—evaporation or consumption by yeasts—involved in the decrease in acetaldehyde concentration in the liquid part during the stationary phase? Furthermore, was the observed inverse correlation between maximum accumulated acetaldehyde and residual acetaldehyde due to a physical or biological phenomenon? The isothermal and anisothermal temperature profiles demonstrated (a) the effect of temperature on the evaporation of the molecule (c.f. Section 3.4.1) and (b) the importance of cell activity and consumption time for acetaldehyde (c.f. Section 3.4.2). Thanks to the online monitoring system, we could access the complete balances for acetaldehyde production (Figure 7), and the biological phenomenon of consumption by yeasts could be distinguished from the physical phenomenon of evaporation of the molecule.

3.4.1. Physical Effect: Evaporation

Figure 8A,B show the cumulative losses of acetaldehyde throughout the alcoholic fermentation as a function of different temperature profiles and for two strains. Molecule losses were enhanced with temperature increases. When fermenting at 18 °C, cumulative losses were about 14.9 mg/L at the end of fermentation compared to 21.8 mg/L and 27.2 mg/L at 24 and 30 °C, respectively (Figure 7A,B). For the rising anisothermal profiles, such as 18–30 °C and 24–30 °C, the progressive increase in temperature generated additional increases in the accumulation of losses of 1.27 mg/L for the 18–30 °C profile and 1.94 mg/L for 24–30 °C compared to the isothermal profiles at 18 and 24 °C, respectively. When the temperature was lowered during the process, for the 24–18 °C profile, the cumulative losses were reduced by 6.68 mg/L at the end of fermentation compared to the isothermal 24 °C profile. On the other hand, different behavior was observed for the 30–18 °C profile. The cumulative losses were similar to the 30 °C isothermal profile, with a total of 30.16 mg/L. Since the temperature decrease was greater at 30–18 °C compared to 24–18 °C, more than half of the fermentation occurred between 30 °C and 26 °C—i.e., above the boiling point of acetaldehyde—and, therefore, evaporation continued to be promoted throughout the alcoholic fermentation.

Figure 8C,D show the rates of losses for all temperature profiles. The maximum rate of loss was systematically obtained after 20% of the fermentation process, which corresponded with the maximum acetaldehyde produced and accumulated in the extracellular medium when the yeast population was at its maximum, as well as with the high CO_2 release rate achieved, which necessarily influences the rate of loss. Therefore, the more acetaldehyde was produced and accumulated at high temperatures, the more acetaldehyde was also lost simultaneously. Using the real-time quantification of losses, online monitoring data for the molecule made it possible to demonstrate that a part of the acetaldehyde was lost in the gas phase through evaporation during alcoholic fermentation [7]. Then, the differences between the physical and biological effects on the consumption of the molecule were demonstrated.

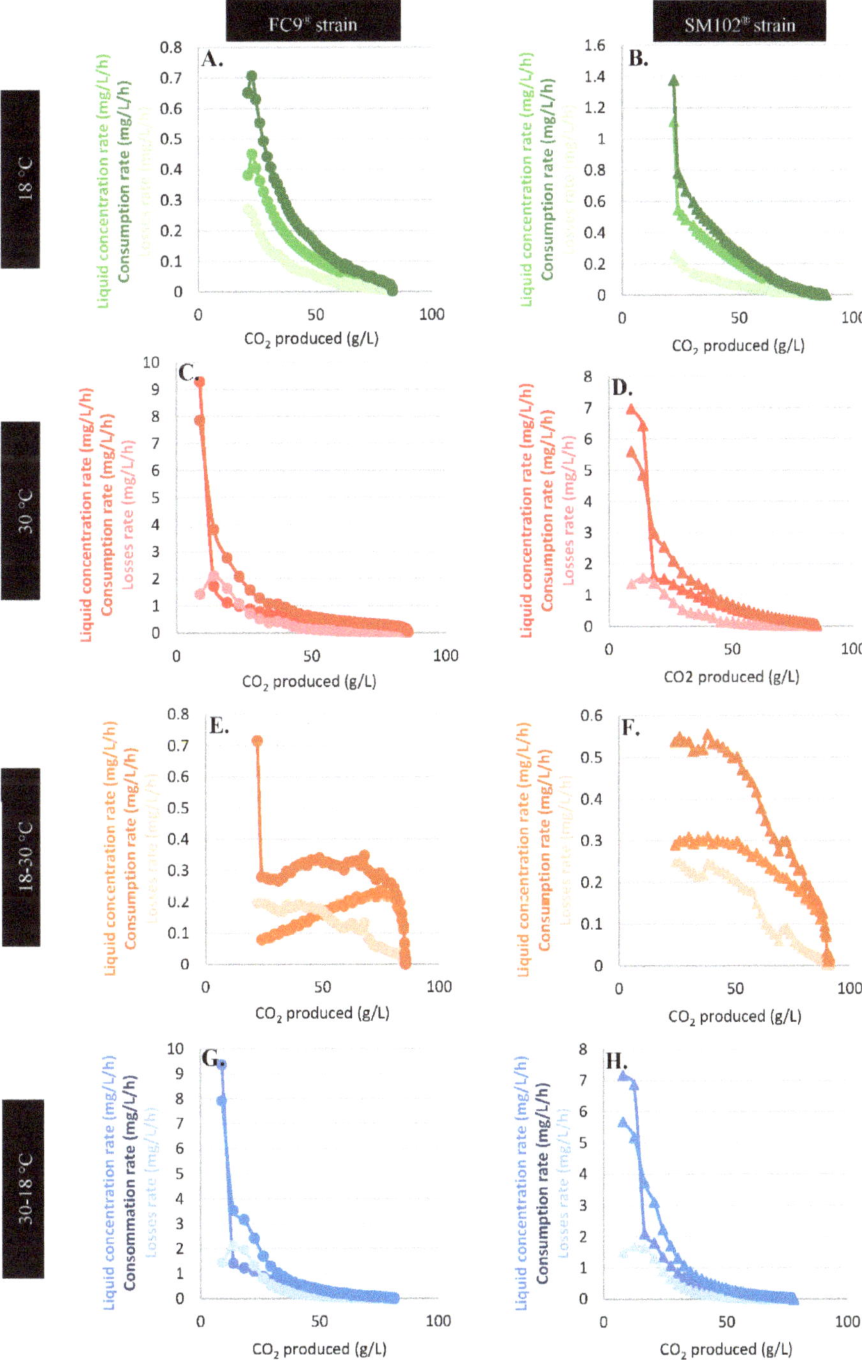

Figure 7. The value of online monitoring, making it possible to produce and visualize complete production balances for the two strains FC9® (**A,C,E,G**) and SM102® (**B,D,F,H**) at 18 °C (**A,B**), 30 °C (**C,D**), 18–30 °C (**E,F**) and 30–18 °C (**G,H**). Assessments carried out during the stationary phase: the decrease in the acetaldehyde concentration was demonstrated by the rate of decrease in the liquid phase (mg/L/h), the rate of physical losses (mg/L/h) and the rate of consumption (mg/L/h).

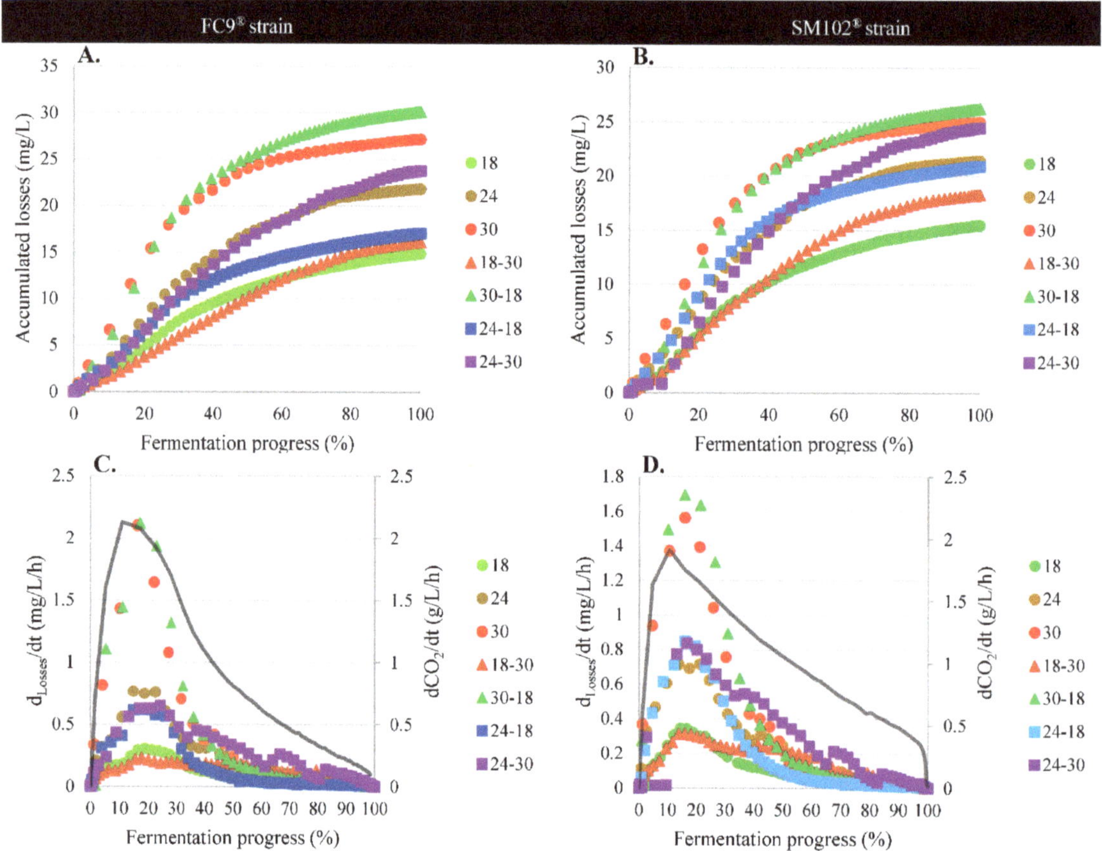

Figure 8. Monitoring of acetaldehyde in the gas phase that provides access to (**A**, **B**) cumulative losses (mg/L) and (**C,D**) loss rates (mg/L/h) for all temperature profiles with the example of a fermentative kinetics (grey) for both strains: (**A,C**) FC9® and (**B,D**) SM102®.

3.4.2. Biological Effects: Consumption by Yeast

The acetaldehyde consumption profiles in the stationary phase were visualized for the two yeast strains used at all the different fermentation temperatures (Figure 9). The same phenomena were observed for both strains. The yeast acetaldehyde catabolism (i.e., the rate of consumption) was higher at the isothermal temperature of 30 °C (Figure 9A,D). For the anisothermal profiles starting at 24 °C (Figure 9B,E), the rate of consumption increased simultaneously with the temperature increase (24–30 °C), whereas, when the temperature was decreased, the rate of consumption was lower. For the 18–30 °C rising temperature profile (Figure 9C,F), the consumption rate was stimulated and plateaus of consumption were maintained at 0.35 mg/L/h and 0.5 mg/L/h for the FC9® and SM102® strains, respectively, in contrast to the isothermal fermentation at 18 °C, for which this rate only decreased. After 70% of the fermentation process with the FC9® strain, yeast consumption was more than three times lower at 18 °C compared to that observed with the 18–30 °C temperature profile.

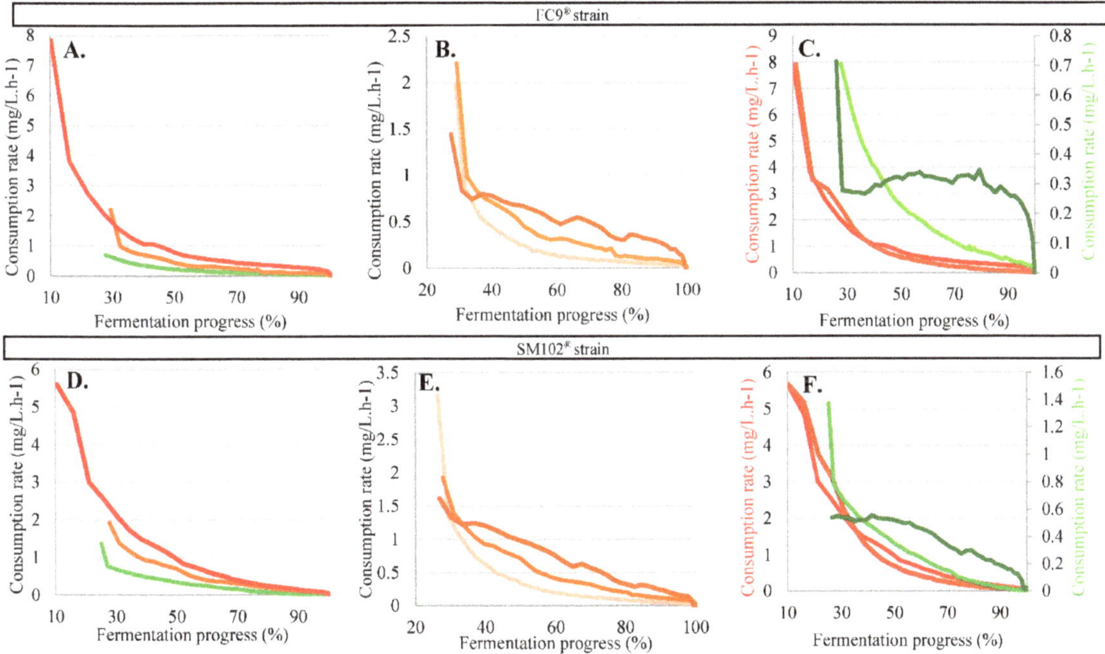

Figure 9. Acetaldehyde consumption rates (mg/L/h) for strains FC9® (**A–C**) and SM102® (**D–F**) with isothermal temperature profiles (**A,D**):18 °C (green), 24 °C (orange) and 30 °C (red)and anisothermal temperature profiles (**B,C,E,F**):24–18 °C (light orange), 24–30 °C (dark orange), 18–30 °C (dark green) and 30–18 °C (dark red).

On the other hand, for the 30–18 °C descending profile, the consumption rate was not slowed down by the decrease in temperature compared to the isothermal profile at 30 °C. Indeed, the amplitude of the temperature difference was important, as it was for the phenomenon of losses, but since half of the fermentation occurred at a temperature higher than 25 °C, yeast metabolic activity remained optimal; this also promoted efficient consumption during half of the fermentation, amounting to between 8 and 1 mg/L/h for the FC9® strain and between 6 and 1 mg/L/h for SM102®. After the accumulation of extracellular acetaldehyde, at the beginning of the stationary phase, the rate of consumption was maximal (Figure 9 and Table 2) and related to temperature: 0.68, 0.72 and 0.99 mg/L/h for FC9®, SM102® and 7013®, respectively, at 18 °C compared to 7.6, 5.38 and 5.64 mg/L/h at 30 °C. Therefore, there was no inhibitory effect from temperature on ADH activity, as previously reported [55]. Thus, high temperatures enabled the promotion of cellular activity with a better catabolism and, even when the fermentation rate was higher, the time necessary to assimilate acetaldehyde was not a limiting factor in obtaining low residual acetaldehyde content, in contrast to the hypothesis posed by [10]. However, acetaldehyde assimilation is lowered by temperature decreases, which entails higher residual contents [7].

Furthermore, at high temperatures, although the yeast accumulated more acetaldehyde, the addition of physical evaporation and biological consumption phenomena led to the lowest residual acetaldehyde levels. While it was the strain that produced the most acetaldehyde at the end of the growth phase, 7013® was the least efficient in catabolizing acetaldehyde (Table 2). The strain effect on acetaldehyde synthesis has been described as a "strain-dependent" trait [9], but so is the strain's ability to consume this metabolite in the medium. These two capacities are, moreover, not correlated: a strain that produces high amounts of acetaldehyde does not necessarily have the highest consuming capacity.

14. Boulton, R.B.; Singleton, V.L.; Bisson, L.F.; Kunkee, R.E. *Principles and Practices of Winemaking*; Springer: Boston, MA, USA, 1996; ISBN 978-1-4613-5718-6.

15. Mouret, J.-R.; Aguera, E.; Perez, M.; Farines, V.; Sablayrolles, J.-M. Study of Oenological Fermentation: Which Strategy and Which Tools? *Fermentation* **2021**, *7*, 155. [CrossRef]

16. Amerine, M.; Ough, C. Studies with Controlled Fermentation. VIII. Factors Affecting Aldehyde Accumulation. *Am. J. Enol. Vitic.* **1964**, *15*, 23–33.

17. Bosso, A.; Guaita, M. Study of Some Factors Involved in Ethanal Production during Alcoholic Fermentation. *Eur. Food Res. Technol.* **2008**, *227*, 911–917. [CrossRef]

18. Li, E.; Mira de Orduña, R. Acetaldehyde Kinetics of Enological Yeast during Alcoholic Fermentation in Grape Must. *J. Ind. Microbiol. Biotechnol.* **2017**, *44*, 229–236. [CrossRef]

19. Romano, P.; Suzzi, G.; Turbanti, L.; Polsinelli, M. Acetaldehyde Production in *Saccharomyces Cerevisiae* Wine Yeasts. *FEMS Microbiol. Lett.* **1994**, *118*, 213–218. [CrossRef] [PubMed]

20. Torija, M. Effects of Fermentation Temperature on the Strain Population of Saccharomyces Cerevisiae. *Int. J. Food Microbiol.* **2003**, *80*, 47–53. [CrossRef]

21. Colombié, S.; Malherbe, S.; Sablayrolles, J.-M. Modeling of Heat Transfer in Tanks during Wine-Making Fermentation. *Food Control* **2007**, *18*, 953–960. [CrossRef]

22. Casalta, E.; Salmon, J.-M.; Picou, C.; Sablayrolles, J.-M. Grape Solids: Lipid Composition and Role during Alcoholic Fermentation under Enological Conditions. *Am. J. Enol Vitic.* **2019**, *70*, 147–154. [CrossRef]

23. Crépin, L.; Nidelet, T.; Sanchez, I.; Dequin, S.; Camarasa, C. Sequential Use of Nitrogen Compounds by Saccharomyces Cerevisiae during Wine Fermentation: A Model Based on Kinetic and Regulation Characteristics of Nitrogen Permeases. *Appl. Environ. Microbiol.* **2012**, *78*, 8102–8111. [CrossRef]

24. Sablayrolles, J.M. Control of Alcoholic Fermentation in Winemaking: Current Situation and Prospect. *Food Res. Int.* **2009**, *42*, 418–424. [CrossRef]

25. Sablayrolles, J.M.; Ball, C.B. Fermentation Kinetics and the Production of Volatiles During Alcoholic Fermentation. *J. Am. Soc. Brew. Chem.* **1995**, *53*, 72–78. [CrossRef]

26. Sablayrolles, J.-M.; Barre, P. Kinetics of Alcoholic Fermentation Under Anisothermal Enological Conditions. I. Influence of Temperature Evolution on the Instantaneous Rate of Fermentation. *Am. J. Enol. Vitic.* **1993**, *44*, 127–133. [CrossRef]

27. Remize, F.; Roustan, J.L.; Sablayrolles, J.M.; Barre, P.; Dequin, S. Glycerol Overproduction by Engineered Saccharomyces Cerevisiae Wine Yeast Strains Leads to Substantial Changes in By-Product Formation and to a Stimulation of Fermentation Rate in Stationary Phase. *Appl. Environ. Microbiol.* **1999**, *65*, 143–149. [CrossRef] [PubMed]

28. Van Dijken, J.P.; Scheffers, W.A. Redox Balances in the Metabolism of Sugars by Yeasts. *FEMS Microbiol Rev.* **1986**, *1*, 199–224. [CrossRef]

29. Roustan, J.L.; Sablayrolles, J.-M. Modification of the Acetaldehyde Concentration during Alcoholic Fermentation and Effects on Fermentation Kinetics. *J. Biosci. Bioeng.* **2002**, *93*, 367–375. [CrossRef] [PubMed]

30. Deroite, A.; Legras, J.-L.; Rigou, P.; Ortiz-Julien, A.; Dequin, S. Lipids Modulate Acetic Acid and Thiol Final Concentrations in Wine during Fermentation by Saccharomyces Cerevisiae × Saccharomyces Kudriavzevii Hybrids. *AMB Expr.* **2018**, *8*, 130. [CrossRef] [PubMed]

31. Guittin, C.; Maçna, F.; Sanchez, I.; Barreau, A.; Poitou, X.; Sablayrolles, J.-M.; Mouret, J.-R.; Farines, V. The Impact of Must Nutrients and Yeast Strain on the Aromatic Quality of Wines for Cognac Distillation. *Fermentation* **2022**, *8*, 51. [CrossRef]

32. Guittin, C.; Maçna, F.; Sanchez, I.; Poitou, X.; Sablayrolles, J.-M.; Mouret, J.-R.; Farines, V. Impact of High Lipid Contents on the Production of Fermentative Aromas during White Wine Fermentation. *Appl. Microbiol. Biotechnol.* **2021**, *105*, 6435–6449. [CrossRef] [PubMed]

33. Scanes, K.T.; Hohrnann, S.; Prior, B.A. Glycerol Production by the Yeast Saccharomyces Cerevisiae and Its Relevance to Wine: A Review. *SAJEV* **2017**, *19*, 17–24. [CrossRef]

34. Gancedo, J.M. The Early Steps of Glucose Signalling in Yeast. *FEMS Microbiol. Rev.* **2008**, *32*, 673–704. [CrossRef] [PubMed]

35. Gombert, A.K.; dos Santos, M.M.; Christensen, B.; Nielsen, J. Network Identification and Flux Quantification in the Central Metabolism of Saccharomyces Cerevisiae under Different Conditions of Glucose Repression. *J. Bacteriol.* **2001**, *183*, 1441–1451. [CrossRef]

36. Bely, M.; Sablayrolles, J.M.; Barre, P. Automatic Detection of Assimilable Nitrogen Deficiencies during Alcoholic Fermentation in Oenological Conditions. *J. Ferment. Bioeng.* **1990**, *70*, 246–252. [CrossRef]

37. Malherbe, S.; Fromion, V.; Hilgert, N.; Sablayrolles, J.-M. Modeling the Effects of Assimilable Nitrogen and Temperature on Fermentation Kinetics in Enological Conditions. *Biotechnol. Bioeng.* **2004**, *86*, 261–272. [CrossRef]

38. Rollero, S.; Bloem, A.; Camarasa, C.; Sanchez, I.; Ortiz-Julien, A.; Sablayrolles, J.-M.; Dequin, S.; Mouret, J.-R. Combined Effects of Nutrients and Temperature on the Production of Fermentative Aromas by Saccharomyces Cerevisiae during Wine Fermentation. *Appl. Microbiol. Biotechnol.* **2015**, *99*, 2291–2304. [CrossRef]

39. Godillot, J.; Sanchez, I.; Perez, M.; Picou, C.; Galeote, V.; Sablayrolles, J.-M.; Farines, V.; Mouret, J.-R. The Timing of Nitrogen Addition Impacts Yeast Genes Expression and the Production of Aroma Compounds During Wine Fermentation. *Front. Microbiol.* **2022**, *13*, 829786. [CrossRef] [PubMed]

40. Ljungdahl, P.O.; Daignan-Fornier, B. Regulation of Amino Acid, Nucleotide, and Phosphate Metabolism in *Saccharomyces Cerevisiae*. *Genetics* **2012**, *190*, 885–929. [CrossRef] [PubMed]

41. Whiting, G.C. Organic acid metabolism of yeasts during fermentation of alcoholic beverages-a review. *J. Inst. Brew.* **1976**, *82*, 84–92. [CrossRef]

42. Thoukis, G.; Ueda, M.; Wright, D. The Formation of Succinic Acid during Alcoholic Fermentation. *Am. J. Enol. Vitic.* **1965**, *16*, 1–8.

43. Camarasa, C.; Grivet, J.-P.; Dequin, S. Investigation by 13C-NMR and Tricarboxylic Acid (TCA) Deletion Mutant Analysis of Pathways for Succinate Formation in Saccharomyces Cerevisiae during Anaerobic Fermentation. *Microbiology* **2003**, *149*, 2669–2678. [CrossRef] [PubMed]

44. Pronk, J.T. Pyruvate Metabolism in Saccharomyces Cerevisiae. *Yeast* **1996**, *12*, 1607–1633. [CrossRef]

45. Saint-Prix, F.; Bönquist, L.; Dequin, S. Functional Analysis of the ALD Gene Family of Saccharomyces Cerevisiae during Anaerobic Growth on Glucose: The NADP+-Dependent Ald6p and Ald5p Isoforms Play a Major Role in Acetate Formation. *Microbiology* **2004**, *150*, 2209–2220. [CrossRef]

46. Beltran, G.; Novo, M.; Guillamón, J.M.; Mas, A.; Rozès, N. Effect of Fermentation Temperature and Culture Media on the Yeast Lipid Composition and Wine Volatile Compounds. *Int. J. Food Microbiol.* **2008**, *121*, 169–177. [CrossRef] [PubMed]

47. Llaurado, J.; Rozes, N.; Bobet, R.; Mas, A.; Constanti, M. Low Temperature Alcoholic Fermentations in High Sugar Concentration Grape Musts. *J. Food Sci.* **2002**, *67*, 268–273. [CrossRef]

48. Neuberg, C.; Hirsch, J. Uber Den Verlauf Der Alkoholischen Garung Bei Alkalischer Reaktion: II. Garung Mit Lebender Hefe in Alkalischen Lösungen. *Biochem. Z* **1919**, *96*, 175–202.

49. Salmon, J.M. Effect of Sugar Transport Inactivation in Saccharomyces Cerevisiae on Sluggish and Stuck Enological Fermentations. *Appl. Environ. Microbiol.* **1989**, *55*, 953–958. [CrossRef]

50. Salmon, J.M.; Vincent, O.; Mauricio, J.C.; Bely, M.; Barre, P. Sugar Transport Inhibition and Apparent Loss of Activity in Saccharomyces Cerevisiae as a Major Limiting Factor of Enological Fermentations. *Am. J. Enol. Vitic.* **1993**, *44*, 56–64. [CrossRef]

51. Roustan, J.L.; Sablayrolles, J.M. Impact of the Addition of Electron Acceptors on the By-Products of Alcoholic Fermentation. *Enzym. Microb. Technol.* **2002**, *31*, 142–152. [CrossRef]

52. Bakker, B.M.; Bro, C.; Kötter, P.; Luttik, M.A.H.; van Dijken, J.P.; Pronk, J.T. The Mitochondrial Alcohol Dehydrogenase Adh3p Is Involved in a Redox Shuttle in Saccharomyces Cerevisiae. *J. Bacteriol.* **2000**, *182*, 4730–4737. [CrossRef]

53. Ciriacy, M. Genetics of Alcohol Dehydrogenase InSaccharomyces Cerevisiac. *Molec. Gen. Genet.* **1975**, *138*, 157–164. [CrossRef] [PubMed]

54. Xu, X.; Niu, C.; Liu, C.; Li, Q. Unraveling the Mechanisms for Low-Level Acetaldehyde Production during Alcoholic Fermentation in *Saccharomyces Pastorianus* Lager Yeast. *J. Agric. Food Chem.* **2019**, *67*, 2020–2027. [CrossRef] [PubMed]

55. Lutstorf, U.; Megnet, R. Multiple Forms of Alcohol Dehydrogenase in S. Cerevisiae. I. Physiological Control of ADH-2 and Properties of ADH-2 and ADH-4. *Arch. Biochem. Biophys.* **1968**, *126*, 933–944. [CrossRef] [PubMed]

fermentation

Article

Metschnikowia pulcherrima in Cold Clarification: Biocontrol Activity and Aroma Enhancement in Verdicchio Wine

Alice Agarbati, Laura Canonico, Maurizio Ciani * and Francesca Comitini *

Department of Life and Environmental Sciences, Polytechnic University of Marche, Via Brecce Bianche, 60131 Ancona, Italy
* Correspondence: m.ciani@univpm.it (M.C.); f.comitini@univpm.it (F.C.)

Abstract: Non-*Saccharomyces* wine yeasts are not only proposed to improve the sensory profile of wine but also for several distinctive promising features. Among them, biocontrol action at different steps of the wine production chain could be a suitable strategy to reduce the use of sulfur dioxide. In this work, the activity of a selected strain of *Metschnikowia pulcherrima* was evaluated as inoculum in cold clarification with the aim to reduce SO_2 and improve the aromatic profile of the wine. Fermentation processes were carried out at the winery level for two consecutive vintages using a *pied de cuve* as the starter inoculum coming from indigenous *Saccharomyces cerevisiae* strains. *M. pulcherrima* revealed an effective bio-protectant action during the pre-fermentative stage even if the timely and appropriate starter inoculum in the two years permitted the effective control of wild yeasts during the fermentation also in the control trials. In general, the main oenological characters did not show differences if compared with an un-inoculated trial, while the inoculum of *M. pucherrima* in cold clarification determined an enhancement of ethyl hexanoate, isobutanol, acetaldehyde, and geraniol even if they are considered in different amounts for each year. Indeed, the analytical and sensory profiles of wines were also influenced by the vintage and variation *pied the cuve* population. Nonetheless, the overall results indicated that *M. pulcherrima* led to biocontrol action and an improvement of the aromatic and sensory profile of the wine.

Keywords: *Metschnikowia pulcherrima*; bioprotectant; aroma profile; Verdicchio wine; winery level

Citation: Agarbati, A.; Canonico, L.; Ciani, M.; Comitini, F. *Metschnikowia pulcherrima* in Cold Clarification: Biocontrol Activity and Aroma Enhancement in Verdicchio Wine. *Fermentation* **2023**, *9*, 302. https://doi.org/10.3390/fermentation9030302

Academic Editor: Niel Van Wyk

Received: 27 February 2023
Revised: 13 March 2023
Accepted: 18 March 2023
Published: 20 March 2023

1. Introduction

Wine consumers are increasingly attentive to the diversification of distinctive styles but also to the negative impact that chemical preservatives have on human health [1]. In the wine industry, the most common chemical additive is sulphur dioxide (SO_2). This additive is considered an essential tool for winemakers due to its low-cost and its combined antioxidant and antimicrobial properties against a wide spectrum of microorganisms [2]. SO_2 is used to decrease undesirable microorganisms, reduce the oxidation of phenolic compounds, and improve the quality and the shelf life of wine during the various stages of the wine production [3]. Sulfites pose a problem for human health, especially for sensitive consumers [4]. Sulphur dioxide is related to headaches, allergic reactions, and breathing difficulties in asthma patients. For these reasons, the use of this additive is strictly controlled by European Union legislation and reductions in all food and beverages products are required.

Several technological approaches were proposed to control wine spoilage microorganisms [5], even if a definitive substitute for SO_2 has not been proposed, especially for wines stored for a long time. In recent years, the use of microorganisms as bioprotective agents or their antimicrobial products was extensively investigated particularly at the prefermentative stage [6–8]. Currently, species such as *Metschnikowia pulcherrima*, *Torulaspora delbrueckii*, and *Lachancea thermotolerans* are the most applied in wine protection [9]. This strategy is based on the inoculum of viable antagonist microorganisms or their antimicrobial products during, at the end, or after the wine fermentation [10–12]. In wine making, bio-protectant strains can be a useful tool to reduce or replace sulfite addition [8,13,14].

In recent years, the increased number of studies has led to a better understanding of the impact of mixed fermentation on overall wine quality. Among non-*Saccharomyces* yeast, *M. pulcherrima* was one of the most investigated species for its positive contribution to wine making, both as a bio-protectant agent and owing to its effect on analytical and sensory wine traits [15–17]. *M. pulcherrima* is a well-characterized species for several positive aspects of wine making. Indeed, its metabolic characteristics are the synthesis of secondary metabolites to improve the volatile profile of the wine and act as a biocontrol agent. In mixed fermentation, *M. pulcherrima* led to a reduction of ethyl acetate compound, favoring the formation of 2-phenylethyl acetate, an increase of acetate esters, β-damascenone, and higher alcohols, particularly isobutanol and 2-phenyl ethanol [18,19]. The action of *M. pulcherrima* manifests as a biocontrol agent due to the production of pulcherrimin, a red pigment with antifungal activity [14] that exhibited wide antimicrobial activity against yeasts such as the *Candida*, *Brettanomyces/Dekkera*, *Hanseniaspora*, and *Pichia* genera as well as filamentous fungi such as the *Botrytis*, *Penicillium*, *Alternaria*, and *Monilia* genera [20–23].

In this work a selected *M. pulcherrima* strain previously characterized for its antimicrobial activity [24] was inoculated at cold clarification stage under winery condition for two consecutive vintages. The impact of *M. pulcherrima* toward the wild yeasts and the contribution on the aroma and sensory profile of wine was evaluated.

2. Materials and Methods

2.1. Fermentation Trials at the Winery Level

2.1.1. *M. pulcherrima* Biomass Production

The biomass of the *M. pulcherrima* selected strain was obtained from pre-cultures in a modified YPD medium (0.5% yeast extract, 0.1% peptone and 2% glucose) and grown for 48 h at 25 °C in an orbital shaker (150 rpm). After this, each pre-culture was used to inoculate a 2 L bench-top bioreactor (Biostat® C; B. Braun Biotech Int., Goettingen, Germany) containing 25 L of modified YPD medium with airflow (1 L/L/min). A feed batch process was used for biomass production. Biomass was collected by centrifugation, washed three times with sterile distilled water, and inoculated at a 1×10^6 cell/mL initial concentration on grape juice in cold clarification. This value was determined using the Thoma-Zeiss counting chamber.

2.1.2. Preparation of Pied de Cuve

In total, 100 kg of undamaged grapes were harvested and soft pneumatic pressed. Thus, the grape juice obtained was subjected to a clarification (10 °C for 24 h) with the addition of enzymes and bentonite (MICROCOL® ALPHA, Laffort) without the addition of SO_2. The clarified grape juice was then separated and left to ferment for two days. After this period, 15 mL/L SO_2 was added and the *pied de cuve* was thus prepared and used to inoculate the steel vessel of the trials.

2.1.3. Inoculation at Cold Clarification Stage and Fermentation

The fermentation trials were carried out for two consecutive vintages (2021–2022). The main analytical parameters of the Verdicchio grape juice used in 2021 were the initial sugar content 265 g/L, pH 3.09, total acidity 5.17 g/L, and yeast assimilable nitrogen (YAN) content 9 mgN/L. In 2022, the main analytical parameters were initial sugar content 270 g/L, pH 3.11, total acidity 5.17 g/L, and yeast assimilable nitrogen (YAN) content 2 mgN/L.

In each vintage, a lot of grape juice (26 hL) was divided into two lots of Verdicchio grape juice to fill two vats of 15 hL. One steel vessel was inoculated with 1×10^6 cell/mL of *M. pulcherrima* and maintained at 10 °C for 24 h in cold static clarification. The other batch was maintained at the same condition without the inoculum of *M. pulcherrima* as a controlled trial. After 24 h of cold static clarification, the YAN was adjusted to 250 mgN/L by the addition of diammonium phosphate and yeast derivative (Genesis Lift® Oenofrance, Bordeaux, France). After this time, the two batches were inoculated with the same *pied de cuve* previously prepared.

2.2. Biomass Evolution

Samples during fermentation were collected to evaluate biomass evolution. A viable cell count was carried out using lysine agar medium (Oxoid, Hampshire, UK) as a selective medium and WL nutrient agar medium (Oxoid, Hampshire, UK) for the detection of colony diversity. The plates were incubated at 25 °C for 48–72 h. The detection and enumeration of inoculated and wild yeasts were evaluated to combine the results of lysine agar, and the analysis of macro- and micro-morphological colonies in WL nutrient agar medium.

2.3. Saccharomyces cerevisiae Typing

Based on the micro- and macro-morphology of the colonies, presumptive *S. cerevisiae* pure cultures derived from *pied de cuves* and samples from 2/3 of respective inoculated fermentations were isolated. DNA was extracted at 95 °C for 10 min, and then it was amplified by PCR using the primers ITS1 (5′-TCCGTAGGTGAACCTCGCG-3′) and ITS4 (5′-TCCTCCGCTTTATTGATATGC-3′) following the procedure reported by Agarbati et al. [25]. A total of 74 strains obtained from the two years were then subjected to genotyping using pairs delta12/21 (delta12: 5′-TCAACAATGGAATCCCAAC-3′; delta21: 5′-CATCTTAACACCGTATATGA-3′) and were used for interdelta sequence analyses, as described by Legras and Karst [26]. The amplification was performed as per the following program: 3 min at 95 °C; followed by 25 s at 94 °C, 30 s at 45 °C, and 90 s at 72 °C for 9 cycles; and 25 s at 94 °C, 30 s at 50 °C, and 90 s at 72 °C for 21 cycles and a final extension at 72 °C for 10 min.

2.4. Analytical Procedures

The Official European Union Methods (2000) were used to determine the total acidity, volatile acidity, pH, and ethanol content. Enzymatic kits (Megazyme International Ireland) were utilized to determine glucose and fructose (K-FRUGL) and malic acid (K-DMAL) following the manufacturer procedures, while the ammonium content was determined using a specific enzymatic kit (kit no. 112732; Roche Diagnostics, Germany) and the free α-amino acids were evaluated following Dukes and Butzke protocol [27]. Acetaldehyde, ethyl acetate, n-propanol, isobutanol, amyl and isoamyl alcohols, and acetoin were quantified by direct injection into a gas chromatography system (GC-2014; Shimadzu, Kyoto, Japan). Each sample was prepared and analyzed as reported by Canonico et al. [28]. The volatile compounds were determined by the solid phase microextraction (HS-SPME) method, preparing the sample as follows: 5 mL of wine was placed into a vial, 1 g of NaCl and 3-octanol as the internal standard (1.6 mg/L) were added, and the vial was closed with a septum-type cap and placed on a magnetic stirrer for 10 min at 25 °C. Then, the sample was heated to 40 °C and extracted with a fiber Divinylbenzene/Carboxen/Polydimethylsiloxane (DVB/CAR/PDMS) for 30 min by insertion into the vial headspace. The compounds were desorbed by inserting the fiber into a Shimadzu gas chromatograph GC injector, in split–splitless modes following the procedure reported by Canonico et al. [29]. The glass capillary column used was 0.25 µm Supelcowax 10, length 60 m, and internal diameter 0.32 mm.

2.5. Sensory Analysis

At the end of the fermentation, the wines were decanted and after three months, were transferred into filled 750 mL bottles, closed with the crown cap, and maintained at 4 °C until sensory analysis. After this period of refinement, they were subjected to sensory evaluation. A group of 15 testers, 10 males and 5 females aged 25–45 years (80% expert and 20% non-expert), used a score scale of 1 to 10, where 10 was the score that quantitatively represented the best judgment (maximum satisfaction) and 1 was the score to be attributed in case of poor satisfaction. The expert testers were composed of oenologists, sommeliers, and wine producers.

2.6. Data Analyses

The analysis of variance (ANOVA) using the statistical software package JMP 11® (statistical discovery from SAS, New York, NY, USA) was used to process all experimental data. The averages obtained were processed significant differences between the averaged data and were determined using the Duncan test. The experimental data were significant with associated p-values < 0.05. The results of the sensory analysis were also subjected to Fisher ANOVA to determine the significant differences with a p-value < 0.05.

3. Results

3.1. Effect of M. pulcherrima Addition in Cold Clarification

The results of the viable wild yeasts (WY) population before and after cold clarification with and without *M. pulcherrima* are reported in Figure 1a,b. Vintages from 2021 (a) and 2022 (b) were considered. In the first year, the initial wild yeast population in grape juice was about 10^5 CFU/mL in both trials. Without the use of *M. pulcherrima* there was an increase of about one log CFU/mL, while the inoculated trial showed a containment of the wild yeast population, which remained almost constant.

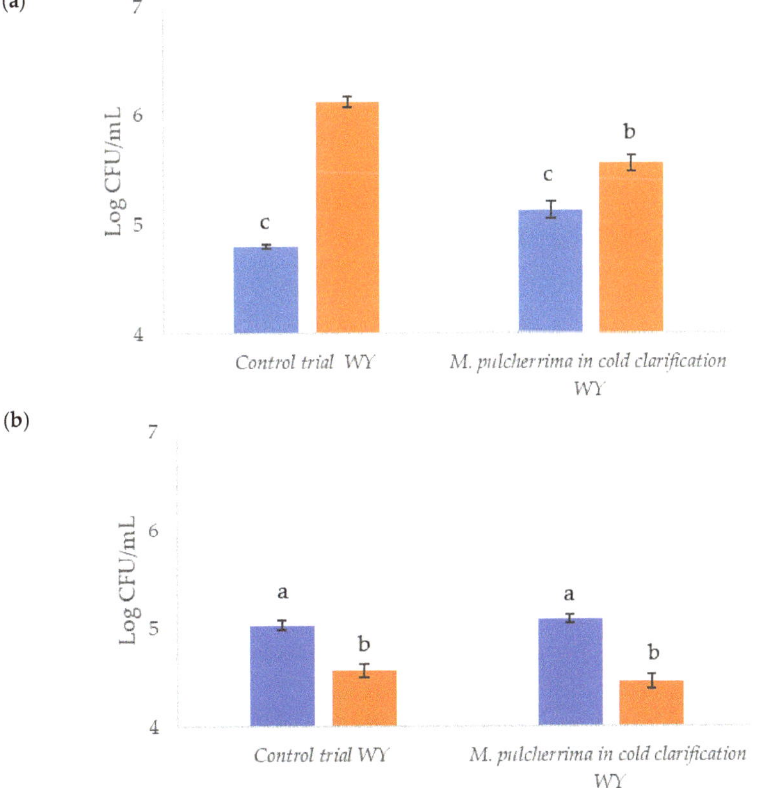

Figure 1. Effect of *M. pulcherrima* addition in cold clarification on wild yeasts (WY) in comparison with control fermentation: (**a**) 2021 vintage and (**b**) 2022 vintage. ■ before clarification; ■ after clarification. Data (n = 3) are the means ± standard deviation. Data with different superscript letters ([a,b,c]) are significantly different (Duncan tests; $p < 0.05$).

During the 2022 vintage (Figure 1b), although the wild population was quantitatively comparable to that of the first year, a significant reduction of WY was shown in both trials with and without the inoculation of *M. pulcherrima* (a reduction of WY of c.a. 0.6 Log CFU/mL and 0.3 Log CFU/mL, respectively). The major containment of WY in the 2022 vintage could also be explained by the presence of wild *M. pulcherrima* (about 10^4 cell/mL) in grape juice.

3.2. Pied de Cuve Inoculum and S. cerevisiae Strains Typing

The two lots of clarified grape juice in both vintages (2021–2022) were inoculated with *pied de cuves* coming from spontaneous fermentation described in the material and methods section. Both *pied de cuves* before the inoculum of vats showed the exclusive presence of *S. cerevisiae* with a viable cell count of 5×10^8 cell/mL. The results of the characterization at the strain level of *S. cerevisiae* population are reported in Table 1. Biotyping results revealed the presence of a total of nine different profiles. The biotypes present in the *pied de cuves* almost dominated the fermentation processes with 90% and 58.3% in the 2021 vintage control and inoculated *M. pulcherrima* trials, respectively, (biotypes I, II, III, and IV) and 60% and 50% in the 2022 vintage control and inoculated *M. pulcherrima* trials, respectively (biotypes II, III, and VIII). Interestingly, 65% of the *S. cerevisiae* population showed the same biotype profile (biotypes I, II, and III) contributing to both 2021 and 2022 vintages.

Table 1. *S. cerevisiae* biotypes found in the spontaneous fermentation of *pied de cuves* and in the fermentation processes.

	2021			**2022**		
	Byotype	**%**	**n. Strains**	**Byotype**	**%**	**Strains**
Pied de cuve	I	12.5	2	II	67.0	12
	II	37.5	6	III	11.0	2
	III	12.5	2	VII	11.0	2
	IV	12.5	2	VIII	11.0	2
	V	25.0	4	-	-	-
Fermentation control	VI	10.0	1	II	40.0	4
	I	20.0	2	III	-	-
	II	30.0	3	VIII	20.0	2
	III	-	-	I	20.0	2
	IV	40.0	4	IX	20.0	2
M. pulcherrima in cold clarification	VI	42.0	5	I	50.0	4
	I	17.0	2	II	50.0	4
	II	17.0	2	III	-	-
	III	8.0	1	-	-	-
	IV	17.0	2	-	-	-

3.3. Biomass Evolution and Saccharomyces cerevisiae Characterization

The biomass evolution after 24 h of cold clarification and the inoculum of *pied de cuve* in the vintage 2021 is reported in Figure 2a. The *S. cerevisiae* evolution in the two fermentations showed a similar trend, therefore the presence of *M. pulcherrima* did not affect the development of *S. cerevisiae*. The initial cell concentration was 10^6 cell/mL, achieving the maximum of growth kinetics at the 5th day of fermentation to remain constant until the end of fermentation. WY in both fermentations exhibited a similar trend: they started with the different amounts found in Figure 1 and disappeared at the 4th of fermentation. After cold clarification, the inoculated *M. pulcherrima* maintained a cell concentration of 10^3 cell/mL until the 8th day and then disappeared on the 12th day. The growth kinetics of trials of the vintage 2022 are reported in Figure 2b. The biomass evolution of *S. cerevisiae* also showed a similar trend to the vintage 2022, although there was a more abundant inoculum (10^7 cell/mL) achieving the maximum biomass on the 3rd day of fermentation (10^8 cell/mL) to remain constant at the end of fermentation. Regarding wild yeast evolution, the two fermentations did not show differences in terms of the cell concentration. The overall results indicated that an adequate starter inoculum is also relevant in the control of WY with and without *M. pulcherrima*.

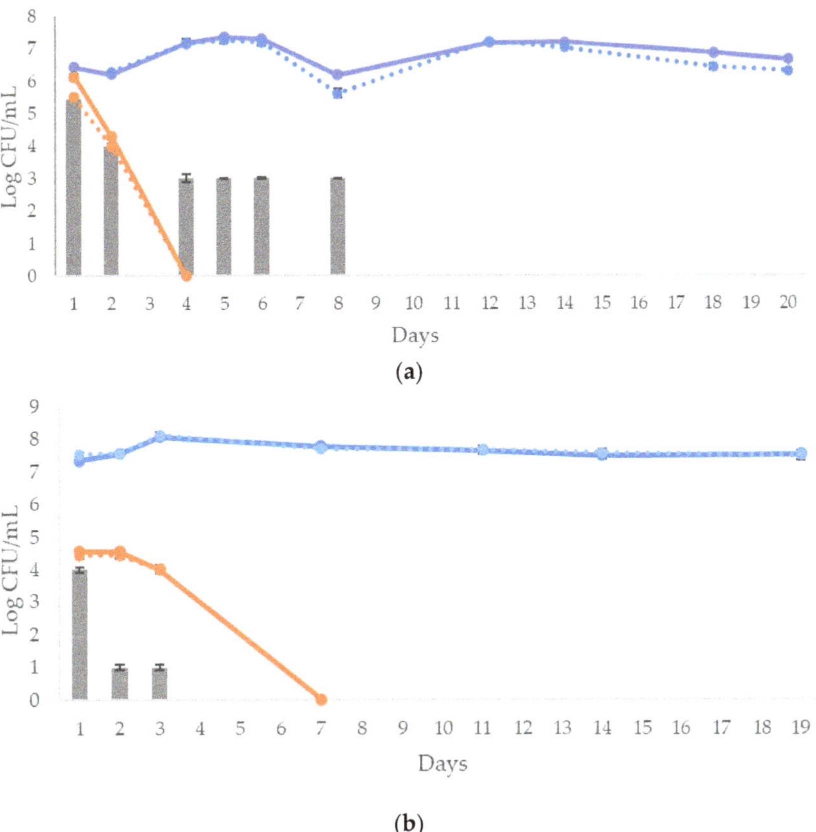

(a)

(b)

Figure 2. Growth kinetics of the yeasts ‧‧●‧‧ population in both (**a**) 2021 vintage and (**b**) 2022 vintage wines. ─●─ *S. cerevisiae* in the control trial; *S. cerevisiae* in *M. pulcherrima* cold clarification; ─●─ WY in the control trial; ‧‧●‧‧ WY in *M. pulcherrima* cold clarification; ▰▰▰ *M. pulcherrima* biomass in the trial with *M. pulcherrima* cold clarification.

3.4. Main Oenological Characters of Wine

The results of the main analytical characters of wines obtained in 2021 and 2022 are shown in Table 2. The presence of *M. pulcherrima* during cold clarification did not generally significantly affect the main parameters analyzed in both vintages 2021 and 2022. Indeed, the resulting wines coming from the two vintages exhibited a similar analytical profile. The only exception was the residual sugar content: the vintage 2021 was significantly higher in the control trial, while in 2022 the data showed an opposite result. Even comparing the wines in the two vintages, the results did not show any differences, except for the higher volatile acidity value in both fermentations in the 2021 vintage.

Table 2. The main analytical characteristics of resulting wine coming from 2021 and 2022 vintages.

	M. pulcherrima in Cold Clarification (2021)	Control Trial (2021)	*M. pulcherrima* in Cold Clarification (2022)	Control Trial (2022)
Ethanol %v/v	15.05 ± 0.02 [a]	14.49 ± 0.01 [b]	14.81 ± 0.15 [a]	15.08 ± 0.02 [a]
Total acidity (g/L tartaric acid)	6.26 ± 0.09 [a]	6.52 ± 0.01 [a]	7.56 ± 0.12 [a]	7.84 ± 0.14 [a]
Sugar residue (g/L)	3.7 ± 0.02 [b]	14.0 ± 0.03 [a]	6.6 ± 0.7 [a]	1.5 ± 0.1 [b]
pH	3.28 ± 0.02 [a]	3.25 ± 0.00 [a]	3.08 ± 0.00 [a]	3.09 ± 0.00 [a]
Volatile acidity (g/L acetic acid)	0.64 ± 0.01 [a]	0.56 ± 0.02 [b]	0.34 ± 0.02 [a]	0.36 ± 0.03 [a]
Total SO$_2$ (mg/L)	25 ± 0.9 [a]	26 ± 0.8 [a]	21 ± 0.9 [a]	21 ± 0.9 [a]
Malic acid (g/L)	0.55 ± 0.03 [a]	0.49 ± 0.02 [a]	0.54 ± 0.05 [a]	0.55 ± 0.05 [a]
Ethanol yield (g/g)	0.45 ± 0.07 [a]	0.45 ± 0.06 [a]	0.44 ± 0.09 [a]	0.44 ± 0.09 [a]

Data (n = 3) are the means ± standard deviation. Data with different superscript letters ([a,b]) within each row and vintage are significantly different (Duncan tests; $p < 0.05$).

3.5. Volatile Compounds of Wine

The resulting volatile compounds of vintage wines coming from 2021 and 2022 are reported in Table 3. The inoculation of *M. pulcherrima* at the start of cold clarification generally led to wine with a different volatile profile. Indeed, a significant increase in ethyl hexanoate with relevant high OAV values in comparison with the control was observed. Moreover, the trials with *M. pulcherrima* exhibited a significant increase in monoterpenes content, in particular with respect to geraniol and nerol (1.5 OAV), and two higher alcohols (n-propanol and isobutanol) in comparison with control trials. The results showed a significant increase in the acetaldehyde content with *M. pulcherrima* but within the limits of the negative threshold. On the contrary, the control trial exhibited a significant increase in the isoamyl acetate and β-phenyl ethanol.

Table 3. The main by-products and volatile compounds in final wines during two vintages 2021–2022.

	M. pulcherrima in Cold Clarification 2021	OAV (Odor Activity Value)	Control Trial 2021	OAV (Odor Activity Value)	*M. pulcherrima* in Cold Clarification 2022	OAV (Odor Activity Value)	Control Trial 2022	OAV (Odor Activity Value)
Esters (mg/L)								
Ethyl butyrate	0.13 ± 0.014 [a]	0.325	0.13 ± 0.00 [a]	0.325	0.187 ± 0.007 [a]	0.467	0.148 ± 0.006 [b]	0.445
Ethyl acetate	35.75 ± 0.41 [a]	2.97	35.55 ± 0.36 [a]	2.96	30.12 ± 0.18 [b]	2.51	31.91 ± 0.35 [a]	2.65
Phenyl ethyl acetate	0.89 ± 0.042 [a]	12.19	0.88 ± 0.034 [a]	12.05	0.821 ± 0.032 [a]	11.24	0.773 ± 0.045 [a]	10.58
Ethyl hexanoate	1.80 ± 0.121 [a]	22.5	0.63 ± 0.142 [b]	7.87	1.470 ± 0.050 [a]	18.37	0.612 ± 0.043 [b]	7.65
Ethyl octanoate	0.00 ± 0.00 [a]	0	0.00 ± 0.00 [a]	0	0.002 ± 0.000 [a]	0.003	0.003 ± 0.001 [a]	0.005
Isoamyl acetate	1.27 ± 0.026 [b]	7.93	2.22 ± 0.147 [a]	13.87	1.484 ± 0.112 [a]	9.27	1.404 ± 0.024 [a]	8.77
Alcohols (mg/L)								
n- propanol	30.88 ± 0.36 [a]	0.100	26.08 ± 0.13 [b]	0.08	36.44 ± 4.92 [a]	0.119	31.64 ± 0.05 [a]	0.103
Isobutanol	20.48 ± 0.29 [a]	0.512	18.30 ± 0.10 [b]	0.45	14.09 ± 0.34 [a]	2.83	11.41 ± 0.24 [b]	0.285
Amyl alcohol	13.00 ± 0.55 [a]	0.203	11.69 ± 0.41 [a]	0.18	10.03 ± 0.29 [a]	0.156	8.64 ± 0.48 [a]	0.135
Isoamyl alcohol	113.94 ± 0.05 [a]	1.89	114.16 ± 0.24 [a]	1.90	89.94 ± 0.12 [a]	1.49	80.95 ± 0.52 [b]	1.34
β-Phenyl Ethanol	55.9 ± 0.130 [b]	3.99	64.5 ± 0.152 [a]	4.60	14.01 ± 0.022 [a]	1	13.32 ± 0.045 [a]	0.951

Table 3. *Cont.*

	M. pulcherrima in Cold Clarification 2021	OAV (Odor Activity Value)	Control Trial 2021	OAV (Odor Activity Value)	*M. pulcherrima* in Cold Clarification 2022	OAV (Odor Activity Value)	Control Trial 2022	OAV (Odor Activity Value)
Carbonyl Compounds (mg/L)								
Acetaldehyde	63.16 ± 0.28 [a]	126.32	25.35 ± 0.37 [b]	70	22.39 ± 0.59 [a]	44.78	16.57 ± 0.1 [b]	33.14
Monoterpenes (mg/L)								
Linalool	0.015 ± 0.001 [a]	0.60	0.012 ± 0.002 [a]	0.48	0.010 ± 0.00 [a]	0.40	0.009 ± 0.00 [a]	0.36
Geraniol	0.010 ± 0.002 [a]	0.33	0.002 ± 0.000 [b]	0.06	0.016 ± 0.00 [a]	0.53	0.005 ± 0.001 [b]	0.17
Nerol	0.023 ± 0.001 [a]	1.53	0.005 ± 0.000 [b]	0.33	0.001 ± 0.000 [a]	0.06	0.001 ± 0.00 [a]	0.06

Data ($n = 3$) are the means ± standard deviation. Data with different superscript letters ([a,b]) within each row and vintage are significantly different (Duncan tests; $p < 0.05$).

The other volatile compounds did not show significant differences. The results of the wines coming from 2022 vintages showed an enhancement of ethyl butyrate and ethyl hexanoate content in the presence of *M. pulcherrima*, while control trials showed a significant increase in the ethyl acetate content.

Considering the two vintages, the results confirmed that the use of *M. pulcherrima* in cold clarification affected the volatile compounds in terms of the terpens, alcohols, and some esters compounds.

The main volatile compounds of the vintage wines obtained from 2021 and 2022 with and without the inoculum of *M. pulcherrima* were elaborated using the Principal Component Analysis (PCA). The total variance explained was 86.6% (PC1 = 66.0%; PC 2 = 26.6%). Figure 3 reports the distribution of fermentations to assess the effect of the yeast and the vintage in the function of the volatile compounds. The fermentation trials were in four different quadrants: PC1 distinguished the trials on the base of the vintage while PC 2 separated the trials in the function of the presence of *M. pulcherrima* during the cold clarification stage. The production of ethyl hexanoate and geraniol is characterized more specifically by the *M. pulcherrima* metabolism of volatile compounds imparting a specific aromatic imprint to the wine.

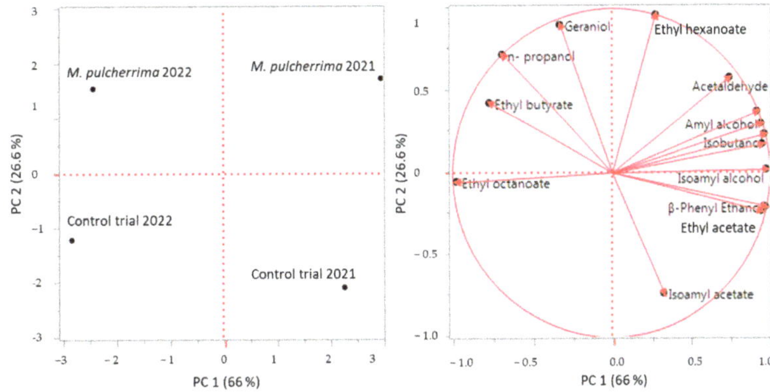

Figure 3. Principal component analysis of the main volatile compounds in wine coming from two consecutive vintages. The variance explained by principal component analysis (PCA) is PC 1 66% X-axis and PC 2 26.6% Y-axis.

3.6. Sensory Analysis

To evaluate the impact of *M. pulcherrima* inoculated in cold clarification on the aroma complexity, the final wines were subjected to sensory analysis. The sensory analysis of wines obtained by the 2021vintage are reported in Figure 4a. The results highlighted a positive judgement of testers regarding each wine, characterized by specific aromatic notes and without defects. The final wine from the control trial exhibited a more pronounced persistence and bitterness. The use of *M. pulcherrima* led to wines with emphasized notes of tropical fruits, sweetness, and more structure. No significant differences were shown regarding the other aromatic descriptors. As for the wines of 2022 (Figure 4b), the use of *M. pulcherrima* led to wine characterized by distinctive descriptors as structure, persistence, herbs, and tropical fruits, in comparison with the control wine. The sensory analysis highlighted the influence of the use of *M. Pulcherrima* in cold clarification on the sensory profile of aroma wine.

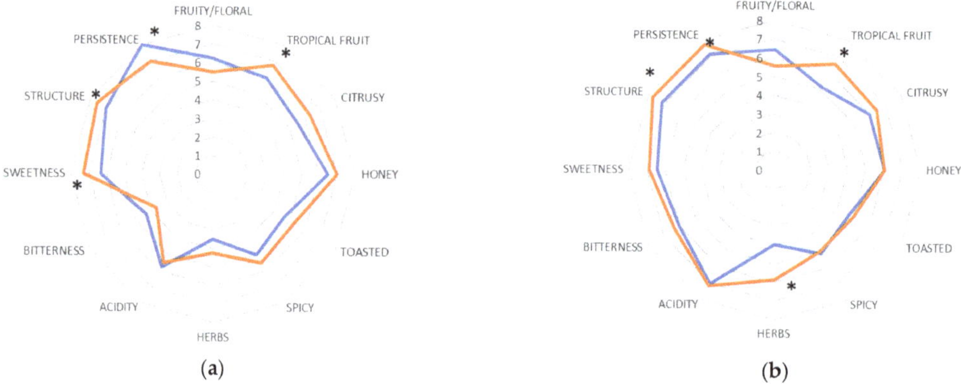

Figure 4. Sensory analysis of Verdicchio wine coming from 2021 vintage (**a**) and 2022 vintage (**b**). —— control trial; —— *M. pulcherrima* in cold clarification. *, significantly different (Fischer ANOVA; p-value < 0.05).

4. Discussion

The term "biocontrol" related to the use of microorganisms as biocontrol agents was defined by Baker and Cook [30] as the reduction of a pathogen or disease activities through an organism. In agri-food, this concept is related to an alternative strategy to the use of chemical products and the use of a microorganism with antagonist action against other microorganisms reducing the use of pesticides and boosting food quality and safety [19,30–32]. The increased consumer demands for safe food and beverages invigorated research on the development of safe and ecofriendly wine, where the protection against undesirable microorganisms before, during, after, or at the end of the fermentation process is required [6,7,14]. Effectively, the use of non-*Saccharomyces* yeasts could be a valid strategy as potential antagonists against phytopathogenic fungi of the genera due to their ability to produce a wide spectrum of secondary metabolites. One of the yeast species best known for its biocontrol potential is *M. pulcherrima*. This characteristic is also linked to the production of the pulcherrimin pigment, which in turn is linked to the availability of iron in the medium which would lead to the deprivation of different yeast species as *Brettanomyces/Dekkera*, *Pichia*, *Hanseniaspora*, spp. *Saccharomycodes ludwigii*, and *Candida* spp. [33]. *M. pulcherrima* can be compatible with the main yeast used for wine production, for example in mixed fermentations, thus resulting in a reduction in the SO_2 dose and hence were usually used as an antimicrobial agent [34]. Moreover, recent investigations conducted using *M. pulcherrima* strains or a mix of *M. pulcherrima* and *T. delbrueckii* in the red wine making process at the prefermentative stage [7,8,14].

In this work carried out at winery conditions, the efficacy of *M. pulcherrima* to control the wild yeast population was found in cold clarification. On the other hand, the timely and appropriate starter inoculum in the two years determined an effective control of wild yeast during fermentation with and without *M. pulcherrima* inoculum. The WY control was particularly evident in the 2022 vintage, where a high concentration of wild *M. pulcherrima* was already present in grape juice before the clarification stage, influencing the control trial. These data corroborate the results reported by other recent studies regarding the use of *M. pulcherrima* in wine making as a bioprotectant agent [6–8]. In particular, the inoculum of a selected *M. pulcherrima* strain would ensure the biocontrol activity and attempt to impart specific aromatic traits to the final wine.

The bioprotectant action of *M. pulcherrima* used in cold clarification was displayed without the addition of SO_2 in Verdicchio grape juice with an inoculum of a *pied de cuve* with indigenous *S. cerevisiae* strains.

The molecular monitoring of biotype profiles that led to the fermentation, including those present in the *pied de cuve*, revealed a rather stable yeast strain consortium in the presence and absence of *M. pulcherrima* (80 %) as well as from a vintage to another (50%), even if same differences can be found. This reflected the results of the PCA analysis indicating that the impact on volatile profile was related not only to the effect of *M. pulcherrima* but also to the different vintages and *S. cerevisiae* yeast strain variation. The presence of *M. pulcherrima* increased ethyl hexanoate, isobutanol, acetaldehyde, and geraniol although with different concentrations in both vintages.

Generally, the use of *M. pulcherrima* affected the esters, alcohol, and monoterpens compounds that contribute to defining the overall sensory characteristics of wines and increasing the perception of the varietal aroma of grapes [1]. Indeed, *M. pulcherrima* was indicated in several works to improve the concentration of volatile compounds [35–37]. This species positively contributes to volatile thiol release in wines, especially during the pre-fermentation stage in wine making [36]. The sensory analysis confirmed the results of main analytical characters and volatile compounds, particularly regarding the descriptors of tropical fruit and structure.

In this work, the multifactorial role of *M. pulcherrima* in wine making was highlighted again. As recently reported [38,39], mixed fermentation with *M. pulcherrima* enhanced the ester profile and increased the final quality of the wine, such as the sensory characteristics and color parameters.

In conclusion, the results obtained indicate that the use of *M. pulcherrima* in the cold clarification stage plays an effective biocontrol action against the wild yeast population. This effect could be a strategy to reduce the use of sulfur dioxide in wine fermentation. Moreover, *M. pulcherrima* was able to enhance the aromatic and sensory profile of the wine.

Author Contributions: A.A., L.C., F.C. and M.C. contributed equally to the manuscript. All the authors participated in the design and discussion of the research. A.A. carried out the experimental part of the work. A.A. and L.C. carried out the analysis of the data. All authors have read and agreed to the published version of the manuscript.

Funding: This research is a part of the project SYSTEMIC "An integrated approach to the challenge of sustainable food systems: adaptive and mitigatory strategies to address climate change and malnutrition", knowledge hub on Nutrition and Food Security, and has received funding from national research funding parties in Belgium (FWO), France (INRA), Germany (BLE), Italy (MIPAAF), Latvia (IZM), Norway (RCN), Portugal (FCT), and Spain (AEI) in a joint action of JPI HDHL, JPIOCEANS, and FACCE-JPI launched in 2019 under the ERA-NET ERA-HDHL (n°696295).

Institutional Review Board Statement: Not applicable.

Informed Consent Statement: Not applicable.

Data Availability Statement: Not applicable.

Acknowledgments: The authors would like to thank the Col di Corte srl Soc. Agricola winery for the samples and the assistance of the winemaker Claudio Calderoni.

Conflicts of Interest: The authors declare no conflict of interest.

References

1. Escribano-Viana, R.; González-Arenzana, L.; Portu, J.; Garijo, P.; López-Alfaro, I.; López, R.; Santamaría, P.; Gutiérrez, A.R. Wine aroma evolution throughout alcoholic fermentation sequentially inoculated with non-*Saccharomyces*/*Saccharomyces* yeasts. *Food Res. Int.* **2018**, *112*, 17–24. [CrossRef] [PubMed]
2. Roullier-Gall, C.; Hemmler, D.; Gonsior, M.; Li, Y.; Nikolantonaki, M.; Aron, A.; Coelho, C.; Gougeon, R.D.; Schmitt-Kopplin, P. Sulfites and the wine metabolome. *Food Chem.* **2017**, *237*, 106–113. [CrossRef]
3. Ubeda, C.; Hornedo-Ortega, R.; Cerezo, A.B.; Garcia-Parrilla, M.C.; Troncoso, A.M. Chemical hazards in grapes and wine, climate change and challenges to face. *Food Chem.* **2020**, *314*, 126222. [CrossRef] [PubMed]
4. Giacosa, S.; Segade, S.R.; Cagnasso, E.; Caudana, A.; Rolle, L.; Gerbi, V. SO₂ in wines: Rational use and possible alternatives. In *Red Wine Technology*; Academic Press: Cambridge, MA, USA, 2019; pp. 309–321.
5. Lisanti, M.T.; Blaiotta, G.; Nioi, C.; Moio, L. Alternative methods to SO₂ for microbiological stabilization of wine. *Compr. Rev. Food Sci. Food Saf.* **2019**, *18*, 455–479. [CrossRef] [PubMed]
6. Chacon-Rodriguez, L.; Joseph, C.L.; Nazaris, B.; Coulon, J.; Richardson, S.; Dycus, D.A. Innovative use of non-*Saccharomyces* in bio-protection: *T. delbrueckii* and *M. pulcherrima* applied to a machine harvester. *Catal. Discov. Into Pract.* **2020**, *4*, 82–90. [CrossRef]
7. Simonin, S.; Alexandre, H.; Nikolantonaki, M.; Coelho, C.; Tourdot-Maréchal, R. Inoculation of *Torulaspora delbrueckii* as a bio-protection agent in winemaking. *Food Res. Int.* **2018**, *107*, 451–461. [CrossRef] [PubMed]
8. Windholtz, S.; Redon, P.; Lacampagne, S.; Farris, L.; Lytra, G.; Cameleyre, M.; Barbe, J.C.; Coulon, J.; Masneuf-Pomarede, I. Non-*Saccharomyces* yeasts as bioprotection in the composition of red wine and in the reduction of sulfur dioxide. *LWT* **2021**, *149*, 111781. [CrossRef]
9. Prior, K.J.; Bauer, F.F.; Divol, B. The utilisation of nitrogenous compounds by commercial non-*Saccharomyces* yeasts associated with wine. *Food Microbiol.* **2019**, *79*, 75–84. [CrossRef] [PubMed]
10. Comitini, F.; Capece, A.; Ciani, M.; Romano, P. New insights on the use of wine yeasts. *Curr. Opin. Food Sci.* **2017**, *13*, 44–49. [CrossRef]
11. Comitini, F.; Agarbati, A.; Canonico, L.; Galli, E.; Ciani, M. Purification and characterization of WA18, a new mycocin produced by *Wickerhamomyces anomalus* active in wine against *Brettanomyces bruxellensis* spoilage yeasts. *Microorganisms* **2020**, *9*, 56. [CrossRef]
12. Oro, L.; Ciani, M.; Comitini, F. Antimicrobial activity of *Metschnikowia pulcherrima* on wine yeasts. *J. Appl. Microbiol.* **2014**, *116*, 1209–1217. [CrossRef]
13. Di Gianvito, P.; Englezos, V.; Rantsiou, K.; Cocolin, L. Bioprotection strategies in winemaking. *Int. J. Food Microbiol.* **2022**, *364*, 109532. [CrossRef]
14. Escribano-Viana, R.; González-Arenzana, L.; Garijo, P.; Fernández, L.; López, R.; Santamaría, P.; Gutiérrez, A.R. Bioprotective effect of a *Torulaspora delbrueckii*/*Lachancea thermotolerans*-mixed inoculum in red winemaking. *Fermentation* **2022**, *8*, 337. [CrossRef]
15. Simonin, S.; Roullier-Gall, C.; Ballester, J.; Schmitt-Kopplin, P.; Quintanilla-Casas, B.; Vichi, S.; Peyron, D.; Alexandre, H.; Tourdot-Maréchal, R. Bio-Protection as an Alternative to Sulphites: Impact on Chemical and Microbial Characteristics of Red Wines. *Front. Microbiol.* **2020**, *11*, 1308. [CrossRef] [PubMed]
16. Pawlikowska, E.; James, S.A.; Breierova, E.; Antolak, H.; Kregiel, D. Biocontrol capability of local *Metschnikowia* sp. isolates. *Antonie Van Leeuwenhoek* **2019**, *112*, 1425–1445. [CrossRef]
17. Kuchen, B.; Vazquez, F.; Maturano, Y.P.; Scaglia, G.J.E.; Pera, L.M.; Vallejo, M.D. Toward application of biocontrol to inhibit wine spoilage yeasts: The use of statistical designs for screening and optimisation. *OENO One* **2021**, *55*, 75–96. [CrossRef]
18. Varela, C.; Sengler, F.; Solomon, M.; Curtin, C. Volatile flavour profile of reduced alcohol wines fermented with the non-conventional yeast species *Metschnikowia pulcherrima* and *Saccharomyces uvarum*. *Food Chem.* **2016**, *209*, 57–64. [CrossRef] [PubMed]
19. Zhang, X.; Li, B.; Zhang, Z.; Chen, Y.; Tian, S. Antagonistic yeasts: A promising alternative to chemical fungicides for controlling postharvest decay of fruit. *J. Fungi.* **2020**, *6*, 158. [CrossRef] [PubMed]
20. Csutak, O.; Vassu, T.; Sarbu, I.; Stoica, I.; Cornea, P. Antagonistic activity of three newly isolated yeast strains from the surface of fruits. *Food Technol. Biotechnol.* **2013**, *51*, 70–77.
21. Kántor, A.; Kačániová, M. Diversity of bacteria and yeasts on the surface of table grapes. *Anim. Sci. Biotechnol.* **2015**, *48*, 149–155.
22. Oro, L.; Feliziani, E.; Ciani, M.; Romanazzi, G.; Comitini, F. Biocontrol of postharvest brown rot of sweet cherries by *Saccharomyces cerevisiae* Disva 599, *Metschnikowia pulcherrima* Disva 267 and *Wickerhamomyces anomalus* Disva 2 strains. *Postharvest Biol. Technol.* **2014**, *96*, 64–68. [CrossRef]
23. Saravanakumar, D.; Ciavorella, A.; Spadaro, D.; Garibaldi, A.; Gullino, M.L. *Metschnikowia pulcherrima* strain MACH1 outcompetes *Botrytis cinerea*, *Alternaria alternata* and *Penicillium expansum* in apples through iron depletion. *Postharvest Biol. Technol.* **2008**, *49*, 121–128. [CrossRef]
24. Agarbati, A.; Canonico, L.; Pecci, T.; Romanazzi, G.; Ciani, M.; Comitini, F. Biocontrol of non-*Saccharomyces* yeasts in vineyard against the gray mold disease agent *Botrytis cinerea*. *Microorganisms* **2022**, *10*, 200. [CrossRef]
25. Agarbati, A.; Canonico, L.; Ciani, M.; Comitini, F. The impact of fungicide treatments on yeast biota of Verdicchio and Montepulciano grape varieties. *PLoS ONE* **2019**, *14*, e0217385. [CrossRef]

26. Legras, J.L.; Karst, F. Optimisation of interdelta analysis for *Saccharomyces cerevisiae* strain characterisation. *FEMS Microbiol. Lett.* **2003**, *221*, 249–255. [CrossRef] [PubMed]
27. Dukes, B.C.; Butzke, C.E. Rapid determination of primary amino acids in grape juice using an o-phthaldialdehyde/N-acetyl-L-cysteine spectrophotometric assay. *Am. J. Enol. Vitic.* **1998**, *49*, 125–134. [CrossRef]
28. Canonico, L.; Comitini, F.; Ciani, M. Influence of vintage and selected starter on *Torulaspora delbrueckii*/*Saccharomyces cerevisiae* sequential fermentation. *Eur. Food Res. Technol.* **2015**, *241*, 827–833. [CrossRef]
29. Canonico, L.; Ciani, E.; Galli, E.; Comitini, F.; Ciani, M. Evolution of aromatic profile of *Torulaspora delbrueckii* mixed fermentation at microbrewery plant. *Fermentation* **2020**, *6*, 7. [CrossRef]
30. Ab Rahman, S.F.S.; Singh, E.; Pieterse, C.M.; Schenk, P.M. Emerging microbial biocontrol strategies for plant pathogens. *Plant Sci.* **2018**, *267*, 102–111. [CrossRef]
31. Rodriguez-Navarro, C.; González-Muñoz, M.T.; Jimenez-Lopez, C.; Rodriguez-Gallego, M. Bioprotection. In *Encyclopedia of Earth Sciences Series*; Finkl, C.W., Ed.; Springer: Berlin, Germany, 2011; pp. 185–189.
32. Biocontrol-Dictionary Agroecology. Available online: https://dicoagroecologie.fr/en/encyclopedia/biocontrol/ (accessed on 2 February 2023).
33. Morata, A.; Loira, I.; Escott, C.; del Fresno, J.M.; Bañuelos, M.A.; Suárez-Lepe, J.A. Applications of *Metschnikowia pulcherrima* in wine biotechnology. *Fermentation* **2019**, *5*, 63. [CrossRef]
34. Türkel, S.; Ener, B. Isolation and characterization of new *Metschnikowia pulcherrima* strains as producers of the antimicrobial pigment pulcherrimin. *Z. Für Nat. C* **2009**, *64*, 405–410. [CrossRef] [PubMed]
35. Zott, K.; Thibon, C.; Bely, M.; Lonvaud-Funel, A.; Dubourdieu, D.; Masneuf-Pomarede, I. The grape must non-*Saccharomyces* microbial community: Impact on volatile thiol release. *Int. J. Food Microbiol.* **2011**, *151*, 210–215. [CrossRef]
36. Ruiz, J.; Belda, I.; Beisert, B.; Navascués, E.; Marquina, D.; Calderón, F.; Rauht, D.; Santos, A.; Benito, S. Analytical impact of *Metschnikowia pulcherrima* in the volatile profile of Verdejo white wines. *App. Microbiol. Biotechnol.* **2018**, *102*, 8501–8509. [CrossRef]
37. Vicente, J.; Ruiz, J.; Belda, I.; Benito-Vázquez, I.; Marquina, D.; Calderón, F.; Santos, A.; Benito, S. The genus *Metschnikowia* Muñoz-in enology. *Microorganisms* **2020**, *8*, 1038. [CrossRef] [PubMed]
38. Redondo, J.M.; Puertas, B.; Cantos-Villar, E.; Jiménez-Hierro, M.J.; Carbú, M.; Garrido, C.; Puiz-Moreno, M.J.; Moreno-Rojas, J.M. Impact of sequential inoculation with the non-*Saccharomyces T. delbrueckii* and *M. pulcherrima* combined with *Saccharomyces cerevisiae* strains on chemicals and sensory profile of rosé wines. *J. Agr. Food Chem.* **2021**, *69*, 1598–1609. [CrossRef] [PubMed]
39. Duarte, F.L.; Egipto, R.; Baleiras-Couto, M.M. Mixed fermentation with *Metschnikowia pulcherrima* using different grape varieties. *Fermentation* **2019**, *5*, 59. [CrossRef]

fermentation

Article

Fermentation of Cocoa (*Theobroma cacao* L.) Pulp by *Laetiporus persicinus* Yields a Novel Beverage with Tropical Aroma

Victoria Klis [1,2], Eva Pühn [2], Jeanny Jaline Jerschow [1,2], Marco Alexander Fraatz [2] and Holger Zorn [1,2,*]

[1] Fraunhofer Institute for Molecular Biology and Applied Ecology, Ohlebergsweg 12, 35392 Giessen, Germany; victoria.klis@ime.fraunhofer.de (V.K.); jeanny.jerschow@ime.fraunhofer.de (J.J.J.)

[2] Institute of Food Chemistry and Food Biotechnology, Justus Liebig University Giessen, Heinrich-Buff-Ring 17, 35392 Giessen, Germany; marco.fraatz@lcb.chemie.uni-giessen.de (M.A.F.)

* Correspondence: holger.zorn@ime.fraunhofer.de; Tel.: +49-641-972-19130

Abstract: Cocoa pulp represents an interesting by-product of cocoa production, with an appealing flavor. We developed a non-alcoholic beverage via the submerged fermentation of 10% pasteurized cocoa pulp in water with *Laetiporus persicinus* for 48 h; the product was characterized by tropical fruity notes such as coconut, mango, passion fruit and peach. The overall acceptance of the beverage compared to the non-fermented medium, as rated by a panel, increased from 2.9 to 3.7 (out of 5.0 points) for odor and from 2.1 to 4.2 for taste. (*R*)-Linalool (flowery, fruity), methyl benzoate (green, sweet), 2-phenylethanol (rose, sweet), 5-butyl-2(5*H*)-furanone (coconut, peach) and (*E*)-nerolidol (flowery, woody) contributed to the overall aroma with odor activity values of >1. During aroma dilution analysis, further substances with coconut, passion fruit and peach-like notes were perceived and structurally assigned to the group of sesquiterpenoids. The fermentation generated a highly interesting beverage using only 10% of the valuable cocoa pulp. The aroma formation via the fungus *L. persicinus* on cocoa pulp is of great interest for further research as an example of the formation of substances not yet described in the literature.

Keywords: cocoa pulp; basidiomycetes; beverage; *Laetiporus persicinus*; aroma dilution analysis

Citation: Klis, V.; Pühn, E.; Jerschow, J.J.; Fraatz, M.A.; Zorn, H. Fermentation of Cocoa (*Theobroma cacao* L.) Pulp by *Laetiporus persicinus* Yields a Novel Beverage with Tropical Aroma. *Fermentation* **2023**, *9*, 533. https://doi.org/10.3390/fermentation9060533

Academic Editor: Niel Van Wyk

Received: 28 April 2023
Revised: 25 May 2023
Accepted: 26 May 2023
Published: 30 May 2023

1. Introduction

Cocoa represents a valuable natural resource with a steadily increasing annual level of production. The forecasted amount of cocoa beans produced in 2022/2023 was over 5.0 million tons [1]. The cocoa value chain offers much potential for improvement as low levels of value added and price fluctuations have major economic and environmental consequences for many smallholders. Several by-products are generated during the production of cocoa beans, which could contribute to more sustainable cocoa farming through upcycling [2]. In addition to cocoa pod husks and cocoa bean shells, by-products include cocoa pulp, which is traditionally used for the fermentation of the cocoa beans and is thus ordinarily lost [3]. However, it has been shown in various studies that a part of the cocoa pulp can be separated prior to fermentation without negatively affecting the flavor of the beans [4,5].

Cocoa pulp has become the focus of recent studies due to its chemical composition and interesting aroma. Depending on the origin, Bickel Haase et al. detected up to 65 different aroma-active substances in the pulp [6]. The substances that typically characterize the aroma include, besides others, 4-vinyl-2-methoxyphenol (clove), δ-decalactone (coconut), linalool (flowery), β-damascenone (fruity, grape) and γ-nonalactone (fruity, coconut). Because of its high sugar content, it represents a suitable starting material for the production of alcoholic beverages like fruit wine or beer via fermentation with yeasts [7,8]. Other fermented products have also been studied, such as the cocoa pulp-based kefir drink [9].

Fermentation employing higher fungi of the division Basidiomycota has been described in the literature, especially due to the potential of these fungi for use in the production of natural-aroma compounds [10]. Well-known examples are the production of vanillin by *Phanerochaete chrysosporium* [11] and benzaldehyde by different *Pleurotus* species [12,13], or the release of a wild strawberry-like flavor caused by the formation of (*R*)-linalool, methyl anthranilate, 2-aminobenzaldehyde and geraniol during the fermentation of black current pomace by *Wolfiporia cocos* [14]. Fermentation by basidiomycetes in submerged cultures has the advantage of a direct use as a beverage after separation of the fungal mycelium [15,16]. A broad spectrum of aroma compounds can be formed by de novo synthesis or by biotransformation, thereby naturally flavoring the beverage. The production of flavoring substances by biotechnological processes is an important alternative to plant-based and chemical sources. It is also advantageous that biotechnologically obtained flavors can be marketed as natural flavors according to current European and US legislation [17–19].

Generating valuable products from cocoa pulp is an important approach for making cocoa farming more sustainable and adding value to the cocoa fruit [2]. Therefore, the aim of the present study was to develop a novel beverage through the fermentation of cocoa pulp with basidiomycetes.

2. Materials and Methods

2.1. Screening of Basidiomycetes

To select a suitable fungus for the fermentation of cocoa pulp, 20 different basidiomycetes were screened in surface cultures for 20 days on cocoa pulp agar plates and on malt extract agar plates. The latter were used as reference media for the comparison of substrate-specific aroma formation. The smell was described and rated every second day for intensity and overall rating by three panelists (all female, 25–27 years old, all non-smokers) as '- -' means very weak/very bad; '-' weak/bad; '0' medium/neutral; '+' intensive/good; and '++' very intensive/very good. Malt extract agar contained 15 g/L agar–agar (Carl Roth GmbH, Karlsruhe, Germany) and 20 g/L malt extract (Carl Roth GmbH, Karlsruhe, Germany). Cocoa pulp agar contained 15 g/L agar–agar and 100 g/L pasteurized cocoa pulp (origin: Ecuador; purchased from Carbosse Naturals AG, Zürich, Switzerland).

2.2. Sterile Control of Pasteurized Cocoa Pulp

Submerged fermentations were carried out with pasteurized cocoa pulp. Sterile controls were performed on LB agar (15 g/L agar–agar, 20 g/L LB-medium (Carl Roth GmbH, Karlsruhe, Germany)). Approximately 1 g cocoa pulp was dispersed in 10 mL sterile, demineralized water and 1 mL of the solution was inoculated on the agar plate and spread with a Drigalski spatula. The plates were incubated at 37 °C for 48 h.

2.3. Fermentation of Cocoa Pulp with Laetiporus persicinus in Submerged Cultures

L. persicinus (CBS 274.92) was obtained from the Westerdijk Fungal Biodiversity Institute (Utrecht, The Netherlands). For strain maintenance, the fungus was kept on malt extract agar plates and transferred to a new plate every nine days using a spatula by cutting a ~1 cm^2 piece of overgrown agar. For the pre-cultures, 100 mL sterilized malt extract media (20 g/L malt extract in drinking water) were placed in an Erlenmeyer flask (250 mL) and inoculated with 1 cm^2 of overgrown agar. Homogenization was performed with an ULTRA-TURRAX (IKA Works Inc., Staufen, Germany) at 10,000 rpm for 30 s. Cultivation took place on a horizontal shaker at 150 rpm at 24 °C in the dark for nine days. Main cultures with cocoa pulp medium (CP-M) were prepared by autoclaving 185 mL tap water in 500 mL Erlenmeyer flasks. After cooling to room temperature, 20 g pasteurized cocoa pulp was added under sterile conditions. The pre-culture was homogenized using an ULTRA-TURRAX as described above and centrifuged (10 min, 3500× *g*, 20 °C). The supernatant was discarded, and the tube was filled with sterile water. This procedure was repeated three times. The washed pre-culture was inoculated with 20 mL (10%) to the main culture medium. Fermentation took place on a horizontal shaker at 150 rpm at 24 °C in the dark for 72 h. The fermentates were harvested after fermentation times of 12, 24, 36, 48, 60 and 72 h by centrifugation (10 min, 3500× *g*, 20 °C). The supernatant was either stored at −20 °C and used for further analysis, or directly subjected to sensory evaluation.

2.4. Sensory Evaluation over the Cultivation Period

For the sensory evaluation of CP-M and fermented beverages, the samples were first examined by a panel in a simple descriptive test (DIN 10964) in order to establish their attributes for odor and taste. Subsequently, a conventional profile test with quantitative descriptions of the intensities of the respective attribute was performed. Therefore, the panelists rated the attributes from 0 (not recognizable) to 5 (very strongly recognizable) (DIN 10967-1). The attributes used were sweet/sweetish, acidic, tropical, passion fruit-, peach-, mango-, coconut-, citrus-, honey-, rhubarb-, pineapple-, apricot-, apple-, and tea-like. The test was followed by an overall evaluation of the acceptability of the respective sample. The beverages were analyzed after 12, 24, 36, 48, 60 and 72 h of fermentation. The cocoa pulp medium (CP–M) served as blank. The panel comprised ten trained panelists (two male, eight female, 21–29 years old, all non-smokers). All sensory descriptions were carried out in a test laboratory according to DIN 10962.

2.5. Aroma Analysis Using Direct Immersion Stir Bar Sorptive Extraction (diSBSE)

A total of 5 mL of the respective fermentate or CP–M was added to a 20 mL GC vial. The stir bars (10 mm with 0.5 mm PDMS coating) were conditioned prior to every analysis in a TubeConditioner TC 2 (GERSTEL, Mülheim an der Ruhr, Germany). diSBSE was carried out at room temperature on a multimagnetic stirring plate (MIXdrive 12, 2 mag, Munich, Germany) at 150 rpm for 30 min. After extraction, the stir bars were rinsed with *dd*H_2O, dried with lint-free tissues and placed back in a conditioned tube.

Gas chromatographic analyses were performed using an Agilent 8890 GC-system (Agilent Technologies, Santa Clara, CA, USA) connected to a 7010B GC/TQ mass spectrometric detector (Agilent Technologies). The system was equipped with a Thermal Desorption Unit 2 (TDU) (GERSTEL), an Olfactory Detection Port 4 (ODP 4) (GERSTEL) and a VF-WAXms column (30 m, i.d. 250 µm, film thickness 0.25 µm) or a DB-5ms column (30 m, i.d. 250 µm, film thickness 0.25 µm; both Agilent Technologies). Desorption started with an initial temperature of 40 °C in the TDU (0.5 min). This temperature increased from 40 °C with 120 °C per min to 250 °C, and this level was maintained for 12 min. Cryogenic focusing started at −70 °C in the CIS (0.5 min), increased from 12 °C/s to 250 °C, and this was maintained for 5 min. Helium 5.0 (Nippon Gases GmbH, Hürth, Germany) served as carrier gas with a constant flow of 1.56 mL/min. The gas flow was split 1:1 between the MSD and the ODP (transferline temperature 250 °C both). The ODP mixing chamber was heated to 150 °C, and N_2 was used as make-up gas. The oven temperature program started at 40 °C (3 min), increased from 5 °C/min to 240 °C, and this level was maintained for 12 min. The MS source temperature was 230 °C; detection was conducted in scan mode with 70 eV (*m/z* 33–300). Splitless measurements were performed with 30 mL/min purge flow to split vent at 2 min in CIS, and a splitless mode was used in TDU.

2.5.1. Aroma Dilution Analysis (ADA)

ADA started at 6.24 mL/min purge flow to the split vent at 0 min in CIS, whereas it began in a splitless mode in TDU. Different split ratios for ADA in CIS and TDU were adapted from Trapp et al. to divide the split ratio in every step by two [20]. ADA was carried out by three trained panelists (all females, 25–27 years old, all non-smokers). In order to determine the flavor dilution factors (FD), the median of the lowest dilution level was chosen at the point at which the odorant could still be perceived by the panelists.

2.5.2. Identification and Quantitation of Selected Aroma Compounds via Standard Addition

Compound identification was carried out via a comparison of their retention indices (RIs) according to the specifications of van den Dool and Kratz [21] and a comparison of their mass spectra (MS), as well as of their odors, with those of authentic standards and/or with literature data. Enantioselective analyses of linalool were performed according to the recommendations of Brescia et al. [22].

2-Nonanone ($\geq$99%, Thermo Fisher Scientific, Waltham, MA, USA), (*R*)-linalool ($\geq$95%, Thermo Fisher Scientific, Waltham, MA, USA), methyl benzoate (99%, Alfa Aesar, Karlsruhe, Germany), 1-phenylethyl acetate ($\geq$99%, Sigma Aldrich, St. Louis, MO, USA), 2-phenylethanol (99%, Acros Organics, Waltham, MA, USA), 5-butyl-2(*5H*)-furanone (76% c.f. Figure S1, synthesized in-house by S.Y. [23]) and (*E*)-nerolidol (100%, Sigma Aldrich, Steinheim, Germany) were quantitated in the final beverage via standard addition in duplicate experiments. A mixed-stock solution was prepared in ddH$_2$O, from which four standard solutions (K1–K4) were prepared. For standard addition, 100 μL of each standard solution was added to the sample (5 mL), respectively (cf. Section 2.5) (S1–S4). For S0, 100 μL ddH$_2$O was added. The concentrations of the stock solution as well as the dilution levels are presented in the Supplementary Material (Table S1). Extraction and measurement were carried out as described above. Quantifier ions were chosen as follows: 2-nonanone (*m/z* 58), (*R*)-linalool (*m/z* 93), methyl benzoate (*m/z* 105), 1-phenylethyl acetate (*m/z* 122), 2-phenylethanol (*m/z* 91), 5-butyl-2(*5H*)-furanone (*m/z* 84) and (*E*)-nerolidol (*m/z* 93).

2.5.3. Odor Threshold and Odor Activity Values (OAV)

Odor thresholds were taken from the literature. For 5-butyl-2(*5H*)-furanone, to the best of our knowledge, no odor threshold in water has been published yet. The determination was thus carried out according to the methods of Czerny et al., 2008 [24]. Starting from a stock-solution of 1 mg/mL in ddH$_2$O, the sample was diluted seven times in 1:3 steps. A total of 10 mL of each dilution was filled into a 35 mL snap lid glass, covered with a small watch glass and equilibrated for 30 min. The odor threshold was tested as a triangle test to determine the difference between descending concentrations. The panel consisted of 21 trained panelists (7 male, 14 female, 24–33 years old, all non-smokers). The evaluation was carried out according to DIN EN ISO 4120:2007 at a significance level of α = 0.05. The odor threshold in water was the mean value between the lowest distinguishable concentration from the references and the highest indistinguishable concentration of the substance. Odor activity values were calculated by dividing the quantitated concentrations by the respective odor threshold [25].

2.5.4. Dynamic Changes of Aroma Compound Formation during Cultivation

In order to investigate the development of selected aroma compounds over the cultivation period, the peak area of selected *m/z* ratios was used. The selected peak areas were 2-pentanone (*m/z* 86), 2-pentanol (*m/z* 45), 1-heptanol (*m/z* 70), 1-octanal (*m/z* 84), **18** (*m/z* 203), **21** (*m/z* 160), **23** (*m/z* 179), **25** (*m/z* 95) and **26** (*m/z* 123). For 2-nonanone, (*R*)-linalool, methyl benzoate, 1-phenylethyl acetate, 2-phenylethanol, 5-butyl-2(*5H*)-furanone and (*E*)-nerolidol, the same mass fragments were chosen as for quantitation via standard addition (cf. Section 2.5.2).

3. Results and Discussion

The results of the screening in surface cultures are summarized in the Supplementary Material (Table S2). *L. persicinus* showed an outstanding aroma formation on cocoa pulp agar plates and was therefore selected for further investigation. In recent years, the fermentation of various substrates with edible fungi of the division Basidiomycota has gained attention from different research groups due to the potential of the method to form a broad spectrum of aroma compounds. However, to the best of our knowledge, *L. persicinus* has thus far not been subjected to in-depth aroma analyses [14–16,26–29], and no data are available on the aroma composition of *L. persicinus*, whether grown in solid-state or in submerged cultures. The present study shows the formation of a highly delicious beverage via the fermentation of cocoa pulp with *L. persicinus* for the first time.

3.1. Sensory Evaluation of Cocoa Pulp Fermented with L. persicinus over the Cultivation Time

The evaluation of the fermented beverage, observed by the sensory panel, revealed differences between the respective cultivation times (Figure 1). Non-fermented CP-M served as a reference and was described mainly by the attributes of acidic, fruity, citrus- and apple-like. During the fermentation process, the highest values for the attributes sweet, tropical, passionfruit, peach, mango, coconut, and apricot were reached after 48 h. The acidic taste was decreased by fermentation and the lowest value was reached after 48 h, while the sweet taste reached its maximum value at this point of time. Longer cultivation periods resulted in the attainment similar sensory profiles, but these displayed lower intensities. The overall acceptance of the beverage in terms of smell and taste showed a maximum after 48 h. The acceptance of smell reached 2.9 out of 5.0 points for CP-M and increased up to 3.7 until a fermentation time of 48 h. Afterwards, the acceptance values decreased gradually. The acceptance of taste started with 2.1 out of 5.0 points for the CP-M and increased up to 4.2 points until 48 h and decreased again afterwards.

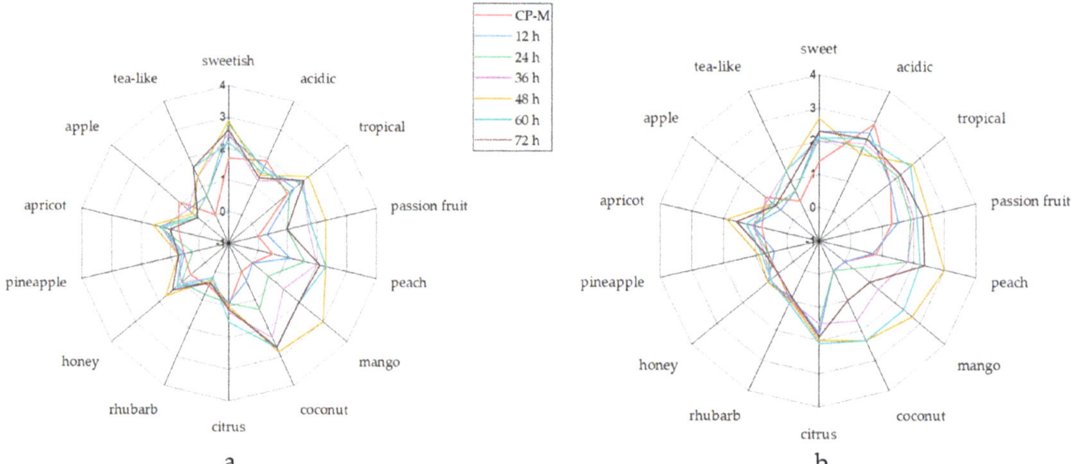

Figure 1. (**a**) Odor evaluation of the cocoa pulp medium (CP-M) and of the fermented beverage at specified cultivation periods; (**b**) taste evaluation (*n* = 10).

The sensory description of CP-M was consistent with those of cocoa pulp in the literature, where it was described as floral, fruity, honey, citrus-like and tropical [6]. An improvement in the overall acceptance of a beverage fermented by basidiomycetes has been shown for other substrates previously. For example, Sommer et al. produced a beverage via submerged fermentation of black current pomace with *Wolfiporia cocos* and the results showed an increase in the overall acceptance of 2.5 to 8.0 out of 10.0 points [27]. Wang et al. reported that the fermentation of okara with edible fungi can improve the flavor quality by decreasing the amount of off-flavor compounds and by forming new aromatic compounds [28]. Different from the fermentation of okara, no off-flavor contents had to be masked in the present study. The beverage was characterized by highly appealing tropical-fruity notes. Cocoa pulp thus represents a side-stream with enormous potential for use in new food products. The proportion of fresh cocoa pulp in cocoa fruits depends on various factors and ranges, according to the literature, from 10.0 to 26.4% [8,30]. Considering that only a part of the fresh cocoa pulp may be used due to the necessity of the fermentation of the beans, the amount of available fresh cocoa pulp is limited. Nevertheless, the development of novel products from cocoa side-streams can significantly contribute to the increased sustainability of the cocoa sector and generate added value for farmers [2].

The use of cocoa pulp in beer and fruit wine production has been reported in the literature. One study reported that, for the development of these alcoholic beverages,

30% cocoa pulp was used in beer production to obtain a high acceptance, and a medium starting from 100% cocoa pulp was used for fruit wine production, whereby the °Brix was subsequently adjusted with sucrose solution [7,8]. Compared to these examples, the beverage fermented by *L. persicinus* required a medium containing only 10% cocoa pulp. In addition, the beverage developed in this study was free of alcohol. The per capita alcohol consumption in Germany is steadily declining, which may be attributed to increased health awareness and a changing age structure [31]. At the same time, the market for non-alcoholic beverages is expanding to a great extent. Fermented non-alcoholic beverages are also increasingly gaining attention of consumers due to advantages regarding enhanced shelf life, improved flavor and the association between fermentation and health benefits [32]. Prior to bringing the novel beverage onto the market, a comprehensive evaluation of the chemical composition, including, e.g., sugars, acids and secondary metabolites potentially formed by the fungus will be required.

3.2. Aroma Compounds in CP-M

The occurrence of aroma substances in cocoa pulp depends on various factors, such as their variety and the origin [6]. In total, 32 aroma compounds were detected olfactometrically by means of GC–MS/MS–O in the CP–M, and seven non-odor-active compounds could be identified additionally (Table 1).

A total 27 of the 32 substances identified here have been described before by Bickel Haase et al., Chetschick et al., Hegmann et al. and Pino et al., who investigated cocoa pulps from different origins and cultivars [6,33–35]. Compounds which have not been described before in the literature on fresh cocoa pulp directly after opening the fruit were ethyl acetate, acetoin, 1-octanol, hexanoic acid and decanoic acid. Ethyl acetate and acetoin are typical flavoring substances produced by yeasts and lactic acid bacteria during fermentation [36]. In contrast to the studies mentioned above, cocoa pulp treated by pasteurization was used in this work. Prior contact with microorganisms could not be excluded and thus may have explained the occurrence of these two aroma compounds. Furthermore, other flavoring substances have been described for cocoa pulp in the literature that could not be detected here, such as β-damascenone (fruity, grape-like), γ- and δ-decalactone (coconut, peach) and *trans*-4,5-epoxy-(*E*)-2-decenal (metallic) [6,33,34].

3.3. Aroma Dilution Analysis of the Beverage after 48 h Fermentation

Aroma dilution analysis (ADA) was performed for the aroma analysis of fermented beverages. A total of 37 substances were olfactometrically perceived with FD factors between 8 and 2048 (Table 2). Linalool (**13**) with sweet, fruity, flowery, and citrus notes, showed the highest FD factor of 2048. The enantioselective analysis showed that (*R*)-linalool was present in the sample, with an *ee* = 98.4%. 5-Butyl-2(*5H*)-furanone (**24**), which has a strong coconut and peach-like odor, was present, as was (*E*)-nerolidol (**27**) with sweet, popcorn, flowery and woody notes, and they had FD factors of 1024. 2-Nonanone (**8**) showed an FD factor of 512 with a fruity, musty, herbaceous, spicy, but also cheesy, odor. Methyl benzoate (**14**) with green, herbaceous, sweet and popcorn notes, as well as 2-phenylethanol (**22**) with sweet, rose, fruity and refreshing notes, showed an FD factor of 128. 1-Phenylethyl acetate (**15**) had an FD factor of 64 and a lavender, flowery, fruity, and tropical odor. Nine other substances with FD factors < 64 were identified: 2-pentanone (**1**), 2-pentanol (**3**), 2-hexanol (**4**), octanal (**6**), 1-octen-3-one (**7**), 1-heptanol (**11**), 2-acetylfuran (**12**) and τ-muurolol (**31**). However, these substances were not quantified due to their low FD factors.

The seven identified substances with FD factors $\geq$ 64 were quantified by means of standard addition. Using odor thresholds extracted from the literature, odor activity values (OAVs) were calculated (Table 3). As no odor threshold has been published so far for 5-butyl-2(*5H*)-furanone, its odor threshold in water was determined for the first time with 62 μg/L. Of the compounds identified and quantified, (*R*)-linalool (**13**), 2-phenylethanol (**22**), 5-butyl-2(*5H*)-furanone (**24**) and (*E*)-nerolidol (**27**) showed OAVs > 1 and thus most likely contributed to the characteristic aroma of the beverage. 2-Nonanone (**8**) and 1-phenylethyl acetate (**15**) had OAVs << 1, indicating no or only a minor contribution to the overall aroma. Methyl benzoate (**14**) had an OAV of 0.8. It is thus difficult to issue a concluding statement on the contribution to the overall aroma. The linear regressions used for quantitation of the aroma compounds by means of standard addition are presented in the Supplementary Material (Table S3).

Table 1. Identified and olfactometrically perceived substances in CP–M with odor impressions and RI indices according to van den Dool and Kratz [21]; n.i. = not identified.

Compound	Odor Impression	$RI_{VF\text{-}Wax}$	$RI_{DB\text{-}5}$	Identification
ethyl acetate	fruity	877 [a]	-	RI, odor, $MS_{VF\text{-}Wax}$
2-pentanone	fruity, sweetish	972 [a]	<700 [a]	RI, odor, MS
2-methyl-3-buten-2-ol	fruity, green	1036 [a]	<700 [a]	RI, odor, MS
2-pentyl acetate	fruity, sweetish	1071 [b]	850 [b]	RI, odor, MS
hexanal	sweetish, caramel, fresh	1080 [a]	801 [a]	RI, odor, MS
n.i.	green, herbaceous, fruity	1103	-	-
2-pentanol	organic solvent, herbaceous	1121 [a]	709 [a]	RI, odor, MS
2-heptanone	-	1182 [a]	892 [a]	RI, MS
2-methyl-1-butanol	-	1216 [a]	735 [a]	RI, MS
2-heptyl acetate	fruity, flowery	1264 [b]	1039 [b]	RI, odor, MS
n.i.	sweetish	1278	-	-
acetoin	sweetish, fatty	1279 [a]	720 [a]	RI, odor, MS
octanal	sweetish, citrus	1290 [a]	1005 [a]	RI, odor, MS
1-octen-3-one	mushroom	1303 [a]	976 [a,c]	RI, odor, MS
2-heptanol	sweetish, coconut	1320 [a]	903 [a]	RI, odor, MS
6-methyl-5-hepten-2-one	-	1339 [a]	986 [a]	RI, MS
1-hexanol	-	1350 [a]	869 [a]	RI, MS
2-nonanone	fruity, herbaceous, cheesy	1390 [a]	1092 [a]	RI, odor, MS
1-heptanol	fruity	1455 [a]	972 [a]	RI, odor, MS
linalool-oxid (isomers)	sweetish, flowery, spicy	1443 [a] / 1471 [a]	1074 [a] / 1089 [a]	RI, odor, MS
acetic acid	acetic acid	1450 [a]	<700 [a]	RI, odor, MS
linalool	sweetish, flowery, citrus	1548 [a]	1101 [a]	RI, odor, MS
1-octanol	sweetish, flowery	1558 [b]	-	RI, odor, $MS_{VF\text{-}Wax}$
acetophenone	sweetish, fruity	1655 [a]	1070 [a]	RI, odor, MS
3-methylbutanoic acid	moldy, cheesy, banana	1681 [a]	842 [a]	RI, odor, MS
α-terpineol	citrus, woody	1698 [a]	1199 [a]	RI, odor, MS
1-phenylethyl acetate	sweetish, fruity, acidic	1704 [a]	1191 [a]	RI, odor, MS
ethylphenyl acetate	sweetish, fruity, flowery	1789 [a]	1245 [a]	RI, odor, MS
hexanoic acid	-	1857 [a]	-	RI, $MS_{VF\text{-}Wax}$
benzyl alcohol	sweetish, flowery	1868 [a]	1036 [a]	RI, odor, MS
2-phenylethanol	sweetish, rose, fruity, coconut	1901 [a]	1116 [a]	RI, odor, MS
γ-nonalactone	coconut	2039 [a]	1361 [a]	RI, odor, MS
octanoic acid	-	2069 [b]	1174 [b]	RI, MS
n.i.	fruity, tropical	2278	-	-
decanoic acid	-	2281 [a]	1370 [a]	RI, MS
n.i.	fruity, acidic	2452	-	-
n.i.	sweetish	2461	-	-
n.i.	fruity, cocoa pulp	2882	-	-

[a] = identified by authentic standard. [b] = identified by comparison with literature data (NIST chemistry webbook 2022). [c] = identified by odor at given RI.

Table 2. Olfactometrically perceived substances in the fermented sample used for ADA with FD factors, odor impressions and RI according to van den Dool and Kratz [21]; n.i. = not identified.

	Compound	FD	Odor Impression	RI$_{VF\text{-}Wax}$	RI$_{DB\text{-}5}$	Identification
1	2-pentanone	32	herbaceous, green, sweetish, flowery	972 [a]	<700 [a]	RI, odor, MS
2	n.i.	32	herbaceous, green	1079	-	-
3	2-pentanol	32	sweetish, flowery	1122 [a]	700 [b]	RI, odor, MS
4	2-hexanol	32	green, herbaceous, fruity, berry, spicy	1220 [a]	800 [a,c]	RI, odor, MS
5	n.i.	16	sweetish, herbaceous	1271	-	-
6	1-octanal	32	sweetish, flowery, citrus	1290 [a]	1000 [a,c]	RI, odor, MS$_{VF\text{-}Wax}$
7	1-octen-3-one	32	mushroom	1303 [a]	976 [a,c]	RI, odor, MS$_{VF\text{-}Wax}$
8	2-nonanone	512	fruity, musty, herbaceous, spicy, cheesy	1390 [a]	1097 [a]	RI, odor, MS
9	n.i.	16	fruity, flowery, citrus, fresh, mushroom	1402	-	-
10	(E)-2-octental	32	herbaceous, green, chocolate, earthy	1431 [a]	-	RI, odor, MS
11	1-heptanol	8	fruity, organic solvent, spicy	1455 [a]	972 [a,c]	RI, odor, MS$_{VF\text{-}Wax}$
12	2-acetylfuran	16	sweetish, citrus, flowery, caramel	1509 [b]	-	RI, odor, MS
13	(R)-linalool	2048	sweetish, fruity, flowery, citrus	1548 [a]	1101 [a]	RI, odor, MS
14	methyl benzoate	128	green, herbaceous, sweetish, popcorn	1626 [a]	1092 [a,c]	RI, odor, MS$_{VF\text{-}Wax}$
15	1-phenylethyl acetate	64	lavender, flowery, fruity, tropical	1704 [a]	1191 [a]	RI, odor, MS
16	n.i.	16	sweetish, popcorn, coconut, fruity, peach	1720	-	-
17	n.i.	64	green, herbaceous, spicy	1750	-	-
18	n.i. (sesquiterpenoid) *	64	fruity, coconut, sweet, passion fruit,	1803	1468	-
19	n.i. (sesquiterpenoid)	32	sweetish, fruity	1842	-	-
20	n.i. (sesquiterpenoid)	64	spicy, herbaceous, sweetish, flowery, fruity, green	1858	-	-
21	n.i. (sesquiterpenoid)	32	fruity, citrus, coconut	1902	1701	-
22	2-phenylethanol	128	sweetish, rose, fruity, refreshing	1910 [a]	1116 [a]	RI, odor, MS
23	n.i. (sesquiterpenoid)	64	sweetish, mushroom-like, fruity, peach	1951	-	-
24	5-butyl-2(5H)-furanone	1024	coconut, peach	1970 [a]	1239 [a]	RI, odor, MS
25	n.i. (sesquiterpenoid)	64	sweetish, coconut, peach	1995	-	-
26	n.i. (sesquiterpenoid)	256	spicy, herbaceous, metallic	2003	-	-
27	(E)-nerolidol	1024	sweetish, popcorn, flowery, woody	2039 [a]	1563 [a]	RI, odor, MS
28	n.i. (sesquiterpenoid)	8	sweetish, fruity, spicy	2056	1603	-
29	n.i. (sesquiterpenoid)	32	fruity, sweetish, caramel	2078	-	-
30	n.i. (sesquiterpenoid)	8	burned, plastic, spicy	2110	1572	-

Table 2. *Cont.*

	Compound	FD	Odor Impression	RI$_{\text{VF-Wax}}$	RI$_{\text{DB-5}}$	Identification
31	τ-muurolol	16	sweetish, Maggi	2199 [b]	1662 [b]	RI, odor, MS
32	n.i. (sesquiterpenoid)	8	sweetish, caramel, peach	2230	1589	-
33	n.i. (sesquiterpenoid)	32	sweetish, citrus, fruity, caramel	2265	1630	-
34	n.i. (sesquiterpenoid)	64	citrus, fruity, popcorn, sweetish	2289	-	-
35	n.i. (sesquiterpenoid)	32	sweetish, fruity	2461	-	-
36	n.i. (sesquiterpenoid)	64	fruity, herbaceous, sweetish, pungent	2491	-	-
37	n.i. (sesquiterpenoid)	32	sweetish, coconut, flowery	2582	1756	-

[a] = identified by authentic standard. [b] = identified by comparison with literature data (NIST chemistry webbook 2022). [c] = identified by odor at given RI. * = assigned to the sesquiterpenes group on the basis of the mass spectrum.

Table 3. Quantitated amounts and calculated OAVs of selected compounds.

Compound	Concentration [µg/L]	Odor Threshold in Water [µg/L]	OAV
2-nonanone (**8**)	1.5 ± 0.1	5.0 [37]	<1
(*R*)-linalool (**13**)	165.0 ± 1.6	0.087 [24]	1897
methyl benzoate (**14**)	0.4 ± 0.1	0.52 [38]	0.8
1-phenylethyl acetate (**15**)	0.6 ± 0.1	19.0 [39]	<1
2-phenylethanol (**22**)	192.8 ± 0.8	140 [24]	1.4
5-butyl-2(*5H*)-furanone (**24**)	457.4 ± 30.6	62	7.4
(*E*)-nerolidol (**27**)	42.4 ± 5.0	0.25 [40]	170

Linalool (**13**) is a well-known monoterpene alcohol that has been detected in many plants and fungi [41]. In other studies, on beverages fermented by different basidiomycetes, linalool almost always had an OAV of >1 [15,16,27,42]. Methyl benzoate (**14**) has been described for a variety of basidiomycetes, such as *Lentinula edodes* and *Grifola frondosa* with concentrations of up to 10 µg/L [10]. It is considered to be a key component of the flavor of mango [38]. Despite its OAV of 0.8, this substance may contribute to the mango-like impression of the fermented beverage. The formation of 5-butyl-2(*5H*)-furanone (**24**) by basidiomycetes of the genus *Laetiporus* was demonstrated by Yalman et al. [23]. They detected 5-butyl-2(*5H*)-furanone in liquid cultures of *Laetiporus montanus* with the highest FD factor of 4096. Little is known about the biosynthesis of this substance. Berger et al. suggested a pathway for biosynthesis, starting from octanoic acid and decanoic acid [23,43]. Both fatty acids were found in CP-M. With an OAV of 7.4, 5-butyl-2(*5H*)-furanone (**24**) contributed to the coconut and peach-like aroma impression of the beverage. (*E*)-Nerolidol (**27**) is a sesquiterpene alcohol that naturally occurs in various plants and is also known to be formed by basidiomycetes like *Polyporus* sp., with up to 260 µg/L [10,44]. Sommer et al., 2023 quantitated a concentration of 0.1 µg/L in submerged cultures of *Wolfiporia cocos* grown on black current pomace where it did not contribute to the overall aroma (OAV < 0.01) [27]. In the present study, (*E*)-nerolidol (**27**) showed the second highest OAV with 170.

3.4. Dynamic Changes of Aroma Compound Formation during Cultivation

The aroma profile of the beverage is composed of aroma compounds already present in CP-M, as well as of new aroma compounds formed during fermentation. Based on the peak areas of a characteristic *m/z* ratio, the concentrations of selected aroma compounds were investigated over a cultivation period of 72 h (Figure 2). The peak areas of the respective substance were related to the highest peak area, which was set to 100%.

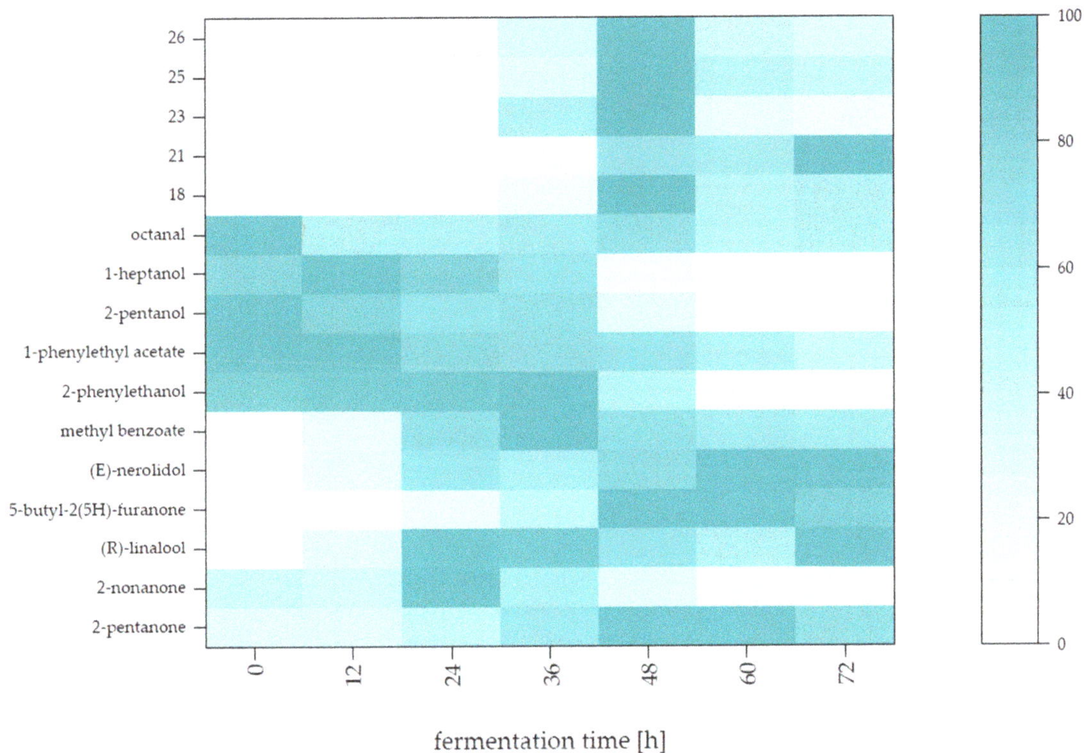

Figure 2. Heat map plot of peak areas of selected aroma compounds in the course of fermentation.

Octanal (**6**), 1-heptanol (**11**), 2-pentanol (**3**), 1-phenylethyl acetate (**15**) and 2-phenylet-hanol (**22**) were already present in the CP-M and showed decreasing intensities during fermentation. It is known that basidiomycetes can also form 2-phenylethanol (**22**) de novo or by biotransformation from asparagine or L-phenylalanine, although this occurs in much lower concentrations than yeasts, for example [18,45]. Özdemir et al. showed the production of 2-phenylethanol by *Lentinula edodes* in wort with an OAV of 1.3 [46]. However, based on the peak area change over time, the contribution to the overall aroma of 2-phenylethanol (**22**) might be attributed to its occurrence in the CP-M. Methyl benzoate (**14**), (*E*)-nerolidol (*27*), 5-butyl-2(*5H*)-furanone (**24**), (*R*)-linalool (**13**), as well as the unidentified substances **18**, **21**, **23**, **25** and **26**, were formed by fermentation. Linalool was also detected in trace amounts in CP-M. However, the content quantified in the beverage could mainly be attributed to the biosynthesis by the fungus. 2-Nonanone (**8**), as well as 2-pentanone (**1**), were already present in CP-M, but the concentrations were increased by fermentation, which indicated that these aroma compounds were formed by the fungus. However, as discussed in Section 3.3, 2-nonanone (**8**) did not contribute to the overall aroma. At the harvest time of 48 h, the highest intensities were detected for 2 pentanone (**1**), 5-butyl-2(*5H*)-furanone (**24**), **18**, **23**, **25** and **26**. For octanal (**6**), 1-phenylethyl acetate (**15**), 2-phenylethanol (**22**), methyl benzoate (**14**), (*E*)-nerolidol (*27*) and (*R*)-linalool (**13**), the intensities were not at the highest level after 48 h, but still at a high level. The formation of methyl benzoate, nerolidol, linalool, 2-nonanone and 2-pentanone by basidiomycetes in submerged fermentations has been described in the literature previously [10,47].

In addition to the identified aroma compounds discussed above, the thus-far-unidentified substances **18**, **21**, **23**, **25** and **26** were also listed in the heat map (Figure 2; related mass spectra cf. Figure S2). These compounds imparted coconut-, passion fruit- and peach-like odor impressions and were assigned to the group of sesquiterpenoids, but they could not be conclusively identified. The supposed sesquiterpenoids showed their maximum concentrations after 48 h and decreased afterwards (except **21**). Compounds **18**, **23** and **25** (all FD 64) exhibited fruity, coconut-, peach- or passion fruit-like notes. The decrease in intensities in the sensory evaluation after 48 h (Figure 1) was accompanied by the decrease in peak intensities. This may indicate that these substances contribute to the coconut, passion fruit and peach notes. Compound **26** showed an FD factor of 256, indicating a contribution to the overall aroma with spicy, herbaceous and metallic notes. Unfortunately, a final identification of these substances was not possible. To the best of our knowledge, no sesquiterpenoids with these aroma notes from *L. persicinus* have been described in the literature to date, which offers an interesting field of research for the future.

The aroma of submerged cultures of *Laetiporus sulphureus* and *Laetiporus montanus*, close relatives of *L. persicinus*, has been described as seasoning-like and meaty, mainly due to the formation of sotolone, (*E*,*E*)-2,4-decadienal, (*E*,*Z*)-2,4-decadienal as well as some sulfur-containing aroma compounds [23,48]. Therefore, fruiting bodies of *Laetipores* are also called chicken of the woods. The aroma achieved here thus offers great research potential, especially as the aroma formation is dependent on the culture medium (cf. Table S2).

4. Conclusions

This study investigated aroma formation during the fermentation of 10% cocoa pulp in water with *L. persicinus* and the development of a non-alcoholic beverage with an outstanding aroma reminiscent of tropical-fruity notes like passion fruit, mango, peach, and coconut. After 48 h of fermentation, the acceptance of the beverage clearly increased. The main contributors to the overall aroma were (*R*)-linalool with an OAV of 1897, (*E*)-nerolidol with an OAV of 170, 5-butyl-2(*5H*)-furanone with an OAV of 7.4 and 2-phenylethanol with an OAV of 1.4. The generation of novel products from cocoa side streams may contribute to increasing the sustainability of the cocoa sector and generate added value for farmers. Aroma formation by *L. persicinus* in submerged cultures has not been described previously and offers an interesting field of research for the future.

Supplementary Materials: The following supporting information can be downloaded at: https://www.mdpi.com/article/10.3390/fermentation9060533/s1, Table S1. Concentrations of stock solution and dilution levels of standard addition. Table S2. Odor impressions during the screening in surface cultures on cocoa pulp agar (CPA) and malt extract agar (MEA), as well as the overall rating R and the intensity of the odor I (–very weak/very bad; - weak/bad; 0 medium/neutral; + intensive/good; ++ very intensive/very good). Table S3. Regression curves and regression coefficient R2 of standard additions (n = 2). Figure S1. Chromatogram for determining the chromatographic purity of 5-butyl-2(5H)-furanone (30.333 min), minus blank measurement of the solvent. Purity = 76%. Figure S2. Mass spectra of 18 (a), 21 (b), **23** (c), **25** (d) and **26** (e).

Author Contributions: V.K.: conceptualization, methodology, investigation, formal analysis, writing—original draft, review, and editing. E.P.: methodology, investigation, formal analysis. J.J.J.: investigation, formal analysis. M.A.F.: methodology, review and editing. H.Z.: review and editing, supervision, funding acquisition, project administration. All authors have read and agreed to the published version of the manuscript.

Funding: The project was funded by the German Federal Ministry of Education and Research (BMBF) under grant number 031B0819. Part of the analytical equipment used was funded by the Deutsche Forschungsgemeinschaft (DFG, German Research Foundation)—463380894. The authors are responsible for the content of this publication.

Institutional Review Board Statement: Not applicable.

Informed Consent Statement: Not applicable.

Data Availability Statement: Not applicable.

Acknowledgments: The authors thank Suzan Yalman for providing the synthesized 5-butyl-2(*5H*)-furanone standard for quantitation. We thank Fabio Brescia for the enantioselective analysis of linalool. We thank all participants for their efforts in the sensory evaluation of the beverages and the determination of the odor threshold in water of 5-butyl-2(*5H*)-furanone.

Conflicts of Interest: The authors declare no conflict of interest.

References

1. The International Cocoa Organization. ICCO Quarterly Bulletin of Cocoa Statistics. 2023. Available online: https://www.icco.org/february-2023-quarterly-bulletin-of-cocoa-statistics/ (accessed on 27 April 2023).
2. Vásquez, Z.S.; Carvalho Neto, D.P.d.; Pereira, G.V.M.; Vandenberghe, L.P.S.; Oliveira, P.Z.d.; Tiburcio, P.B.; Rogez, H.L.G.; Góes Neto, A.; Soccol, C.R. Biotechnological Approaches for Cocoa Waste Management: A Review. *Waste Manag.* **2019**, *90*, 72–83. [CrossRef] [PubMed]
3. Anvoh, K.; Zoro Bi, A.; Gnakri, D. Production and Characterization of Juice from Mucilage of Cocoa Beans and its Transformation into Marmalade. *Pak. J. Nutr.* **2009**, *8*, 129–133. [CrossRef]
4. Schwan, R.F.; Wheals, A.E. The Microbiology of Cocoa Fermentation and its Role in Chocolate Quality. *Crit. Rev. Food Sci. Nutr.* **2004**, *44*, 205–221. [CrossRef] [PubMed]
5. Amanquah, D.T. *Effect of Mechanical Depulping on the Biochemical, Physicochemical and Polyphenolic Constituents during Fermentation and Drying of Ghanaian Cocoa Beans*; University of Ghana: Accra, Ghana, 2013.
6. Bickel Haase, T.; Schweiggert-Weisz, U.; Ortner, E.; Zorn, H.; Naumann, S. Aroma Properties of Cocoa Fruit Pulp from Different Origins. *Molecules* **2021**, *26*, 7618. [CrossRef]
7. Nunes, C.S.O.; Da Silva, M.L.C.; Camilloto, G.P.; Machado, B.A.S.; Hodel, K.V.S.; Koblitz, M.G.B.; Carvalho, G.B.M.; Uetanabaro, A.P.T. Potential Applicability of Cocoa Pulp (*Theobroma cacao* L) as an Adjunct for Beer Production. *Sci. World J.* **2020**, *2020*, 3192585. [CrossRef]
8. Dias, D.R.; Schwan, R.F.; Freire, E.S.; Serôdio, R.d.S. Elaboration of a Fruit Wine from Cocoa (*Theobroma cacao* L.) Pulp. *Int. J. Food Sci. Tech.* **2007**, *42*, 319–329. [CrossRef]
9. Puerari, C.; Magalhães, K.T.; Schwan, R.F. New Cocoa Pulp-based Kefir Beverages: Microbiological, Chemical Composition and Sensory Analysis. *Food Res. Int.* **2012**, *48*, 634–640. [CrossRef]
10. Abraham, B.G.; Berger, R.G. Higher Fungi for Generating Aroma Components through Novel Biotechnologies. *J. Agric. Food Chem.* **1994**, *42*, 2344–2348. [CrossRef]
11. Barbosa, E.d.S.; Perrone, D.; Vendramini, A.L.d.A.; Leite, S.G.F. Vanillin Production by *Phanerochaete Chrysosporium* Grown on Green Coconut Agro-Industrial Husk in Solid State Fermentation. *BioResources* **2008**, *3*, 1042–1050.
12. Beltran-Garcia, M.J.; Estarron-Espinosa, M.; Ogura, T. Volatile Compounds Secreted by the Oyster Mushroom (*Pleurotus ostreatus*) and Their Antibacterial Activities. *J. Agric. Food Chem.* **1997**, *45*, 4049–4052. [CrossRef]
13. Liu, J.; Vijayakumar, C.; Hall, C.A.; Hadley, M.; Wolf-Hall, C.E. Sensory and Chemical Analyses of Oyster Mushrooms (*Pleurotus sajor-caju*) Harvested from Different Substrates. *J. Food Sci.* **2005**, *70*, S586–S592. [CrossRef]
14. Sommer, S.; Fraatz, M.A.; Büttner, J.; Salem, A.A.; Rühl, M.; Zorn, H. Wild Strawberry-like Flavor Produced by the Fungus *Wolfiporia cocos*—Identification of Character Impact Compounds by Aroma Dilution Analysis after Dynamic Headspace Extraction. *J. Agric. Food Chem.* **2021**, *69*, 14222–14230. [CrossRef]
15. Zhang, Y.; Fraatz, M.A.; Müller, J.; Schmitz, H.-J.; Birk, F.; Schrenk, D.; Zorn, H. Aroma Characterization and Safety Assessment of a Beverage Fermented by *Trametes versicolor*. *J. Agric. Food Chem.* **2015**, *63*, 6915–6921. [CrossRef]
16. Rigling, M.; Liu, Z.; Hofele, M.; Prozmann, J.; Zhang, C.; Ni, L.; Fan, R.; Zhang, Y. Aroma and Catechin Profile and in vitro Antioxidant Activity of Green Tea Infusion as Affected by Submerged Fermentation with *Wolfiporia cocos* (Fu Ling). *Food Chem.* **2021**, *361*, 130065. [CrossRef]
17. Janssens, L.; Pooter, H.L.d.; Schamp, N.M.; Vandamme, E.J. Production of Flavours by Microorganisms. *Process. Biochem.* **1992**, *27*, 195–215. [CrossRef]
18. Lomascolo, A.; Stentelaire, C.; Asther, M.; Lesage-Meessen, L. Basidiomycetes as New Biotechnological Tools to Generate Natural Aromatic Flavours for the Food Industry. *Trends Biotechnol.* **1999**, *17*, 282–289. [CrossRef]
19. European Parliament and the Council of December 2008 on Flavourings and certain food ingredients with flavouring properties for use in and on foods, amending Regulation (EC) No. 1334/2008 and Council Regulation (EEC) No 1601/91, Regulations (EC) No 2232/96 and (EC) No 110/2008 and Directive 2000/13/EC. *Off. J. Eur. Union* **2008**.
20. Trapp, T.; Jäger, D.A.; Fraatz, M.A.; Zorn, H. Development and Validation of a Novel Method for Aroma Dilution Analysis by Means of Stir Bar Sorptive Extraction. *Eur. Food Res. Technol.* **2018**, *244*, 949–957. [CrossRef]
21. Van den Dool, H.; Kratz, P.D. A Generalization of the Retention Index System Including Linear Temperatur Programmed Gas—Liquid Partition Chromatography. *J. Chromatogr. A* **1963**, *11*, 463–471. [CrossRef]
22. Brescia, F.F.; Pitelas, W.; Yalman, S.; Popa, F.; Hausmann, H.G.; Wende, R.C.; Fraatz, M.A.; Zorn, H. Formation of Diastereomeric Dihydromenthofurolactones by *Cystostereum murrayi* and Aroma Dilution Analysis Based on Dynamic Headspace Extraction. *J. Agric. Food Chem.* **2021**, *69*, 5997–6004. [CrossRef]

23. Yalman, S.; Trapp, T.; Vetter, C.; Popa, F.; Fraatz, M.A.; Zorn, H. Formation of a meat-like flavor by submerged cultivated *Laetiporus Montanus*. *J. Agric. Food Chem.* **2023**. [CrossRef]
24. Czerny, M.; Christlbauer, M.; Christlbauer, M.; Fischer, A.; Granvogl, M.; Hammer, M.; Hartl, C.; Hernandez, N.M.; Schieberle, P. Re-Investigation on Odour Thresholds of Key Food Aroma Compounds and Development of an Aroma Language based on Odour Qualities of Defined Aqueous Odorant Solutions. *Eur. Food Res. Technol.* **2008**, *228*, 265–273. [CrossRef]
25. Grosch, W. Evaluation of the Key Odorants of Foods by Dilution Experiments, Aroma Models and Omission. *Chem. Senses* **2001**, *26*, 533–545. [CrossRef] [PubMed]
26. Rigling, M.; Yadav, M.; Yagishita, M.; Nedele, A.-K.; Sun, J.; Zhang, Y. Biosynthesis of Pleasant Aroma by Enokitake (*Flammulina velutipes*) with a Potential Use in a Novel Tea Drink. *LWT* **2021**, *140*, 110646. [CrossRef]
27. Sommer, S.; Hoffmann, J.L.; Fraatz, M.A.; Zorn, H. Upcycling of Black Currant Pomace for the Production of a Fermented Beverage with *Wolfiporia cocos*. *J. Food Sci. Technol.* **2023**, *60*, 1313–1322. [CrossRef]
28. Wang, Z.; Gao, T.; He, Z.; Zeng, M.; Qin, F.; Chen, J. Reduction of Off-Flavor Volatile Compounds in Okara by Fermentation with Four Edible Fungi. *LWT* **2022**, *155*, 112941. [CrossRef]
29. Zhang, Y.; Fraatz, M.A.; Horlamus, F.; Quitmann, H.; Zorn, H. Identification of Potent Odorants in a Novel Nonalcoholic Beverage Produced by Fermentation of Wort with Shiitake (*Lentinula edodes*). *J. Agric. Food Chem.* **2014**, *62*, 4195–4203. [CrossRef]
30. Tịnh, N.T.T.; An, N.T.; Hòa, H.T.T.; Tươi, N.T. A Study of Wine Fermentation from Mucilage of Cocoa Beans (*Theobroma cacao* L.). *DLU JOS* **2016**, *6*, 387. [CrossRef]
31. Statista Research Department. Konsum von alkoholischen Getränken. 2023. Available online: https://de.statista.com/themen/22/alkohol/#topicOverview (accessed on 24 February 2023).
32. Şanlier, N.; Gökcen, B.B.; Sezgin, A.C. Health Benefits of Fermented Foods. *Crit. Rev. Food Sci. Nutr.* **2019**, *59*, 506–527. [CrossRef]
33. Chetschik, I.; Kneubühl, M.; Chatelain, K.; Schlüter, A.; Bernath, K.; Hühn, T. Investigations on the Aroma of Cocoa Pulp (*Theobroma cacao* L.) and Its Influence on the Odor of Fermented Cocoa Beans. *J. Agric. Food Chem.* **2018**, *66*, 2467–2472. [CrossRef]
34. Hegmann, E.; Niether, W.; Rohsius, C.; Phillips, W.; Lieberei, R. Besides Variety, also Season and Ripening Stage have a Major Influence on Fruit Pulp Aroma of Cacao (*Theobroma cacao* L.). *J. Appl. Bot. Food Qual.* **2020**, *93*, 266–275. [CrossRef]
35. Pino, J.A.; Ceballos, L.; Quijano, C.E. Headspace Volatiles of *Theobroma cacao* L. Pulp from Colombia. *J. Essent. Oil Res.* **2010**, *22*, 113–115. [CrossRef]
36. Vuyst, L.d.; Leroy, F. Functional Role of Yeasts, Lactic Acid Bacteria and Acetic Acid Bacteria in Cocoa Fermentation Processes. *FEMS Microbiol. Rev.* **2020**, *44*, 432–453. [CrossRef]
37. Teranishi, R.; Buttery, R.G.; Guadagni, D.G. Odor Quality and Chemical Structure in Fruit and Vegetable Flavors. *West. Reg. Res. Lab.* **1974**, *237*, 209–216. [CrossRef]
38. Pino, J.A.; Mesa, J. Contribution of Volatile Compounds to Mango (*Mangifera indica* L.) Aroma. *Flavour. Fragr. J.* **2006**, *21*, 207–213. [CrossRef]
39. Carunchia Whetstine, M.E.; Cadwallader, K.R.; Drake, M. Characterization of Aroma Compounds Responsible for the Rosy/Floral Flavor in Cheddar Cheese. *J. Agric. Food Chem.* **2005**, *53*, 3126–3132. [CrossRef]
40. Yin, P.; Wang, J.-J.; Kong, Y.-S.; Zhu, Y.; Zhang, J.-W.; Liu, H.; Wang, X.; Guo, G.-Y.; Wang, G.-M.; Liu, Z.-H. Dynamic Changes of Volatile Compounds during the Xinyang Maojian Green Tea Manufacturing at an Industrial Scale. *Foods* **2022**, *11*, 2682. [CrossRef]
41. Breheret, S.; Talou, T.; Rapior, S.; Bessière, J.-M. Monoterpenes in the Aromas of Fresh Wild Mushrooms (Basidiomycetes). *J. Agric. Food Chem.* **1997**, *45*, 831–836. [CrossRef]
42. Rigling, M.; Heger, F.; Graule, M.; Liu, Z.; Zhang, C.; Ni, L.; Zhang, Y. A Robust Fermentation Process for Natural Chocolate-like Flavor Production with *Mycetinis scorodonius*. *Molecules* **2022**, *27*, 2503. [CrossRef]
43. Berger, R.G.; Neuhäuser, K.; Drawert, F. Biosynthesis of Flavor Compounds by microorganisms: 6. Odorous Constituents of *Polyporus durus* (Basidiomycetes). *Z. Für Nat.* **1986**, *41*, 963–970. [CrossRef]
44. Chan, W.-K.; Tan, L.T.-H.; Chan, K.-G.; Lee, L.-H.; Goh, B.-H. Nerolidol: A Sesquiterpene Alcohol with Multi-Faceted Pharmacological and Biological Activities. *Molecules* **2016**, *21*, 529. [CrossRef]
45. Chreptowicz, K.; Wielechowska, M.; Główczyk-Zubek, J.; Rybak, E.; Mierzejewska, J. Production of Natural 2-Phenylethanol: From Biotransformation to Purified Product. *Food Bioprod. Process.* **2016**, *100*, 275–281. [CrossRef]
46. Özdemir, S.; Heerd, D.; Quitmann, H.; Zhang, Y.; Fraatz, M.; Zorn, H.; Czermak, P. Process Parameters Affecting the Synthesis of Natural Flavors by Shiitake (*Lentinula edodes*) during the Production of a Non-Alcoholic Beverage. *Beverages* **2017**, *3*, 20. [CrossRef]
47. Gallois, A.; Gross, B.; Langlois, D.; Spinnler, H.-E.; Brunerie, P. Influence of Culture Conditions on Production of Flavour Compounds by 29 Ligninolytic Basidiomycetes. *Mycol. Res.* **1990**, *94*, 494–504. [CrossRef]
48. Lanfermann, I.; Krings, U.; Schopp, S.; Berger, R.G. Isotope Labelling Experiments on the Formation Pathway of 3-Hydroxy-4,5-dimethyl-2(5H)-furanone from 1-Isoleucine in Cultures of *Laetiporus sulphureus*. *Flavour. Fragr. J.* **2014**, *29*, 233–239. [CrossRef]

MDPI AG
Grosspeteranlage 5
4052 Basel
Switzerland
Tel.: +41 61 683 77 34

Fermentation Editorial Office
E-mail: fermentation@mdpi.com
www.mdpi.com/journal/fermentation